T0074714

Seafood Safety
and Quality

Books Published in *Food Biology* series

Food Biology Series

Seafood Safety and Quality

Editors

Md. Latiful Bari

University of Dhaka
Dhaka 1000
Bangladesh

and

Koji Yamazaki

Hokkaido University
3-1-1, Minato, Hakodate
Hokkaido 041-8611
Japan

CRC Press
Taylor & Francis Group
Boca Raton London New York

CRC Press is an imprint of the
Taylor & Francis Group, an **informa** business

A SCIENCE PUBLISHERS BOOK

CRC Press
Taylor & Francis Group
6000 Broken Sound Parkway NW, Suite 300
Boca Raton, FL 33487-2742

First issued in paperback 2020

© 2018 by Taylor & Francis Group, LLC
CRC Press is an imprint of Taylor & Francis Group, an Informa business

No claim to original U.S. Government works

ISBN-13: 978-1-138-54300-3 (hbk)
ISBN-13: 978-0-367-78085-2 (pbk)

Library of Congress Cataloging-in-Publication Data

Names: Bari, Md. Latiful (Mohammad Latiful), editor. | Yamazaki, Kåoji, 1961-
    editor.
Title: Seafood safety and quality / editors, Md. Latiful Bari, University of
    Dhaka, Dhaka 1000, Bangladesh, and Koji Yamazaki, Hokkaido University,
    3-1-1, Minato, Hakodate, Hokkaido 041-8611, Japan.
Description: Boca Raton, FL : Taylor & Francis Group, [2018] | Series: Food
    biology series | "A Science Publishers book." | Includes bibliographical
    references and index.
Identifiers: LCCN 2018016965 | ISBN 9781138543003 (hardback : acid-free paper)
Subjects: LCSH: Seafood--Microbiology. | Seafood industry--Safety measures. |
    Seafood--Contamination. | Foodborne diseases.
Classification: LCC QR118 .S43 2018 | DDC 664/.94--dc23
LC record available at https://lccn.loc.gov/2018016965

Visit the Taylor & Francis Web site at
http://www.taylorandfrancis.com

and the CRC Press Web site at
http://www.crcpress.com

Preface to the Series

Food is the essential source of nutrients (such as carbohydrates, proteins, fats, vitamins, and minerals) for all living organisms to sustain life. A large part of daily human efforts is concentrated on food production, processing, packaging and marketing, product development, preservation, storage, and ensuring food safety and quality. It is obvious therefore, our food supply chain can contain microorganisms that interact with the food, thereby interfering in the ecology of food substrates. The microbe-food interaction can be mostly beneficial (as in the case of many fermented foods such as cheese, butter, sausage, etc.) or in some cases, it is detrimental (spoilage of food, mycotoxin, etc.). The *Food Biology* series aims at bringing all these aspects of microbe-food interactions in form of topical volumes, covering food microbiology, food mycology, biochemistry, microbial ecology, food biotechnology and bio-processing, new food product developments with microbial interventions, food nutrification with nutraceuticals, food authenticity, food origin traceability, and food science and technology. Special emphasis is laid on new molecular techniques relevant to food biology research or to monitoring and assessing food safety and quality, multiple hurdle food preservation techniques, as well as new interventions in biotechnological applications in food processing and development.

The series is broadly broken up into food fermentation, food safety and hygiene, food authenticity and traceability, microbial interventions in food bio-processing and food additive development, sensory science, molecular diagnostic methods in detecting food borne pathogens and food policy, etc. Leading international authorities with background in academia, research, industry and government have been drawn into the series either as authors or as editors. The series will be a useful reference resource base in food microbiology, biochemistry, biotechnology, food science and technology for researchers, teachers, students and food science and technology practitioners.

Ramesh C. Ray
Series Editor

Preface

Seafood Safety and Quality continues to be a major public health issue and its importance has escalated to unprecedented levels in recent years. In this book, major seafood borne diseases and key safety issues are reviewed. In addition, emerging microbial agents, fish toxins and other contaminants including heavy metal; allergy, water safety and related topics are discussed. It also addresses the challenges faced by both developed and developing countries to ensure seafood safety in new seafood products and processing technologies, seafood trade, safety of foods derived from biotechnology, microbiological risks, emergence of new and antibiotic-resistant pathogens, particularly from emerging pathogens, directing research to areas of high-risk, focus intervention and establishment of target risk levels and target diseases or pathogens. The book serves as a comprehensive resource on the seafood borne diseases and a wide variety of responsible etiologic agents, including bacteria, viruses, parasites, seafood toxins, and environmental toxins. It has been written in a simple manner and should promote the efforts of the scientific community to deliver safe seafood for a better health and environment. The book is intended to highlight key challenges in food safety and existing control measures in seafood sectors; to provide scientifically non-biased perspectives on these issues; and to provide assistance to the reader in understanding these issues.

The editors would like to acknowledge people who provided valuable input and assistance and to express our sincere appreciation for their efforts. This appreciation is especially extended to R.C. Ray, M. Mostafizur Rahman, Sonia Shahnaz, Dominic Kasujja Bagenda, Toshiyuki Suzuki, Masataka Satomi, Hiroki Saeki, Yoshihiro Ohnishi, Md. Ali Reza Faruk, Shigenobu Koseki, Hisae Kasai and Shotaro Nishikawa, Kenichi Watanabe, Shogo Yamaki, Chihiro Ohshima, Hajime Takahashi, Chiraporn Ananchaipattana and Pongtep Wilaipun, for their enthusiasm and assistance in preparing this volume, and to the authors of all chapters.

Md. Latiful Bari
Koji Yamazaki

Contents

Seafood Poisoning
Overview

M.A. Khaleque,[1] *Sonia Shahnaz*[2] *and Md. Latiful Bari*[3,*]

1. General Overview

Seafood is any form of sea life regarded as food by humans. Seafood prominently includes fish and shellfish. Shellfish include various species of mollusks, crustaceans, and echinoderms. Seafood-borne illness, or seafood poisoning, occurs because of human consumption that includes, but is not limited to, finfish and shellfish. The incidence of seafood poisoning is on the rise due to high consumption of seafood, as well as increase in international trade and travel. In the United States, one in six food poisoning outbreaks are due to seafood, various fish, and other marine organisms (Mowry et al. 2014). A wide variety of etiologic agents are responsible, including toxins, bacteria, viruses, and parasites.

2. Raw and other Seafood

Raw seafood is usually consumed as sushi and sashimi, as well as clams or oysters on the half shell, and ceviche. Seafood salads, sandwiches and

[1] Department of Biochemistry & Microbiology, North South University, Bashundhara, Dhaka-1229, Bangladesh.
  E-mail: abdul.khaleque@northsouth.edu
[2] Japanese School at Dhaka under the occupancy of Japan Embassy in Bangladesh.
  E-mail: Shahnaz.sonia@gmail.com
[3] Center for Advanced Research in Sciences, University of Dhaka, Dhaka-1000, Bangladesh.
* Corresponding author: latiful@du.ac.bd

cocktails are considered ready-to-eat, which are not usually cooked before they are consumed. Raw seafood contains bacteria, such as "spoilage bacteria" and "food poisoning" bacteria that can grow and multiply rapidly if the food is left for several hours at room temperature. Pathogenic bacteria are the primary food safety concern regarding seafood consumption. Other seafood products such as lightly smoked fish (e.g., salmon lox) are also partially cooked. The raw or partially cooked fish and shellfish in these products may contain pathogens that would normally be killed if these products were fully cooked before they are eaten. Some types of fish may also have naturally occurring parasites. If these fish are used in raw or partially cooked products like sushi, ceviche, or lightly smoked fish, they must be frozen prior to serving to kill any parasites that may be present. When seafood is properly handled and cooked, the risk of food-borne illness from pathogens or parasites is minimal. Poor handling practices, such as failure to prevent raw foods from contact with cooked or ready-to-eat foods (cross contamination), and lack of proper temperature control are significant factors that can lead to pathogen growth and food-borne illness (Clark et al. 1999).

3. Seafood Toxins

Natural toxins are produced by minute organisms called dinoflagellates and diatoms. These phytoplankton move up in the food chain into shellfish and other carnivorous marine organisms, where they concentrate in viscera, affecting those who consume them. The toxins are not only tasteless and odorless, but also heat and acid stable (Isbister and Kiernan 2005).

Seafood-toxin-borne diseases can be categorized into two groups depending on their vectors: shellfish and fish. Shellfish harbors the toxins that produce paralytic shellfish poisoning (PSP), neurotoxic shellfish poisoning (NSP), diarrhetic shellfish poisoning (DSP), and amnesic shellfish poisoning (ASP) (See Table 1.1). On the other hand, fish carry the toxins responsible for ciguatera (Dickey and Plakas 2010, Chan 2013) and tetrodotoxin (fugu or pufferfish) poisoning. The shellfish-associated diseases generally occur in association with algal blooms or "red tides", that is characterized by patches of discolored water due to the dead fishes. Most seafood toxins usually target our nervous system and gastrointestinal tract.

3.1 Finfish poisoning

3.1.1 Ciguatera poisoning

Ciguatera poisoning results from the ingestion of fish contaminated with several marine toxins, all produced by small sea algae, which sticks on to dead coral and multiplies rapidly with the disturbances of coral reefs by

Table 1.1 Marine toxin diseases.

Disease	Ciguatera poisoning	Scombroid poisoning	Puffer fish poisoning	PSP paralytic shellfish poisoning	DSP diarrheal shellfish poisoning	ASP amnestic shellfish poisoning	NSP neurotoxic shellfish poisoning
Vector	carnivorous reef fish (barracuda, grouper, muray eel, parrot fish, red snapper, surgeon fish, trigger fish, amberjack, wrass, mullet)	scombroid fish (tuna mackerel, skipjack bonito)	puffer fish (served as Fugu in Japanese restaurants)	shellfish	shellfish	shellfish	shellfish
Organism	dinoflagellate	surface bacteria	? bacteria	red tide dinoflagellate	red tide dinoflagellate	red tide dinoflagellate	red tide dinoflagellate
Distribution	worldwide tropical and subtropical	worldwide	worldwide	worldwide temperate	worldwide temperate	North America	Gulf of Mexico, New Zealand
Incubation	hours	minutes-hours	5–30 minutes		hours	hours	30 mins-hours
Duration	months-years	< 24 hours	days	days	days	years	days
Symptoms	GI neurological	cutaneous	neurological (respiratory paralysis)	neurological (respiratory paralysis)	GI	neurological	GI neurological
Mortality	< 1%	0%	60%	1–14%	0%	3%	0%
Treatment	supportive mannitol	antihistamines H2-blockers	supportive	supportive (respiratory)	supportive (respiratory)	supportive	supportive
Prevention (avoid)	large carnivorous fish	poorly preserved fish	puffer fish	shellfish	shellfish	shellfish	shellfish

Adapted from Shoreland © 2001 Shoreland, Inc.

different activities such as underwater dredging or storms (Lehane and Lewis 2000). The toxin is concentrated in the viscera, liver, and gonads of the affected fish. The toxin is also tasteless, odorless, and heat stable that makes it hard to neutralize by cooking (Schnorf et al. 2002, Stewart et al. 2010). Successful treatment of ciguatera fish poisoning with intravenous mannitol was reported recently (Schwarz et al. 2008).

3.1.2 Scombroid poisoning

Scombroid is a fish-borne illness reported worldwide, and is particularly common in countries where fish consumption is high, and severe scombroid fish poisoning may result in dermatologic emergency (Jantschitsch et al. 2011). It results from the eating of spoiled fishes from *Scombroidea* or *Scomberesocidae* families including tuna, mackerel, skipjack, and bonito (Etkind et al. 1987). The phenomenon occurs with the consumption of both fresh and canned fish. However, non-scombroid fishes, for example, mahi-mahi (dolphin fish), bluefish, amberjack, swordfish, herring, sardines, anchovies, salmon, and trout produce scombrotoxism (Hungerford 2010).

Scombroid is usually caused by bacterial overgrowth associated with unsatisfactory storage conditions, for instance storing at 20° to 30°C that favors bacterial growth. These surface bacteria contain decarboxylate histidine to produce high levels of histamine, the source of "histamine toxicity (Taylor 1986). If the bacterial overgrowth and decarboxylation have not taken place, refrigeration or cooking both can easily prevent scombroid (Taylor et al. 1984).

3.2 Puffer fish poisoning

Puffer fish poisoning results from the ingestion of tetrodotoxin (TTX) and, while rare, is the most common lethal seafood poisoning. TTX is found in many puffer fish species as well as in the blue-ringed octopus, newts, gobies, frogs, starfish, and the horseshoe crab (Fukuda and Tani 1941, Chen et al. 2016). Several incidences of fatalities have been reported after ingestion of horseshoe crab eggs containing TTX (Kanchanapongkul 2008). In the puffer fish, TTX concentration and distribution depend on the species and typically the ovary, liver, and skin frequently contain the highest amount. The minimum lethal dose in humans for TTX is estimated to be 2 mg (Noguchi and Arakawa 2008), but this number can vary based on the age and existing comorbidities. The CDC has reported symptoms of TTX toxicity with consumption as minimal as 0.25–1.5 oz of incorrectly prepared fugu (Narahashi 2008). Puffer fishes are most commonly found in the tropical water of the Indo-Pacific region, and poisoning occurs predominantly in

Southeast Asia. Japan accounts for the most poisonings worldwide, which is approximately 30–50 percent per year (CDC 1996). The incidence of puffer fish poisoning is decreasing, presumably because of increased awareness and proper preparation of fish (Cohen et al. 2009).

3.3 Sardine poisoning (clupeotoxism)

Sardine poisoning, or clupeotoxism, is a rare form of seafood poisoning caused by the ingestion of clupeiformes, which include sardines, herrings, and anchovies. The exact identity of the toxin is still unknown, although it is assumed to be a palytoxin (Randall 2005, Onuma et al. 1999).

3.4 Hallucinogenic fish poisoning

Hallucinogenic fish poisoning, or ichthyoallyeinotoxism, is a rare form of seafood poisoning related to ingestion of several species, which are also implicated in ciguatera poisoning. It is mainly found in reef fish from the Mediterranean Sea, and the Indian and the Pacific Oceans. The exact toxin is unknown, although indole compounds formed by macro algae have been implicated in many instances (de Haro and Pommier 2006, Helfrich and Banner 1960). Unlike ciguatera poisoning, it has pronounced CNS involvement.

4. Shellfish Poisoning

Shellfish poisonings are associated with hepatisis A virus, Norwalk virus, and proteobacteria family members, such as genus *Vibrio* species. Although several types of toxic ingestions have been reported, the four classic ones are paralytic, neurologic, diarrheal, and amnestic shellfish poisonings. Toxins are found in microscopic diatoms and dinoflagellates, with concentrations occurring in filter feeding bivalves, such as clams or mollusks (James et al. 2010). Most dinoflagellate toxins are neurotoxins, often acting on voltage-sensitive ion channels affecting humans (Isbister and Kiernan 2005).

Red tides

Contrary to common belief, red tides are not well correlated to outbreaks of shellfish poisoning. Red tide gets its name from the phenomenon by which pigmented phytoplankton reproduce in such a high concentration that it turns the water red or dark brown, also termed as harmful algae bloom (HAB) for toxic red tides (Meier and White 1995). The incidence of shellfish poisoning has been declining, most likely because of careful monitoring, beach closures, and increased public awareness.

4.1 Paralytic shellfish poisoning

Paralytic shellfish poisoning (PSP) is not only the most common form of shellfish poisoning, but it is also the deadliest (mortality rate 6%) (Lehane 2001, Meier and White 1995). The causative agent is saxitoxin, which is produced by dinoflagellates (microalgae) ingested by filter-feeding bivalve shellfish (e.g., oysters, mussels, and clams) that concentrate the toxin, and are subsequently consumed by predators, including humans. Saxitoxin is one of the most potent neurotoxins known (0.5–1 mg fatal to human) (James et al. 2010, Clark et al. 1999). It acts similarly to TTX in that it blocks voltage-sensitive sodium channels. The PSP toxins are not destroyed by heat, marinating, or freezing. Contaminated seafood smells, tastes, and appears to be normal (Isbister and Kiernan 2005, Xu et al. 2015).

Potential vectors for PSP include bivalve mollusks (e.g., cockles, salt- and fresh-water mussels, or butter/little neck clams), gastropod mollusks (e.g., whelk, moon snails, or abalone), crustaceans (e.g., Dungeness crabs, shrimp, or lobsters), puffer fish (saxitoxin puffer fish poisoning), and zooplanktivorous fish (e.g., Atlantic salmon, herring, and mackerel) (Mowry et al. 2014). Major outbreaks of PSP are associated with salt-water bivalve mollusks, especially mussels or clams.

In the United States, PSP primarily occurs in seafood harvested from the Northeast, Pacific Northwest, and Alaskan waters (CDC 2011). Commercial harvests of shellfish are routinely monitored for PSP toxins prior to consumption (Isbister and Kiernan 2005).

4.2 Neurotoxic shellfish poisoning

Neurotoxic shellfish poisoning (NSP) is a rare form of shellfish poisoning. The causative agents are brevetoxins, which are produced by dinoflagellates (microalgae) ingested by filter-feeding bivalve shellfish (e.g., oysters, mussels, clams) that concentrate the toxin and are subsequently consumed by predators, including humans. Like many marine toxins, the brevetoxins are tasteless, odorless, and heat stable. They act as ciguatoxins as they are sodium channel openers that cause neuroexcitatory effects (Isbister and Kiernan 2005). Symptoms are generally mild.

4.3 Diarrheal shellfish poisoning

Diarrheal shellfish poisoning (DSP) is a form of mild seafood poisoning. Mainly okadaic acid which is produced by dinoflagellates is ingested by filter-feeding bivalve shellfish (e.g., oysters, mussels, clams), that concentrate the toxin and are subsequently consumed by humans, leading to poisoning. Symptoms are generally mild, causing nausea, vomiting, diarrhea, and abdominal pain (Wu et al. 2005).

4.4 Amnestic shellfish poisoning (domoic acid poisoning)

Amnestic shellfish poisoning (ASP) is a serious form of seafood poisoning (Perl et al. 1990). Diatoms produce domoic acid (DA), which enters the food chain via contaminated mussels is a glutamate agonist. ASP can cause nausea, vomiting, diarrhea, which is unlike other shellfish poisonings due to its profound effects on the CNS (Isbister and Kiernan 2005). The only reported human outbreak of ASP occurred in 1987 in Canada's eastern province of Prince Edward Island, and affected more than 100 people (James et al. 2010).

4.5 Azaspiracid poisoning

Azaspiracid poisoning (AZP) is one of the more recently discovered seafood poisonings. It was identified following cases of severe GI illness from the consumption of contaminated mussels from Ireland, and recently contamination has been confirmed throughout the western coastline of Europe. The implicated toxins, azaspiracids, accumulate in bivalve mollusks that feed on toxic microalgae. Toxicological studies have indicated that azaspiracids can induce widespread organ damage in mice, and that they are probably more dangerous than previously known classes of shellfish toxins (James et al. 2004, Furey et al. 2010).

5. Allergens

Both Finfish and crustacean poisoning cause an allergic reaction in humans. Current regulations require that all seafood that contains food allergens be properly labeled so that individuals with an allergy to a specific food can avoid it. Individuals are often only allergic to a certain species of fish or shellfish, and thus can safely eat other types of seafood. Fish and crustaceans, like crab, shrimp, and lobsters, are commonly consumed seafood that account for all known seafood allergies. Allergy to one type of seafood does not mean the individual is allergic to all seafood; so it is highly recommended to test distinct allergens (Prester 2015).

6. Environmental Contaminants

Large-scale industrial activities release several heavy metals that are washed into both fresh and salt water reservoirs, where they accumulate through the food chain and return to humans in the form of contaminated seafood. Exposure of fish to toxins like methyl mercury, the cause of Minamata Disease in Japan; and other heavy metals like PCBs; organochlorides; pesticides; and radioactive waste can be avoided by removing the skin and trimming the fat (Vandermeersch et al. 2015). Mercury occurs in the environment and is transformed by bacteria in water to an organic form called methyl mercury (MeHg). It is the most toxic form of mercury, and it can accumulate in larger and older fish. In special groups of populations including pregnant women, breastfed women, and young children, the poisoning occurs primarily. Large marine fishes, like tuna, shark, swordfish, tilefish, and king mackerel are suggested to be consumed less to reduce mercury exposure. Most of the commonly consumed seafood is low in mercury; smaller and short lived species such as salmon, pollock, shrimp, catfish, or shellfish are all popular low mercury choices (Cano-Sancho et al. 2015).

6.1 Bacteria

Historically, the genus *Vibrio* is implicated in seafood poisoning. *Vibrio cholerae*, the most infamous, is often acquired from unhygienic food handling at home and on the street. Other species include *V. parahaemolyticus*, an organism that causes diarrhea, and *V. vulnificus*, which can cause fatal septicemia in a person with liver disease or an immunocompromised host (Janda et al. 2015). *Listeria monocytogenes*, on the surface of fish, may cause severe illness to a neonate born through an infected mother. *Salmonella* and *Aeromonas* species are also a risk through the consumption of seafood. In Latin America, ceviche (raw fish) is a common source of seafood poisoning. Since the organism requires an alkaline medium to grow, the addition of citrus juice (an acid) will reduce, but not eliminate, the risk (Iwamoto et al. 2010).

Seafood-borne diseases caused by bacterial toxins are generally associated with improper food preparation and storage. Scombroid poisoning is the most common phenomenon. *Clostridium botulinum* type E produces a toxin on smoked fish, fish eggs, and uneviscerated and salted whitefish (Reddy et al. 1997) *Staphylococcus aureus* incorporates a toxin on improperly stored seafood, especially if the fish is garnished with cream sauces or mayonnaise (Vázquez-Sánchez et al. 2012).

6.2 Viruses

Seafood harvested in contaminated water is a major source for seafood-borne sickness. Shellfish harvested in waters contaminated with inadequately treated sewage are extremely efficient vectors of seafood pathogens, such as, Hepatitis A. It is the most common cause of seafood-associated hepatitis. Small, round, structured viruses (such as Norwalk-like viruses), which is the causative agent for gastroenteritis, are not eliminated completely by cooking of shellfish (Hall et al. 2014).

6.3 Parasites

Many parasite infections occur from ingestion of inadequately cooked fish. *Anisakiasis*, a roundworm, is acquired through the consumption of raw fish, especially cod, herring, mackerel, and salmon. It causes abdominal pain and eosinophilia in symptomatic individuals. Diphyllobothriasis due to fish tapeworm has been reported with the ingestion of raw Pacific salmon and gefilte fish (Vidaček 2014).

References

Cano-Sancho, G., Sioen, I., Vandermeersch, G., Jacobs, S., Robbens, J., Nadal, M. et al. 2015. Integrated risk index for seafood contaminants (IRISC): Pilot study in five European countries. Environ. Res. 2015 Nov 143(Pt B): 109–15.

CDC. 1996. From the Centers for Disease Control and Prevention. Tetrodotoxin poisoning associated with eating puffer fish transported from Japan—California. JAMA. 1996 Jun 5. 275(21): 1631.

CDC. 2011. Paralytic shellfish poisoning southeast Alaska, May–June 2011. MMWR Morb. Mortal. Wkly. Rep. 2011 Nov 18. 60(45): 1554–6.

Chen, Tai-Yuan, Cheng-Hong Hsieh, Deng-Fwu Hwang. 2016. Development of standardized methodology for identifying toxins in clinical samples and fish species associated with tetrodotoxin-borne poisoning incidents. Journal of Food and Drug Analysis 24: 9–14.

Chan, T.Y. 2013. Severe bradycardia and prolonged hypotension in ciguatera. Singapore Med. J. 2013 Jun 54(6): e120–2.

Clark, R.F., Williams, S.R., Nordt, S.P. and Manoguerra, A.S. 1999. A review of selected seafood poisonings. Undersea Hyperb. Med. 26(3): 175–84.

Cohen, N.J., Deeds, J.R., Wong, E.S., Hanner, R.H., Yancy, H.F. and White, K.D. 2009. Public health response to puffer fish (Tetrodotoxin) poisoning from mislabeled product. J. Food Prot. 2009 Apr 72(4): 810–7.

de Haro, L. and Pommier, P. 2006. Hallucinatory fish poisoning (ichthyoallyeinotoxism): two case reports from the Western Mediterranean and literature review. Clin. Toxicol. (Phila) 44(2): 185–8.

Dickey, R.W. and Plakas, S.M. 2010. Ciguatera: a public health perspective. Toxicon. 2010 Aug 15. 56(2): 123–36.

Etkind, P., Wilson, M.E., Gallagher, K. and Cournoyer, J. 1987. Bluefish-associated scombroid poisoning. An example of the expanding spectrum of food poisoning from seafood. JAMA. 1987 Dec 18. 258(23): 3409–10.

Fukuda, A. and Tani, A. 1941. Records of puffer poisonings. Report 3. Nippon Igaku Oyobi Kenko Hoken (3528): 7–13.

Furey, A., O'Doherty, S., O'Callaghan, K., Lehane, M. and James, K.J. 2010. Azaspiracid poisoning (AZP) toxins in shellfish: toxicological and health considerations. Toxicon. 2010 Aug 15. 56(2): 173–90.

Hall, A.J., Wikswo, M.E., Pringle, K., Gould, L.H. and Parashar, U.D. 2014. Vital signs: foodborne norovirus outbreaks d-United States, 2009–2012. Morb. Mortal. Wkly. Rep. June 2014 75(22): 491–495.

Helfrich, P. and Banner, A.H. 1960. Hallucinatory mullet poisoning: a preliminary report. J. Trop. Med. Hyg. 1960 Apr 63: 86–9.

Hungerford, J.M. 2010. Scombroid poisoning: a review. Toxicon. 2010 Aug 15. 56(2): 231–43.

Isbister, G.K. and Kiernan, M.C. 2005. Neurotoxic marine poisoning. Lancet. Neurol. 2005 Apr 4(4): 219–28.

Iwamoto, M., Ayers, T., Mahon, B.E. and Swerdlow, D.I. 2010. Epidemiology of seafood associated infections in the United States. Clin. Microbiol. Rev. 23: 399–411.

James, K.J., Fidalgo Sáez, M.J., Furey, A. and Lehane, M. 2004. Azaspiracid poisoning, the food-borne illness associated with shellfish consumption. Food Addit. Contam. 2004 Sep 21(9): 879–92.

James, K.J., Carey, B., O'Halloran, J., van Pelt, F.N. and Skrabáková, Z. 2010. Shellfish toxicity: human health implications of marine algal toxins. Epidemiol. Infect. 2010 Jul 138(7): 927–40.

Janda, J. Michael, Anna E. Newton and Cheryl A. Bopp. 2015. Vibriosis. Clin. Lab. Med. 35: 273–288.

Jantschitsch, C., Kinaciyan, T., Manafi, M., Safer, M. and Tanew, A. 2011. Severe scombroid fish poisoning: an underrecognized dermatologic emergency. J. Am. Acad. Dermatol. 2011 Jul 65(1): 246–7.

Kanchanapongkul, J. 2008. Tetrodotoxin poisoning following ingestion of the toxic eggs of the horseshoe crab Carcinoscorpius rotundicauda, a case series from 1994 through 2006. Southeast Asian J. Trop. Med. Public Health 2008 Mar 39(2): 303–6.

Lehane, L. and Lewis, R.J. 2000. Ciguatera: recent advances but the risk remains. Int. J. Food Microbiol. 2000 Nov 1. 61(2-3): 91–125.

Lehane, L. 2001. Paralytic shellfish poisoning: a potential public health problem. Med. J. Aust. 2001 Jul 2. 175(1): 29–31.

Meier, J. and White, J. 1995. Handbook of Clinical Toxicology of Venoms and Poisons. 1st Ed. Boca Raton: CRC press.

Mowry, J.B., Spyker, D.A., Brooks, D.E., McMillan, N. and Schauben, J.L. 2014. Annual report of the American Association of Poison Control Centers' National Poison Data System (NPDS): 32nd Annual Report. Clin. Toxicol. (Phila) 2015 Dec 53(10): 962–1147.

Narahashi, T. 2008. Tetrodotoxin: a brief history. Proc. Jpn. Acad. Ser. B Phys. Biol. Sci. 84(5): 147–54.

Noguchi, T. and Arakawa, O. 2008. Tetrodotoxin—distribution and accumulation in aquatic organisms, and cases of human intoxication. Mar. Drugs 6(2): 220–42.

Onuma, Y., Satake, M., Ukena, T., Roux, J., Chanteau, S. and Rasolofonirina, N. 1999. Identification of putative palytoxin as the cause of clupeotoxism. Toxicon. 1999 Jan 37(1): 55–65.

Perl, T.M., Bedard, L., Kosatsky, T., Hockin, J.C., Todd, E.C. and Remis, R.S. 1990. An outbreak of toxic encephalopathy caused by eating mussels contaminated with domoic acid. N. Engl. J. Med. 1990 Jun 21. 322(25): 1775–80.

Prester, L. 2015. Seafood allergy, toxicity, and intolerance: A review. J. Am. Coll. Nutr. 2015 Aug 7: 1–13.

Randall, J. 2005. Review of clupeotoxism, an often fatal illness from the consumption of clupeoid fishes. Pacific Science 2005 Jan 59(1): 73–77.

Reddy, N.R., Solomon, H.M., Yep, H., Roman, M.G. and Rhodehamel, E.J. 1997. Shelf life and toxin development by *Clostridium botulinum* during storage of modified atmosphere-packaged fresh aquaculture salmon fillets. J. Food Prot. 9: 1055–1063.

Schnorf, H., Taurarii, M. and Cundy, T. 2002. Ciguatera fish poisoning: a double-blind randomized trial of mannitol therapy. Neurology 2002 Mar 26. 58(6): 873–80.

Schwarz, E.S., Mullins, M.E. and Brooks, C.B. 2008. Ciguatera poisoning successfully treated with delayed mannitol. Ann. Emerg. Med. 2008 Oct 52(4): 476–7.

Stewart, I., Lewis, R.J., Eaglesham, G.K., Graham, G.C., Poole, S. and Craig, S.B. 2010. Emerging tropical diseases in Australia. Part 2. Ciguatera fish poisoning. Ann. Trop. Med. Parasitol. 2010 Oct 104(7): 557–71.

Taylor, S.L. et al. 1984. Toxicology of Scombroid poisoning. Seafood Toxins. American Chemical Society. Washington DC 417–430.

Taylor, S.L. 1986. Histamine food poisoning: toxicology and clinical aspects. Crit. Rev. Toxicol. 17(2): 91–128.

Vandermeersch, G., Lourenço, H.M., Alvarez-Muñoz, D., Cunha, S., Diogène, J., Cano-Sancho, G. et al. 2015. Environmental contaminants of emerging concern in seafood—European database on contaminant levels. Environ. Res. 2015 Nov 143(Pt B): 29–45.

Vázquez-Sánchez, D., López-Cabo, M., Saá-Ibusquiza, P. and Rodríguez-Herrera, J.J. 2012. Incidence and characterization of *Staphylococcus aureus* in fishery products marketed in Galicia (Northwest Spain). Int. J. Food Microbiol. 157(2): 286–296.

Vidaček, S. 2014. Seafood. Risk and controls in the food supply chain. pp. 195–197. *In*: Yasmine Motarjemi and Huub Lelieveld (eds.). Book, Food Safety Management: A Practical Guide for the Food Industry. Elsevier Inc.

Wu, J.-Y., Zheng, L. and Wang, J.-H. 2005. Contamination of shellfish from Shanghai seafood markets with paralytic shellfish poisoning and diarrhetic shellfish poisoning toxins determined by mouse bioassay and HPLC. Food Additives & Contaminants 22(7).

Xu, X.M., Yu, X.W., Lu, M., Huang, B.F. and Ren, Y.P. 2015. Study of the matrix effects of tetrodotoxin and its content in cooked seafood by liquid chromatography with triple quadrupole mass spectrometry. J. Sep. Sci. 2015 Oct 38(19): 3374–82.

2

Microbial Safety and Quality Control on Seafood

Chiraporn Ananchaipattana[1],* and *Pongtep Wilaipun*[2]

1. Introduction

Seafood includes finfish, crustaceans (shrimp, lobster, crabs), and mollusks harvested from marine waters. In general, they are rich in protein and non-protein nitrogenous compounds (NPN); their fat content varies with type and season. Except for mollusks, they are very low in carbohydrates; mollusks contain about 3% glycogen (Biber and Bhunia 2008). The microbial population in these products varies greatly with the pollution level and temperature of the water. Bacteria from many groups, as well as viruses, parasites, and protozoa, can be present in the raw materials. Muscles of fish and shellfish are sterile, but scales, gills, and intestines harbor microorganisms. Finfish and crustaceans can have $10^{3–8}$ bacterial cell/g. During feeding, mollusks filter large volumes of water and can thus concentrate bacteria and viruses. Products harvested from marine environments can have halophilic vibrios, as well as *Pseudomonas*, *Alteromonas*, *Flavobacterium*, *Enterococcus*, *Micrococcus*, coliforms, and pathogens such as *Vib. parahaemolyticus*, *Vib. vulnificus*, and *Cl. botulinum*

[1] Rajamangala University of Technology Thanyaburi, 39 Village No. 1 Rangsit-Nakornnayok Road, Klong 6, Thanyaburi, Pathumthani, 12110 Thailand.
[2] Faculty of Fisheries, Kasetsart University, Chatuchak, Bangkok 10900, Thailand.
  E-mail: ffisptw@ku.ac.th
* Corresponding author: chiraporn_a@rmutt.ac.th

type E. Fish and shellfish harvested from water polluted with human and animal waste can contain *Salmonella*, *Shigella*, *Cl. perfringens*, *Vib. cholera*, and hepatitis A and Norwalk-like viruses. They can also contain opportunistic pathogens such as *Aeromonas hydrophila* and *Plesiomonas shigelloides*. As many of the bacterial species are psychrotrophs, they can grow at refrigerated temperature. Pathogens can remain viable for a long time during storage. Microbial loads are greatly reduced during their subsequent heat processing to produce different products.

Seafood consumption has increased worldwide in recent decades, reaching a high during the past decade (Iwamoto et al. 2010). However, along with the nutrients and benefits derived from seafood consumption come the potential risks of eating contaminated seafood. Chemicals, metals, marine toxins, and infectious agents have been found in seafood. Infectious agents associated with food-borne illnesses include bacteria, viruses, and parasites, and the illnesses caused by these agents range from mild gastroenteritis to life-threatening syndromes. Seafood is responsible for an important proportion of food-borne illness and outbreaks in EU, Asia, USA, and worldwide. Although seafood is also an important vehicle for marine toxins and chemical contamination, they have been described in Chapters 9 to 13 in this book.

2. Microorganisms in Seafood

2.1 Spoilage microorganisms

Fish: Fish harvested from saltwater are susceptible to spoilage through autolytic enzyme actions, oxidation of unsaturated fatty acids, and microbial growth. Protein hydrolysis by autolytic enzymes (proteinases) is predominant if the fish are not gutted following catch. Microbial spoilage is determined by the microbial types, their level, fish environment, fish types, methods used for harvest, and subsequent handling. Fish tissues have high levels of NPN compounds (free amino acids trimethylamine oxide, and creatinine), peptides, and proteins, but almost no carbohydrates; the pH is generally above 6.0. Gram-negative aerobic rods, such as *Pseudomonas* spp., *Acinetobacter*, *Moraxella*, and *Flavobacterium*, and facultative anaerobic rods, such as *Shewanella*, *Alcaligenes*, *Vibrio*, and coliforms, are the major spoilage bacteria. However, because of the relatively shorter generation time, spoilage by psychrotrophic *Pseudomonas* spp. predominates under aerobic storage at both refrigerated and slightly higher temperature. On the other hand, fish stored under vacuum or CO_2 conditions, lactic acid bacteria (including *Enterococcus*) can become predominant because growth of aerobic spoilage bacteria is prevented. However, anaerobic and facultative anaerobic bacteria can grow, including lactic acid bacteria. Under refrigeration, products have a relatively long shelf life due to slower growth of spoilage bacteria.

Gram-negative rods initially metabolize the NPN compounds by decay (oxidation), followed by putrefaction to produce different types of volatile compounds such as NH_3, trimethylamine, histamine, putrescine, cadaverine, indoles, H_2S, mercaptans, dimethyl, and volatile fatty acids. Proteolytic bacterial species also produce extracellular proteinases that hydrolyze fish proteins and supply peptides and amino acids for further metabolism by spoilage bacteria. The volatile compounds produce different types of off-odors such as stale, fishy (due to trimethylamine), and putrid. Bacterial growth is also associated with slime production, discoloration of grills and eyes (in whole fish), and loss of muscle texture (soft due to proteolysis). Oxidation of unsaturated fatty acids is also high in fatty fish (Biber and Bhunia 2008). Salted fish, especially lightly salted fish, are susceptible to spoilage by halophilic bacteria, such as *Vibrio* (at low temperature) and *Micrococcus* (at higher temperature). Smoked fish, especially with lower aW, inhibit growth of most bacteria. However, molds can grow on the surface.

Minced fish flesh, surimi, and seafood analogs prepared from fish tissues generally have high initial bacterial levels due to extensive processing (ca. 10^{5-6}/g). The type include those present in fish and those that get in during processing. These products, such as fresh fish, can be spoiled rapidly by Gram-negative rod unless frozen quickly or used soon after thawing. Canned fish (tuna, salmon, and sardines) are given heat treatment to produce commercially sterile products. They can be spoiled by thermophilic sporeformers unless proper preservation and storage conditions are maintained (Biber and Bhunia 2008).

Crustaceans: Microbial spoilage of shrimps is more prevalent than that of crabs and lobsters. Whereas crabs and lobsters remain alive until they are processed, shrimps die during harvest. The flesh of crustaceans is rich in NPN compounds (amino acids, especially arginine, trimethylamine oxide), contains ca. 0.5% glycogen, and has a pH above 6.0. The predominant microflora is *Pseudomonas* and several Gram-negative rods. If other necessary factors are present, the nature of spoilage is quite similar to that in fresh fish. Microbial spoilage of shrimp is dominated by odor changes due to production of volatile metabolites of NPN compounds, slime production, and loss of texture (soft) and color. If the shrimps are processed and frozen rapidly, the spoilage can be minimized. Lobsters are frozen following processing or sold live, and thus are not generally exposed to spoilage conditions. Crabs, lobster, and shrimps are also cooked to extend their shelf life. However, they are subsequently exposed to conditions that cause post-heat contamination and then stored at low temperature (refrigerated and frozen). Blue crabs are streamed under pressure, and the meat is picked and marketed as fresh crab meat. To extend shelf life (and safety), the meat is also heat processed (85°C for 1 min) and stored at refrigerated temperature. Under refrigerated conditions, they have a limited shelf life because of the growth of surviving bacteria and post-harvest contaminants (Biber and Bhunia 2008).

Mollusks: As compared with fish and crustaceans, oyster, clam, and scallop meats are lower in NPN compounds, but higher in carbohydrates (glycogen 3.5–5.5%), with pH normally above 6.0. The mollusks are kept alive until processed (shucked); thus, microbiological spillage occurs only after processing. The resident microflora is predominantly *Pseudomonas* and several other Gram-negative rods. During refrigerated storage, microorganisms metabolize both NPN compounds and carbohydrates. Carbohydrate can be metabolized to produce organic acids by lactic acid bacteria (*Lactobacillus* spp.), enterococci, and coliforms, thereby lowering the pH. Breakdown of nitrogenous compounds primarily by *Pseudomonas* and *Vibrio*, especially at refrigerated temperature, results in production of NH_3, amines, and volatile fatty acids (Biber and Bhunia 2008).

3. Pathogenic Bacteria Associated with Seafood

In 2010, an epidemiological study (1973 to 2006) described that there were 188 seafood-borne outbreaks occurred in the United States, of which 76.1% were bacterial, 21.3% were viral, and 2.6% were parasitic (Iwamoto et al. 2010). In recent years, outbreaks of infections were reported in seafood harvested from warm waters that had not been reported previously. Water quality, feeding habits, and diseases can change the normal microbial types and level (Elbashir et al. 2018). Pathogens such as *Vibrio parahaemolyticus*, *Vibrio Vulnificus*, and *Vibrio Cholera* are of major concern from these sources. *Vibrio vulnificus* and *Vibrio parahaemolyticus* are ubiquitous bacterial pathogens found naturally in marine and estuarine waters. *V. cholerae*, unlike most other vibrios, can survive in freshwater environments (CA 2010). The incidents and levels of vibrios present in marine organisms are greatly affected by water temperature, as they multiply rapidly between 20°C and 40°C. Due to the halophilic nature and the marine source of these pathogens, raw seafood is naturally contaminated and is the main vehicle of infection. *V. parahaemolyticus* and *V. vulnificus* occupy a similar ecological niche, but have different disease symptoms, growth temperatures, and salt tolerances. *V. vulnificus* does not tolerate low temperatures or high salinity (FAO/WHO 2005, Martinez-Urtaza et al. 2010).

Contamination of water and sediments is due to the presence of naturally occurring pathogens such as *Vibrio* spp., some species of *Aeromonas*, spore of *C. botulinum* type F, or enteric bacteria such as non-typhi *Salmonella* and *Campylobacter*. Cross contamination of seafood and seafood production are caused by processing, transportation, and storage. The major concern of bacterial agents in seafood include *Vibrio* and *Salmonella* and minor concern agents include *Shigella, Listeria monocytogenes, Clostridium botulinum,* and staphylococcal enterotoxin C (SEC) and staphylococcal enterotoxin A (SEA) (Hadler 1991).

4. Sources and Transmission of Pathogens in Seafood

Seafood includes mollusks (e.g., oysters, clams, and mussels), finfish (e.g., salmon and tuna), marine mammals (e.g., seal and whale), fish eggs (roe), and crustaceans (e.g., shrimp, crab, and lobster). Some seafood commodities are naturally more risky than others owing to many factors, including the nature of the environment from which they come, their mode of entry, the harvesting season, and how they are prepared and consumed. Fish, mollusks, and crustaceans can acquire pathogens from various sources. All seafood can be susceptible to surface or tissue contamination originating from the marine environment. Bivalve mollusks feed by filtering large volumes of seawater. During this process, they can accumulate and concentrate pathogenic microorganisms that are naturally present in harvest waters, such as Vibrios. Contamination of seafood by pathogens with a human reservoir can occur when growing areas are contaminated with human sewage. Outbreaks of seafood-associated illness linked to polluted waters have been caused by calicivirus, hepatitis A virus, and *Salmonella enterica* serotype Typhi (Iwamoto et al. 2010). Identified sources of seafood contamination have included overboard sewage discharge into harvest areas, illegal harvesting from sewage-contaminated waters, and sewage runoff from points inland after heavy rains or flooding. Additionally, seafood may become contaminated during handling, processing, or preparation. Contributing factors may include storage and transportation at inappropriate temperatures, contamination by an infected food handler, or cross-contamination through contact with contaminated seafood or seawater. Adequate cooking kills most pathogens; however, unlike other foods, such as meat and poultry, that are usually fully cooked, seafood is often consumed raw or prepared in ways that do not kill organisms.

5. Bacterial Pathogens of Major Concern

***Vibrio* spp.:** *Vibrio* spp. are Gram-negative bacteria that are rod-shaped and curved, halophilic, facultative anaerobic, non-spore forming, motile with polar flagella and sheath, and oxidase positive. Their natural habitat is in estuarine and coastal areas, where they are found free living in water, sediments, plankton, and nearly all flora and fauna found in coastal environments (Scharer et al. 2011, Thompson et al. 2005). The major human pathogenic *Vibrio* species are *V. parahaemolyticus, V. vunificus*, and *V. cholera*

(Parveen and Tamplin 2013, Nishibuchi and DePaola 2005). They constitute a significant human health hazard causing outbreak and sporadic food-borne illnesses associated with the consumption of raw or undercooked contaminated seafood (Scallan et al. 2011, Nishibuchi and DePaola 2005, Elbashir et al. 2018).

V. parahaemolyticus is recognized as one of the leading etiologic pathogens that caused sporadic infections and outbreaks of gastroenteritis in the U.S. in 1997 (CDC 1998). The pathogenicity of *V. parahaemolyticus* is usually associated with the presence of two principal virulence genes; the thermostable direct hemolysin (*tdh*), and/or TDH-related hemolysin (*trh*) (Scharer et al. 2011). The presence of predisposing factors, such as use of antacid medications, absence of gastric hydrochloric acid (achlorhydria), and partial or complete gastrectomy, increase the risk of illness (Nishibuchi and DePaola 2005, Dennis and Gage 2004). Consumption of raw or undercooked oysters harvested from the Pacific Northwest and Texas caused outbreaks of *V. parahaemolyticus* between 1997 and 1998 (CDC 1998). In 2013, the CDC reported a multistate, shellfish-associated *V. parahaemolyticus* (Pacific North West strain) outbreak, in which 28 illnesses were detected in nine states. Other outbreaks induced by the same strain were reported along the U.S. Atlantic Coast, causing illnesses in 104 persons from 13 states during the summer (CDC 2013a) (Elbashir et al. 2018).

Vibrio vulnificus biotype 1 is pathogenic to humans through seafood consumption or wound, however, biotypes 2 and 3 were responsible for only a few direct wound infections (Nishibuchi and DePaola 2005). Multiple virulence factors such as ability to acquire iron, capsule (encapsulated phase variant), proteins in type IV pilus and type II, the hemolysin encoded by the *vvh* (cytotoxin-hemolysin), zinc metalloprotease, and transmembrane regulatory protein (*Tox*R) may contribute to *V. vulnificus* infection (Nishibuchi and DePaola 2005). Metalloprotease has a significant role in skin lesions produced by *V. vulnificus* infections. It is a cardinal pathogenic factor, and is assigned as *V. vulnificus* protease (*VVP*) (Miyoshi 2006). Liver diseases, hemochromatosis, and immune-compromised patients are predisposed to severe *V. vulnificus* illness. Infections may develop through open wounds during seafood processing, fishing, and leisure swimming (Oliver 2006, 2005). Septicemia caused by *V. vulnificus* is often fatal, with an average mortality rate of 37% (Scallan et al. 2011). Two hundred seventy four illnesses of *V. vulnificus* in the U.S. were transmitted through consumption of undercooked seafood (Oliver 2005).

Vibrio cholerae and its close relative *V. mimicus*, are found in fresh and brackish waters. Both are typical Vibrio species because they are capable of growing in laboratory media without adding sodium chloride (Parveen and Tamplin 2013, Nishibuchi and DePaola 2005). The disease is characterized by profuse diarrhea, dehydration, and electrolyte imbalance (Parveen and Tamplin 2013). Different virulent agents are associated with their pathogenicity; of them the cholera toxin and toxin coagulated pilus (TCP) are the most important pathogenicity factors associated with epidemic cholera (Rivera et al. 2001). Only toxigenic strains of serogroups O1 and O139 have caused widespread epidemics and are reportable to the World Health Organization (WHO) as "cholera" (WHO 2012). No major outbreaks of this disease have occurred in the U.S. since 1911. The sporadic cases between 1973 and 1991 were caused by a non O1 *Vibrio cholerae*, and were associated with the consumption of raw shellfish or of shellfish either improperly cooked or re-contaminated after proper cooking (FDA 2012b, Elbashir et al. 2018).

Antimicrobial drug resistance in *Vibrio* spp. can develop through mutation or through acquisition of resistance genes transferred between bacteria (Wozniak et al. 2009, Burrus et al. 2006). In Louisiana, Han et al. (2007) found that among all tested isolates recovered from Louisiana Gulf oysters (168 *V. parahaemolyticus* and 151 *V. vulnificus*), nearly all isolates were susceptible to the antimicrobials tested. However, *V. parahaemolyticus* showed weak resistance for ampicillin and exerted significantly higher minimal inhibitory concentrations (MICs) for cefotaxime, ciprofloxacin, and tetracycline, compared to *V. vulnificus*. However, both *V. parahaemolyticus* and *V. vulnificus* did demonstrate susceptibility patterns to therapeutic antimicrobials such as: doxycycline, tetracycline, aminoglycides, and cephalosporins (Shaw et al. 2014). Baker-Austin et al. (2009) reported multiple antimicrobial resistance of *V. vulnificus* to doxycycline, tetracycline, aminoglycosides, and cephalosporins (Elbashir et al. 2018).

Salmonella: Salmonellae are Gram-negative, small, rod-shaped, facultative anaerobic bacteria, usually motile with peritrichous flagella, and are catalase positive, oxidase negative, and produce gas from glucose. Salmonella is a non-lactose fermenter (Clark and Barret 1987). Salmonella produces enterotoxins, and causes inflammatory reaction and diarrhea. Symptoms often start 12–72 hours after the ingestion of contaminated food. Onset of acute symptoms may last for 1–2 days or more, depending on the individual host, ingested dose, and strains (Jay et al. 2005). Salmonella virulence depends on different determinants, such as *sod*C1 which is responsible for resisting killing by macro phages, *mgt*BC that allows the organism to nourish intracellularly, *spv*B that interferes with actin polymerization, and SPI-2 that alters vesicular trafficking (Andrews and Baumler 2005, Elbashir et al. 2018).

Salmonella infections from consumption of seafood products are most commonly associated with raw, undercooked, and/or poorly cooked finfish and crustaceans (NACMCF 2008). Salmonella cross-contamination of seafood may take place during processing and storage. However, this can be prevented by good manufacturing practices (GMPs) and HACCP (CDC 2014a, Elbashir et al. 2018).

Salmonellosis is a worldwide health hazard and is the second leading cause of food-borne illness in the U.S. (Scallan et al. 2011, McCoy et al. 2011). Between 1990 and 1998; the U.S. Food and Drug Administration (FDA) investigated the incidence of Salmonella in 11,312 imported and 768 domestic seafood. The incidence of the pathogen was 7.2% and 1.3%, respectively (CDC 2014a). The incidence was higher in raw seafood tested in an FDA study (10% and 2.8%, respectively) (Heinitz et al. 2000). The risk of seafood-borne salmonellosis can be reduced by using specific control measures focused on: monitoring harvest water pollution levels; employing best management hygienic production practices; establishing biosecurity measures in production areas; and ensuring appropriate cooking temperatures, appropriate storage temperatures, and prevention of cross-contamination during harvesting, handling, and processing (FDA 2011). *Salmonella Bareilly* was confirmed to cause outbreaks involving over 140 illnesses in Washington, D.C. and 20 states across the U.S. on April 19, 2012. On April 11, 2012 another outbreak caused by the same *Salmonella* sp. was also recognized in 21 states including Washington, D.C. and involved 190 illnesses. A third outbreak was also confirmed, with products coming from a different distributor. Frozen, raw yellow fin-tuna from India, labeled as Nakaochi Scrape grades AA or AAA imported from India, was implicated as the possible sources of all outbreaks. *Salmonella* Nchanga was confirmed to be the cause of an outbreak in 5 states involving 10 illnesses and the source was the same imported Indian tuna (FDA 2012a, Elbashir et al. 2018). Distribution trends of *Salmonella* serovars in various countries and predominant *Salmonella* serovars in seafood are shown in Tables 2.1 and 2.2, respectively.

Reports on antimicrobial resistance in *Salmonella* spp. started in the early 1960s. Salmonellae resistant to erythromycin and penicillin, and to some degree to nitrofurantoin was discovered in fish from China (Broughton and Walker 2009). Resistance to sulfamethizol and carbenicillin, and a relatively moderate resistance to nalidixic acid and oxytetracycline was identified in fish from India (Kumar 2009). In 2001, the FDA conducted a study of *Salmonella* isolates recovered from imported foods, including seafood, and found that some of the isolates were resistant to tetracycline, sulfamethoxazole streptomycin, nalidixic acid, and trimethoprim/sulfamethoxazole. Others were resistant to ampicillin, chloramphenicol, gentamicin, kanamycin, nalidixic acid, sulfamethoxazole, tetracycline, and trimethoprim/sulfamethoxazole (Zhao et al. 2006).

Table 2.1 Epidemiological data about salmonellosis outbreaks and Salmonella occurrence in seafood.

Geographic area/country	Salmonellosis outbreaks linked to seafood (% of total salmonellosis outbreaks)	Product type with prevalence (%)	Reference
US	Mollusks (4 outbreaks, 32 cases) Crutaceans (4 outbreaks, 81 cases) Finfish (10 outbreaks, 261 cases) Total: 18 outbreaks, 374 cases of salmonellosis between 1973 and 2006	Not reported	Iwamoto et al. (2010)
EU	Fish and fish products (5 outbreaks, 1%) Crustaceans, shellfish, mollusks and products (7 outbreaks, 1.4%)	Fish and fishery products (0.3) Crustaceans (0.5), live bivalve mollusks (0.9), molluskan shellfish (1.1)	EFSA (2010)
India		Clam (342) Mussel (31) Finfish (28.2) Shimp (26.7) Squid (17.3) Octopus (16.6) Oyster (12.5) Crab (9.6) Lobster (4.7) Total seafood (23)	Kumar (2009)
Japan	Fish and shellfish (1 outbreak, 1.5%)		Japan Food Poisoning Statistics Report (2009)
China		Seafood (20.8)	Yan et al. (2010)
Thailand		Open market shrimp (53) Seafood (24)	Minami et al. (2010) Ananchaipattana et al. (2012)
Morocco		Seafood (1.9)	Bouchrif et al. (2009)

Modified from: Amagliani et al. 2012.

6. Bacterial Pathogen of Minor Concern

Listeria monocytogenes: *Listeria monocytogenes* is a Gram-positive, rod shaped non-acid fast, non-spore-forming, catalase positive bacterium that ferments glucose producing lactic acid. The bacterium is ubiquitous in soil, water, animal excreta, and plants. The ability of the bacterium to accommodate a vast range of temperatures (1–45°C), and pHs (4.1–9.6),

Table 2.2 Predominant *Salmonella* serovars in seafood.

Geographic area/country	Product type	Serovar (number or % of isolates)	Reference
US	All tested products	*S.* Enteritidis (16%), *S.* Typhimurium (15%), *S.* Newport (10%), *S.* I 4, [5], 12: i: – (5%), *S.* Javiana (5%), *S.* Heidelberg (4%), and *S.* Montevideo (3%)	CDC (2009)
EU	Fish and fish products Crustaceans, shellfish, mollusks and products thereof	*S.* Enteritidis (1.7% of human case) *S.* Enteritidis (1.37%) and *S.* Typhimurium (0.15%)	EFSA (2010)
India	Seafood	*S.* Worthington (18 isolates), *S.* Weltevreden (13), *S.* Typhimurium (9), *S.* Enteritidis (9), *S.* Bareilly (7), *S.* Gallinarum (4) and *S.* Infantis (3) *S.* Weltevreden (22), *S.* Rissen (20), *S.* Typhimurium (17) and *S.* Derby (16)	Kumar (2009)
Thailand	Seafood Water Open markets shrimp Seafood	*S.* Weltevreden (26%) *S.* Weltevreden (26%) *S.* Stanley *S.* Corvallis (5%) *S.* O4: i:- (5%)	Bangtrakulnonth et al. (2004) Minami et al. (2010) Ananchaipattana et al. (2012)

Modified from: Amagliani et al. 2012.

form biofilms, and remain viable for long periods promotes its broad distribution, colonization, and adaptation to various environments (Jay et al. 2005). Listeriosis caused by *Listeria* spp. is recognized as a food-borne disease of increasing public health and food safety concern since early 1981. Ingestion is the main route of transmission to humans, mainly through the consumption of ready-to-eat foods (Miya et al. 2010). Data indicates that 93% of raw seafood is contaminated with less than 1 CFU/serving. However, raw and processed crab meats are the foremost incriminated seafood that transmits *Listeria*. Approximately 7.2% of raw seafood contains $1–10^3$ CFU/ serving, 1.2% contain $10^3–10^6$ CFU/serving and < 0.3% contain $> 10^6$ CFU/ serving (NACMCF 2008). Different *L. monocytogenes* serotypes were isolated from ready-to-eat minced tuna and fish roe products in Japan (Miya et al. 2010). Little information is available on the prevalence and sources of *L. monocytogenes* in ready-to-eat (RTE) seafood products, which passively affects the seafood industry. Pagadala et al. (2011) reported the presence of *Listeria* species and *L. monocytogenes* in raw crab (6%), cooked crab meat (0.1%), and processing environments (10%), and concluded that contamination may be caused during processing. The incidence of presumptive *Listeria* detection was 19.5% in raw crab, 10.8% in crab meat, and 65.9% in the processing environment; among them *L. monocytogenes* incidences were 4.5%, 0.2%, and 2.8%, respectively (Pagadala et al. 2012). Markkula et al. (2005) reported a 4% incidence of *L. monocytogenes* in raw rainbow trout in Finland. Numerous studies have shown the antimicrobial resistance of *Listeria monocytogenes* to numerous compounds. *L. monocytogenes* isolated from popular seafood products sampled at the retail market and processing facilities were found to be resistant to penicillin, ampicillin, tetracycline, and vancomycin (Fallah et al. 2013). Yan et al. (2010) reported that *L. monocytogenes* showed antimicrobial resistance for ciprofloxacin, tetracycline, and streptomycin. Pagadala et al. (2012) also reported multimicrobial resistance among the tested *L. monocytogenes*. Their study showed that isolates were resistant to erythromycin, ciprofloxacin, and tetracycline, and were susceptible to gentamicin, sulphamethoxazole/trimethoprim, and kanamycin. In Mexico, Rodas-Suarez et al. (2006) reported multi-antimicrobial resistance of *L. monocytogenes* isolated from oysters, fish, and estuarine water to ampicillin (60.3%), erythromycin (30.9%), penicillin (57.4%), cefuroxime and cephalothin (13.4%), ceftazidime (67.6%), tetracycline (16.7%), dicloxacillin (9.7%), and trimethoprim-sulfamethoxazole (37.4%) (Elbashir et al. 2018).

***Shigella* spp.:** *Shigella* species are Gram-negative, rod-shaped, non-motile, oxidase-negative, and non-lactose fermenting bacteria, except *Shigella sonnei* which can ferment lactose after prolonged incubation. According to their somatic antigen O; *Shigella* species are serologically grouped into *Shigella dysenteriae* (group A), *Shigella flexneri* (group B), *Shigella boydii* (group C), and *Shigella sonnei* (group D) (Lampel 2005, O'Connel et al. 1995). Clinical features of shigellosis vary from watery loose stool to more severe symptoms, including fever, abdominal pain, tenesmus, and bloody diarrhea. Children may exhibit serious complications, such as toxic megacolon, Reiter's syndrome, and hemolytic uremic syndrome. *Shigella* species produce three toxins that cause bloody diarrhea in infected humans. *Shigella* enterotoxin 1 and 2 induce watery diarrhea (Scallan et al. 2011). *Shigella* species possess multi-virulence factors that allow them to adhere to and invade intestinal cells, survive stomach acid, overcome immune defenses, and secrete toxins. Human feces, not animal feces, are the fundamental source of food contamination (Iwamoto et al. 2010). Wang et al. (2011) reported that 32% of the seafood (shrimp, tilapia, and salmon) samples PCR tested positive for *Shigella*; however, none was recovered by culture. *Shigella dysenteriae* was isolated from two edible fish (*Priacanthus hamrur* and *Megalaspis cordyla*) in India (Sujatha et al. 2011). These species (39.6%) were also isolated from Nile tilapia in Kenya (David et al. 2009).

Clostridium botulinum: *Clostridium botulinum* is a Gram-positive, anaerobic, spore forming, bacillus. Foodborne botulism is an intoxication caused by the consumption of food contaminated by a neurotoxin produced by *C. botulinum*. Signs begin within 4 to 8 days, and symptoms include fatigue, dizziness, followed by double vision, and progressive difficulty speaking and swallowing, dyspnea, and muscle weakness (Aberoumand 2010). Four transmission categories including wound, infant, old, and immunocompromised people have been indicated as seafood or foodborne human botulism (CDC 2014b). Seven serologically classified toxins are produced by this organism; among them A, B, E, and F are responsible for most human food-borne botulism (Iwamoto et al. 2010). The majority of illnesses in seafood products have been due to temperature abuse, and/ or an inadequate preservation process (NACMCF 2008). Improper heating during processing may increase the incidence of the intoxication. *Clostridium botulinum* is relatively heat resistant, however, internal heating to 78°C for 1 min kills the vegetative state of the bacterium; the spore state is highly heat resistant. In North America, botulism outbreaks were linked to the ingestion of canned and fermented or salted seafood. *Clostridium botulinum* type E spores and toxins were commonly found in fish such as white fish, flounder, cod, rock fish, smoked fish, etc. (Iwamoto et al. 2010). It is capable of producing toxins even at low temperatures, such as 3.3°C. Modified atmosphere packaging of seafood (salmon, tilapia, etc.) has extended shelf life, especially when kept at 4°C. High temperatures increase the possibility of spoilage and augment bacterial spores' vegetation (Reddy et al. 1997, Elbashir et al. 2018).

Staphylococcus aureus: *Staphylococcus aureus* is a Gram-positive coccus and catalase positive chemotrophic bacterium. It produces Staphylococcus food poisoning (food intoxication syndrome), although it is not considered a problem with raw seafood (NACMCF 2008). *S. aureus* enterotoxins (SEs) are responsible for the pathogenicity and virulence of the bacterium. Nine different serological types of SEs (SEA to SEE and SEG to SEJ) have been shown to have emetic activity. SEA is the foremost Staphylococcus-related food poisoning (Pinchuk et al. 2010). In a recent survey, Consumer Reports tested 342 samples of frozen shrimp (284 raw and 58 cooked). Seven raw shrimp samples tested positive for methicillin resistant *S. aureus* (MRSA), and 11 samples of raw shrimp tested positive for illegal antibiotic residues. The 11 samples represented about 5% of the imported farmed samples tested. Residues of tetracycline, enrofloxacin, sulfamethoxazole, and sulfamethazine were found in the different samples (Consumer Reports 2015). Albuquerque et al. (2007) reported a multidrug resistance of *S. aureus* isolated from fish and attendants in a fish-stall in Brazil included resistance to ampicillin, cephalothin and ampicillin, chloramphenicol and ampicillin, lincomycin and erythromycin, lincomycin and ampicillin, and

Table 2.3 Contamination of food-borne pathogenic bacteria in some fresh fish and shellfish in Thailand.

Fish and shellfish	Food poisoning bacteria
Black tiger shrimp	*Vibrio alginolyticus* *Vibrio damsela* *Vibrio vulnificus* *Vibrio parahaemolyticus* *Aeromonas* spp.
Estuary grouper	*Vibrio parahaemolyticus*
Hybrid catfish (Hybrid clarias fish)	*Pseudomonas* spp. *Vibrio cholerae* *Aeromonas hydrophila* *Aeromonas sobria* *Salmonella* spp. *Staphylococcus* spp. *Edwardsiella tarda*
Tilapia	*Aeromonas hydrophila*
Blood cockle (Blood clam)	*Vibrio parahaemolyticus*
Oyster	*Vibrio parahaemolyticus*

Modified from: Virunhakul et al. 2004.

oxacillin. Intermediate resistance was reported for streptomycin, oxacillin, erythromycin, lincomycin, and ampicillin (Elbashir et al. 2018).

7. Emerging Significant Bacterial Species

Delftia acidovorans: *Delftia acidovorans* is an aerobic, nonfermenting Gram-negative bacillus. It is usually a nonpathogenic environmental organism, and is rarely clinically significant. *D. acidovorans*, formerly known as *Comamonas acidovorans* or *Pseudomonas acidovorans*, is found in soil, water, and the hospital environment. *Delftia acidovorans*, while common in the environment, is variously described as an opportunistic pathogen of humans, having been isolated from AIDS patients with pneumonia, and cases of nosocomial bacteraemia (Smith and Gradon 2003). In one noteworthy case, *D. acidovorans* was attributed to a bacteraemia acquired from inhalation of aerosols from a tropical fish tank in which the fish were dying (Smith and Gradon 2003).

A further possible risk associated with *D. acidovorans* as a zoonotic pathogen, as well as a potential aquatic animal pathogen, is its ability to degrade herbicide chemicals and use them as a carbon source (Müller et al. 1999). In areas of intensive rice culture where these chemicals are in use (Köck et al. 2010), there may be an unexpected enrichment for this bacterium in the soils and water of the rice fields, which are a common habitat for eels.

This may present an added risk for the health of field workers, as well as the aquatic animals present, when untreated water harboring potentially harmful bacteria is released from the rice paddies into the surrounding environment. *D. acidovorans* has been described as a normal member of the gut microbiota of fish (Sun et al. 2009), but it is not normally a strict pathogen of fish. However, the handling and overcrowding that the glass eels experienced in preparation for shipment provided ample stress to leave them immune compromised and enhanced their susceptibility to opportunistic infections caused by potential pathogens already present in the population. Bacteraemia due to *Pseudomonas* spp. possibly associated with exposure to tropical fish was described by Smith and Gradon (2003).

Edwardsiella tarda: *Edwardsiella tarda* is a motile, facultatively anaerobic, Gram-negative rod that is categorized as a member of the family Enterobacteriaceae. *E. tarda* is typically isolated from fresh or brackish water environments such as river mouths. It has also been isolated from the intestines of humans (after eating fresh water food sources such as catfish (Wyatt et al. 1979) or eels (Joh et al. 2011)) and from animals, including reptiles and freshwater fish. *E. tarda* rarely causes infections in humans. The colonization rate in humans ranges from 0.0073% in the Japanese (Onogawa et al. 1976) to 1% in Panamanians (Kournay et al. 1977). Approximately 80% of *E. tarda* infections in humans are characterized as gastroenteritis, and *E. tarda* is primarily isolated from stool samples. *E. tarda* should be categorized as a severe food- and water-borne infection, which results in high mortality for patients with liver cirrhosis (Hirai et al. 2015).

Legionella pneumophila: Legionnaire's disease is a respiratory illness transmitted solely by water and aerosols. The bacterium Legionella grows in natural waters, but also in distribution systems and water and air conditioning systems, and is inhaled through contaminated aerosols (Rose et al. 2001). Suzuki et al. (2002) described eight cases of community-acquired legionellas' pneumonia, and in one case the legionnaire disease was associated with the work of the patient at a fish market.

Plesiomonas shigelloides: Pl. *shigelloides* is a facultative anaerobic bacterium from the family Vibrionaceae which is known as a causative agent in water-borne diseases. *Plesiomonas* spp. was isolated from 1.5% of freshwater fishes investigated in Okayama prefecture from 1987 to 1990 (Nakajima et al. 1991). The symptoms of Pl. *shigelloides* infection are diarrhoea (watery or bloody), fever, vomiting, and abdominal pain (Wong et al. 2000).

Shigella **spp.:** The genus *Shigella* is specific host-adapted to humans and higher primates, and its presence in the environment is associated with faecal contamination. *Shigella* spp. isolates have been reported to survive for up to 6 months in water (Wachsmuth and Morris 1989). *Shigella* spp. is the cause of shigellosis (earlier name was bacillary dysentery), which is an infection of the gut. The great majority of cases of shigellosis are caused by direct person-to-person transmission of the bacteria via the oral-faecal route. Water-borne transmission is also important, especially where hygiene standards are low. However, food, including seafood (shrimp-cocktail, tuna salads) has also been the cause of a number of outbreaks of shigellosis. This has nearly always been as a result from contamination of raw or previously cooked foods during preparation by an infected, asymptomatic carrier with poor personal hygiene.

8. Viral Agents of Major Concern

Norovirus: Norovirus is a highly contagious RNA calicivirus causing infection in humans. Its clinical symptoms include nausea, vomiting, abdominal cramps, diarrhea, fever, and headache (Iwamoto et al. 2010). Besides bacteria, shellfish can also accumulate large numbers of viruses from sewage waters. The most important are noroviruses and hepatitis A. The most significant outbreak of hepatitis A infection occurred in Shanghai, China, in 1988, in which almost 300,000 cases were caused by consumption of clams harvested from a sewage-polluted area. The genus Norovirus, belonging to the family Caliciviridae, is considered the leading cause of nonbacterial human gastroenteritis in developed countries (Croci et al. 2012). As stated above, government authorities have programs for monitoring fecal pollution in water, but it is not a reliable means of determining the extent of viral contamination of shellfish. Current treatment regimens for placing live mollusks on the market (depuration and relaying), as is commonly practiced, do not effectively reduce noroviruses. The most effective measure to control infection by norovirus from raw mollusk consumption is to produce them in areas which are not fecally contaminated (EFSA 2012b).

Forty eight percent of human norovirus outbreaks were due to the consumption of contaminated shellfish (Alfano-Sosbey et al. 2012). Although norovirus genogroup I (GI) and genogroup II (GII) are infectious to humans (Woods et al. 2016, Bazzardi et al. 2014), genogroup II infection is the dominant type (Woods et al. 2016, Terio et al. 2010). Norovirus infections are considered the foremost cause of non-bacterial gastroenteritis due to the consumption of raw or partially-cooked shellfish, especially oysters, and exposure to contaminated water (Woods et al. 2016). Infected seafood handlers are potential sources of infection for both sporadic cases and outbreaks (Hall et al. 2012, Hall et al. 2014). Raising fish and shellfish in the vicinity of sewage-contaminated waters will increase the possibility of pathogen transmission to humans (Iwamoto et al. 2010). Richards (2014), reviewed that environment is a possible cause of such transmission. Mussels, oysters, and clams (Bazzardi et al. 2014, Terio et al. 2010), oysters (Woods et al. 2016, Webby et al. 2007), clams, crab, prawns, finfish (Anbazhagi and Kamatchiammal 2010) spider crabs, gooseneck barnacles, and blue crab (Sala et al. 2008) were the main reservoirs of norovirus due to their eating habits as filter feeders and their ability to concentrate the virus from contaminated water (Woods et al. 2016). Generally, 50% of the food-borne gastroenteritis in the U.S. and European Union (EU) were caused by norovirus infection (CDC 2012). Scallan et al. (2011) reported that 58% of the human gastroenteritis were linked to noroviruses in the U.S. Eight norovirus outbreaks were reported in the U.S. from 2009 to 2014. Seven of those outbreaks were associated with contaminated raw oysters, while the eighth was associated with partially cooked oysters. Outbreaks were in Tennessee, Mississippi, and Alaska (2009), Mississippi (2010), Washington and Pennsylvania (2011), Louisiana (2013), Massachusetts (2013), Louisiana (2014), and Washington (2014) (Woods et al. 2016, CDC 2012). Norovirus genogroup II was detected in the majority of the outbreaks (Woods et al. 2016). Among the 364 outbreaks in the U.S. from 2001 to 2008 that were caused by the consumption of a single food, mollusks were found to be the cause of 13% of them (Hall et al. 2012). Many norovirus outbreaks were reported worldwide due to the consumption of contaminated food. Vidal et al. (2005) reported 14 outbreaks caused by caliciviruses including norovirus, and were associated with seafood between 2000 and 2003 in Chile. Multiple seafood-related outbreaks of different strains of norovirus were reported in Australia (Webby et al. 2007). Norovirus was reported as the main etiologic agent of travelers' diarrhea in Mexico. Nine percent of stool samples collected from 320 U.S. students in Mexico were positive for this virus (Ajami et al. 2010). A study in Italy reported that 4.1% of shellfish samples over a 9 year period (2003 to 2011) were positive for both genogroups (Pavoni et al. 2013). High prevalence of norovirus was detected in different batches of shellfish and fishery products processed by a Belgian company (Dan et al. 2014). Norovirus is resistant to freezing, not highly resistant to free chlorine disinfection, and resistance to commercial disinfectants varies depending on the chemical nature of the disinfectant. Norovirus is sensitive to autoclaving, but stable in water and shellfish. It displays stability in depurated shellfish for 7 days (Cook et al. 2016, Elbashir et al. 2018).

9. Viral Agent of Minor Concern

Hepatitis A virus: Hepatovirus A or *Hepatitis A Virus* (HAV) is the non-enveloped virus species that belongs to the genus Hepatovirus and the family Picornaviridae. Vertebrates including human are its natural habitat (ICTV 2012). It transmitted through the fecal-oral route and causes acute, self-limiting liver disease (hepatitis). The incubation period of illness ranges between 14 and 28 days (WHO 2016), and sometimes reaches 45 days (Richards 2014). Symptoms vary from asymptomatic, unnoticeable in children less than 6 years, mild in most patients, to severe in old or immunocompromised groups. Patients may show clinical signs of fever, malaise, loss of appetite, abdominal pain, diarrhea, and jaundice (Ghasemian et al. 2016). The CDC's Advisory Committee on Immunization Practice, in 1996, recommended immunization and vaccination against HAV to risk groups (CDC 2007). HAV is an acute illness with low fatality. It is transmitted through ingestion of fecal contaminated seafood and/ or water, inadequate sanitation, and poor personal hygiene (WHO 2016, Richards 2014). Hepatitis A is a worldwide pandemic; about 1.4 million cases were reported worldwide annually. Fifty percent of the cases were reported in Asia (Hadler 1991). Raw clams were the seafood incriminated in the Shanghai's largest epidemic of HAV in 1988 in China. This epidemic was responsible for 292,301 cases and 47 deaths (Halliday et al. 1991). In the period between 1986 and 2012; forty six outbreaks of HAV were reported and linked to seafood vehicles such as oysters, clams, mussels, and cockles, worldwide (Bellou et al. 2013). Consumption of raw oysters was linked to an outbreak in Sweden in 1995 (629 cases) (La Rosa et al. 2012). Between 1996 and 1997, there were two major outbreaks of HAV in Puglia, Southern Italy. The numbers of cases per each outbreak were 5,673 and 5,382 cases; respectively (Lopalco et al. 1997, Malfait et al. 1996). Raw shellfish consumption was the seafood vehicle in both outbreaks (Chironna et al. 2002). In the United States, HAV ranged between intermediate and low outbreaks in the period between 1900 and 1950 (CDC 2007, Bell et al. 1998). In 1994, an outbreak of HAV infection with 26,796 reported cases was reported in the U.S. (CDC 2007, Elbashir et al. 2018).

10. Parasites of Concern

A large number of parasites infect fish, but only a few cause illnesses in humans: *Opisthorchiidae* and *Heterophyidae* (Class Trematodea, Subclass Digenea), *Anisakidae* and *Gnathostomidae* (Phylum Nematoda) and *Diphyllobothridae* (Class Cestoda) (Lima dos Santos and Howgate 2011). Humans acquire the fish-borne parasitic diseases through the consumption of infected raw, undercooked, or inadequately preserved fish (Vidaček 2014). Although many aquatic organisms may carry these parasites, human infections are mainly related to fish consumption. These parasites have different life cycles, with different organisms as primary, intermediate, and definitive hosts. Therefore, routes of human infections by these parasites are different. Infections by nematodes are caused by consumption of marine fish and cephalopods, mainly from open marine waters (only one case of farmed salmon infected by Anisakis has been reported). Infections by trematodes are related to consumption of farmed freshwater fish and crustaceans. More than 100 species of freshwater fishes belonging to 13 families, especially the Cyprinidae, and three species of freshwater shrimp can serve as a second intermediate host of liver flukes, which are trematodes of the highest public health concern. For cestodes, wild and farmed freshwater and marine fish living in cold water habitats can be intermediate hosts. In the worldwide picture of fish-borne parasitoses, cestodiasis (diphyllobothriasis) is considered a mild disease. Nematodiasis is also not as severe as trematodiases. However, the parasite borne illnesses can be severe, and the incidences are high in endemic areas (Lima dos Santos and Howgate 2011). The number of people currently infected with fish-borne trematodes exceeds 18 million (WHO 1995), but worldwide the number of people at risk, including those in developed countries, is more than half a billion. Infections by the liver flukes (trematodes Clonorchis and Opisthorchis) are a major health concern specifically in the Far East, Eastern Europe, and Southeast Asia. The public health significance of these diseases is increasing because of intensification of aquaculture, environmental damage, a lack of appropriate tools for control, links with poor sewage treatment and poverty (90% of world aquaculture is situated in Asia), and cultural traditions of eating raw or minimally processed fishery products (WHO 1995, 2004).

In Europe, most of the parasitic infestations of fish are related to nematodes in marine fish. In 2010, a 41% increase of parasitic infestation with Anisakis of fish (one case of squid) has been reported in the EU (compared to 2009). The reports were mostly for chilled fish and, in some cases, for frozen fish (RASFF 2011). The other health-related issue regarding Anisakis is allergy. Anisakis is the only parasite known to cause allergic-type reactions to sensitive individuals. The responsibility of the industry is to provide fish that have no visible parasitic larvae and to ensure that the fish do not pose a health risk to humans. Control strategies for industry to reduce the risk of helminthic infections include visually inspecting fish, by means such as candling, for parasites that are large enough to be detected visually (Huss et al. 2003). Freezing or heat treatments are the most effective processes used for the killing of parasitic larvae. However, only for nematodes temperature–time conditions of freezing and heating are well defined. The only data available for trematodes would seem to indicate a higher heat resistance of trematodes compared to nematodes. More research is also required on the survival of trematodes in edible fish tissues during traditional processing and preparation (WHO 1999). For the killing of A, simplex larvae, requirements include freezing at −20°C for not less than 24 hours at the core of the fishery products (or treatments which provide an equivalent level of health protection, like freezing at −35°C for at least 15 hours or at −15°C for at least 96 hours) and heat treatment at > 60°C for at least 1 minute (EFSA 2010). The freezing of fish to be consumed either raw or after mild processing (cold smoking, marinating) is compulsory in many European countries (EU 2004). For prevention and control of liver flukes, education campaigns are important for communicating the risk to consumers, who should be advised to consume only cooked fish. Environmental sanitation is also important because efficient control should only produce parasite-free fish. More recently, mass chemotherapy of people at risk in endemic areas was recommended as the most effective control strategy (WHO 2004).

11. Seafood-borne Diseases

Infection: The shellfish-borne bacterial infections from sewage waters may include infections with pathogenic bacteria such as *Salmonella* spp., *Shigella* spp. or *Escherichia coli*. Hazards also present *Yersinia enterocolitica* and *Campylobacter* spp. In the EU in 2010, the sources of the highest number of notifications reported for *Escherichia coli* were live bivalve mollusks (RASFF 2011). To control the risk of developing seafood-borne infections from sewage waters, regulatory authorities have monitoring programs for classifying the waters where shellfish are harvested according to a number of bacteria. When the number of bacteria exceeds the set criteria, depuration of bivalve mollusks is required prior to marketing to ensure that they are safe for consumption (Huss et al. 2003).

Poisoning: Seafood poisoning occurs as a result of human consumption of food harvested from the sea. This includes, but is not limited to, finfish and shellfish. Raw seafood contains bacteria, such as "spoilage bacteria" and "food poisoning" bacteria that can grow and multiply rapidly if the food is left for several hours at room temperature. Natural toxins are produced by minute organisms called dinoflagellates and diatoms. These phytoplankton move up in the food chain into shellfish and other carnivorous marine organisms, where they concentrate in viscera, affecting those who consume them. The toxins are not only tasteless and odorless, but also heat and acid stable. Seafood-toxin-borne diseases can be categorized into two groups depending on their vectors: shellfish and fish. Shellfish harbors the toxins that produce paralytic shellfish poisoning (PSP), neurotoxic shellfish poisoning (NSP), diarrheic shellfish poisoning (DSP), and amnesic shellfish poisoning (ASP) (See Chapter 1 for detailed information). On the other hand, fish carries the toxins responsible for ciguatera and tetrodotoxin (fugu or pufferfish) poisoning. The shellfish-associated diseases generally occur in association with algal blooms or "red tides", that is characterized by patches of discolored water due to the dead fishes. Most seafood toxins usually target our nervous system, and gastrointestinal tract.

Intoxication: The seafood outbreaks were more linked to intoxication illnesses than infection. As stated before, bacteria causing intoxications need to be present in a sufficient number before producing toxins, and therefore they do not present risks at the time of catch. Growth conditions of these bacteria are different and can be found in Huss et al. (2003) and FDA (2011).

The main preventive measure is control of growth. Spore-forming bacteria that produce toxins and are associated with seafood outbreaks—*Clostridium* spp. and *Bacillus* are commonly found in soil, involved in organic matter decay, and are natural inhabitants of the gastrointestinal tract of insects and many warm-blooded animal species. Most of them are bacteria from the general environment (except some types of *C. botulinum*) (Huss et al. 2003) that contaminate products during processing. *Clostridium botulinum* is ubiquitous in nature and its spores are naturally present in soil and water. The bacterium produces a neurotoxin—botulin—under anaerobic, low-acid conditions. The types of bacteria pathogenic to humans (types A, B, E, and F) can be divided into two groups. Group I strains (proteolytic types A, B, and F), the spores of which are highly heat resistant, mesophilic, NaCl tolerant, and have the general environment as the natural habitat, are frequently related to insufficiently processed home-preserved foods such as canned vegetables and cured meats. Group II strains (non-proteolytic types B, E, and F) are heat sensitive, NaCl sensitive, and have the aquatic environment as their natural habitat, and owing to their ability to grow at refrigerated temperatures, they are a safety risk in modern industrially processed foods. These foods are processed with mild heat treatments that may allow the survival of group II spores (Huss et al. 2003, Lindström et al. 2006). Most seafood-associated botulism cases in the USA are caused by toxin type E. Implicated seafood has been fermented under anaerobic conditions that favor the germination of *C. botulinum*. The main preventive measures are control of growth of the bacteria by controlling the temperature, pH, oxygen or salt, or by adding preservatives. Unlike biotoxins and histamine, botulism toxin is sensitive to heat, so cooking for a sufficient time can inhibit the toxin, which can be an added measure to ensure safety. *Clostridium perfringens* and *Bacillus cereus* are also spore-forming bacteria and are ubiquitous. *C. perfringens* is an anaerobe commonly found in mammalian feces and soil. *B. cereus* is an aerobic bacterium that is commonly found in soil, on vegetables, and in many raw and processed foods. Spores may survive cooking, and rapid growth may occur if the food is not chilled promptly. Outbreaks are usually associated with food left at inappropriate temperatures for prolonged periods, allowing multiplication of the bacteria. Only a few reports of illness due to the presence of these microorganisms in seafood have been published (Iwamoto et al. 2010). An incident in 2007 in Spain was caused by *B. cereus* in ready-to-eat tuna (Doménech-Sánchez et al. 2011): several vomiting episodes were reported a few hours after the tuna fish consumption in a beach club. Microbiological analyses detected high levels of *B. cereus* in ready-to-eat fish samples, indicating inappropriate cooking procedures. The important non-spore-forming, toxin-producing pathogenic bacterium in seafood is *S. aureus*. Although *S. aureus* is an ubiquitous organism, the largest reservoir of enterotoxin-producing staphylococci is human nasal passages, but they are also found on skin, hands, wounds, and cutaneous abscesses. The presence of staphylococci in cooked or processed foods can serve to indicate poor hygiene among food handlers. Freshly caught fish is generally free from this bacterium. A recent study conducted during 2008 and 2009 showed high incidence of *S. aureus* (~ 25%) in fish products in Spain. The incidence was highest in fresh (43%) and frozen (30%) products, but it was high in salted and smoked fish, ready-to-cook products, and ready-to-eat products (Vázquez-Sánchez et al. 2012). Unlike botulin, enterotoxins produced by *S. aureus* are heat resistant. The main preventive measure to control *S. aureus* during processing is an effective prerequisite program (hygienic handling).

12. Microbial Quality Control on Seafood

Traditionally, three principal means have been used by governmental agencies and food processors to control microorganisms in food, as listed by ICMSF (1988). These are (a) education and training, (b) inspection of facilities and operations, and (c) microbiological testing. These programmes have been directed toward developing an understanding of the causes and consequences of microbial contamination and to evaluate facilities, operations and adherence to good handling practices. Although these are essential parts in any food control program, they have certain limitations and shortcomings. Beyond ICMSF, in order to comply with more stringent quality & safety standards, prevention practices based on HACCP principle need to be adopted in controlling seafood hazards.

12.1 Hygiene and sanitation for seafood processing

Pre-harvest hazards in seafood are difficult or impossible to manage using ordinary control measures. On the other hand, contamination, as well as recontamination or survival of microbes during processing are more easily managed and controlled by implementing and following Good Manufacturing Practice (GMP). Seafood undergoes processing (washing, cleaning surfaces, etc.) after catch from the sea, and these processing facilities should be located, constructed, and maintained according to sanitary design principles. This plant sanitation has been defined as the controlling of all condition or practices within the plant so that the seafood processed is free from disease producing microorganisms and foreign matter. Each segment of the processing facility must provide the conditions necessary to protect food while the food is under that segment's control. This protection has traditionally been accomplished through the application of good manufacturing practices (GMP), good hygiene practice (GHP), and Hazard Analysis and Critical Control Points (HACCP) programs (Elbashir et al. 2018).

Table 2.4 Summarization of biological hazards, chemical hazards, and physical hazards in fish and fishery products in Thailand.

Fish and fishery products	Hazards		
	Biological	**Chemical**	**Physical**
1. Fresh fish and shellfish	- *V. cholera* - *V. parahaemolyticus* - *Salmonella* spp. - Pathogenic *E. coli* - *S. aureus* - Parasites	- Tetrodotoxin PSP (marine puffer fish) - PSP (freshwater puffer fish) - PSP, DSP, NSP (bivalve mollusks) - Histamine (Scombroid fish) - Antibiotic residues - Pesticide - Heavy metal - Illegal food preservative	
2. Dried and salted fish	- *E. coli* - *S. aureus*	- Histamine (Scombroid fish) - Heavy metal - Pesticide - Illegal synthetic color	
3. Fermented fish	- Parasites - *Salmonella* spp.	- Histamine (fish sauce) - Dimethylnitrosamine (fermented fish with rice bran) - Borax (low salt fermented fish)	
4. Canned fish	- Staphylococcal toxin - *C. botulinum*	- EDTA and sulfur dioxide (canned shrimp and crab) - Histamine (canned tuna)	- Foreign matter - Metal - Fish bone and hard shell of shellfish
5. Chilled fish	- *V. parahaemolyticus* - *Salmonella* spp.	- Borax (raw minced fish and value added products from minced fish) - Sodium benzoate	
6. Frozen fish	- *Salmonella* spp. - *V. parahaemolyticus* - *S. aureus* - *Shigella* spp.	- Antibiotic residues (cultured shrimp)	

Modified from: Virunhakul et al. 2004.

12.2 Sanitary indicator bacteria

Sources of seafood contamination have been identified as overboard sewage discharge into harvest areas, illegal harvesting from sewage-contaminated waters, and sewage runoff from points inland after heavy rains or flooding. Traditionally, indicator micro-organisms have been used to suggest the presence of pathogens (FDA 2011). Today, however, we understand a myriad of possible reasons for indicator presence and pathogen absence, or vice versa. In short, there is no direct correlation between numbers

Table 2.4A

Group	Definition
Process indicator	A group of organisms that demonstrates the efficacy of a process, such as total heterotrophic bacteria or total coliforms for chlorine disinfection.
Faecal indicator (such as *E. coli*)	A group of organisms that indicate the presence of faecal contamination, such as the bacterial groups thermotolerant coliforms or *E. coli*. Hence, they only infer that pathogens may be present.
Index and model organisms	A group/or species indicative of pathogen presence and behavior respectively, such as *E. coli* as an index for *Salmonella* and F-RNA coliphages as models of human enteric viruses.

of any indicator and enteric pathogens (Grabow 1996). To eliminate the ambiguity in the term 'microbial indicator', the following three groups are now recognized and their definition has been presented in Table 2.4A.

Seafood vehicles were categorized into the following three commodity groups: (i) mollusks (e.g., oysters, clams, and scallops); (ii) crustaceans (e.g., shrimp, crab, and lobster); and (iii) finfish (e.g., salmon, tuna, and finfish eggs), and other aquatic vertebrates (e.g., whale and seal), or "fish" for brevity. Notably, in recent years, outbreaks of infections due to some pathogens usually associated with seafood harvested from warm waters were reported from more northerly areas of the country that had not previously reported outbreaks. However, in developing countries the three groups of bacteria as indicators are still using: (1) Presence of coliforms is used as an indicator of sanitary quality of water or as a general indicator of sanitary condition in the fish-processing environment, (2) Fecal coliforms remain the standard indicator of choice for shellfish and shellfish harvest waters, and (3) *E. coli* is used to indicate recent fecal contamination or unsanitary processing.

12.3 Hurdle technology for controlling microorganism in seafood

The microbial safety and stability of most food are based on an application of preservative factors called hurdles. Each hurdle implies putting microorganisms in a hostile environment, which inhibits their growth or causes their death (Leistner 2000). Some of those hurdles have been empirically used for years to stabilize meat, fish, milk, and vegetables. This sometimes leads to completely different product with its own new taste characteristics. Examples of hurdles in marine products are salt (salted cod, klipfish), smoke (cold or hot smoked salmon, herring), acids (marinated products, pickles), temperature (high or low), fermentative microorganisms (traditional Asian sauces) and more recently redox potential (vacuum-packed products). Those preservative factors have been studied for years, but a large amount of potential hurdles for food have already been described, including organic acids, bacteriocins, chitosan, nitrate, lactoperoxidase, essential oil, modified atmosphere packaging, etc., as well as novel decontamination technologies, such as microwave and radio frequency, ohmic and inductive heating, high pressure, pulsed electric field, high voltage arc discharge, pulsed light, oscillation magnetic field, ultraviolet light, ultrasound, X-ray, electrolyze NaCl water, ozone (Kim et al. 1999, Weber 2000, Mahmoud et al. 2006). Hurdles that have a positive effect by inhibiting microorganisms may have a negative one on other parameters such as nutritional properties or sensory quality, depending on their intensity. As an example, salt content in food must be high enough to inhibit pathogens and spoilage microorganisms, but not so high as to impair taste. In order to lower the preservative level, the hurdle technology concept has been developed (Leistner 1985), consisting of using combined hurdles to establish an additive antimicrobial effect, and sometimes even a synergetic one, thus improving the safety and the sensory quality of food. Preservation implies putting microorganisms in a hostile environment, in order to inhibit their growth or shorten their survival or cause their death. The feasible responses of microorganisms to this hostile environment determine whether they may grow or die. More research is needed in view of these responses; however, recent advances have been made by considering the homeostasis, metabolic exhaustion, and stress reactions of microorganisms in exhaustion, and stress reactions of microorganisms inducing the novel concept of multi-target preservation for a gentle but most effective preservation of hurdle-technology foods (Leistner 1995).

13. Conclusion

The concepts of safety and quality differ. Safety dictates the non-negotiable characteristics that must be present for a product to be fit for human consumption, whereas quality defines other characteristics that determine what a product is fit for, thus, although healthful, seafood consumption is not risk-free. Consumers should be aware of the potential health risks associated with eating seafood. Seafood-borne infections can be prevented by cooking seafood thoroughly, storing foods properly, and avoiding cross-contamination after cooking. However, some seafood is commonly consumed raw or minimally cooked. Persons with underlying medical conditions such as liver disease, diabetes, or immune-suppressing conditions are at higher risk of acquiring severe infection, and should be especially careful. Multiple outbreaks of seafood-associated outbreaks of *Vibrio* infections continue to occur every year worldwide, suggesting that existing control strategies have not been optimally effective. Thus, preventive measures in seafood-associated infections requires an understanding not only of the etiologic agents and seafood commodities associated with illness, but also of the mechanisms of contamination that are amenable to control.

References

Aberoumand, A. 2010. Occurrence of Clostridium in fish and fishery products in retail trade, a review article. World J. Fish Mar. Sci. 2: 246–250.

Ajami, N.J., Koo, H.L., Darkoh, C., Atmar, R.L., Jiang, Z.D. and DuPont, H.L. 2010. Characterization of norovirus-associated Travelers' diarrhea. Clin. Infect. Dis. 51(2): 123–130.

Albuquerque, W.F., Macrae, A., Sousa, O.V., Vieira, G.H.F. and Vieira, R.H.S.F. 2007. Multiple drug resistant *Staphylococcus aureus* strains isolated from a fish market and from fish handlers. Braz. J. Microbiol. 38: 131–134.

Alfano-Sosbey, E., Sweat, D., Hall, A., Breedlove, F., Rodríguez, R., Greene, S., Pierce, A., Sobsey, M., Davies, M. and Ledford, S.I. 2012. Norovirus outbreak associated with undercooked oysters and secondary household transmission. Epidemiol. Infect. 140: 276–282.

Amagliani, G., Brandi, G. and Schiavano, G.F. 2012. Incidence and role of seafood safety. Food Res. Inter. 45: 780–788.

Ananchaipattana, C., Hosotani, Y., Kawasaki, S., Pongswat, S., Latiful, M.B., Isobe, S. and Inatsu, Y. 2012. Prevalence of foodborne pathogens in retailed foods in Thailand. Journal of Foodborne Pathogen and Disease 9(9): 835–840.

Anbazhagi, S. and Kamatchiammal, S. 2010. Contamination of seafood by norovirus in India. Int. J. Virol. 6(3): 138–149.

Andrews, H.L. and Baumler, A.J. 2005. *Salmonella* species. pp. 327–339. *In*: Fratamico, P.M., Bhunia, A.K. and Smith, J.L. (eds.). Foodborne Pathogens Microbiology and Molecular Biology. Caister Academic Press, Norfolk, UK.

Baker-Austin, C., McArthur, J.V., Lindell, A.H., Wright, M.S., Tuckfield, R.C., Gooch, J., Warner, L., Oliver, J. and Stepanauskas, R. 2009. Multi-site analysis reveals widespread antibiotic resistance in the marine pathogen *Vibrio vulnificus*. Microb. Ecol. 57: 151–159.

Bangtrakulnonth, A., Pornreongwong, S., Pulsrikarn, C., Sawanpanyalert, P., Hendriksen, R.S., LoFoWong, D.M.A. et al. 2004. *Salmonella* serovars from humans and other sources in Thailand, 1993–2002. Emerging Infectious Diseases 10: 131–137.

Bazzardi, R., Fattaccio, M.C., Salza, S., Canu, A., Marongiu, E. and Margherita Pisanu, M. 2014. Preliminary study on Norovirus, hepatitis A virus, *Escherichia coli* and their potential seasonality, Italy. Italian J. Food Saf. 3(1601): 125–130.

Bell, B.P., Shapiro, C.N., Alter, M.J. et al. 1998. The diverse patterns of hepatitis A epidemiology in the United States—implications for vaccination strategies. J. Infect. Dis. 178: 1579–1584.

Bellou, M., Kokkinos, P. and Vantarakis, A. 2013. Shellfish-borne viral outbreaks: a systematic review. Food Environ. Virol. 5: 13–23.

Biber Pay and Arun Bhunia, Fundamental Food Microbiology, 4th Edition 2008, CRC Press Taylor and Francis Group.

Bouchrif, B., Paglietti, B., Murgia, M., Piana, A., Cohen, N., Ennaji, M.M. et al. 2009. Prevalence and antibiotic-resistance of Salmonella isolated from food in Morocco. Journal of Infections in Developing Countries 3: 35–40.

Broughton, E.I. and Walker, D.G. 2009. Prevalence of antibiotic resistant *Salmonella* in fish in Guangdong, China. Foodborne Pathogens Dis. 6: 519–521.

Burrus, V., Marrero, J. and Waldor, M.K. 2006. The current ICE age: biology and evolution of SXT-related integrating conjugative elements. Plasmid J. 55: 173–183.

CA, Codex Alimentarius. 2010. Guidelines on the application of general principles of food hygiene to the control of pathogenic *Vibrio* species in seafood. CAC/GL 73-2010.

Centers for Disease Control and Prevention (CDC). 1998. Outbreak of *Vibrio parahaemolyticus* Infections Associated with Eating Raw Oysters-Pacific Northwest, 1997. MMWWR 1998. Available at: http://www.cdc.gov/mmwr/preview/ mmwrhtml/00053377.htm (Accessed 9 August 2016).

Centers for Disease Control and Prevention (CDC). 2007. Viral hepatitis surveillance United States, 2007. MMWR Surveill. Summ. 2009 58(SS-3).

Centers for Disease Control and Prevention (CDC). 2009. Food Net 2007 surveillance report. Atlanta: U.S. Department of Health and Human Services.

Center for Disease Control and Prevention (CDC). 2010a. Surveillance for foodborne disease outbreaks-United States, 2007. Morb. Mortal. Wkly. Rep. 59: 973–979.

Centers for Disease Control and Prevention. 2012. Notes from the field: norovirus infections associated with frozen raw oysters-Washington, 2011. Morb. Mortal. Wkly. Rep. (MMWR) 307(14): 1480. Available at: http://www.cdc.gov/mmwr/preview/mmwrhtml/mm6106a3.htm (Accessed 10 October 2016).

Centers for Disease Control and Prevention (CDC). 2013a. Increase in *Vibrio parahaemolyticus* illnesses associated with consumption of shellfish from several Atlantic coast harvest areas, United States, 2013. Available at: http://www.cdc. gov/vibrio/investigations/index.html (Accessed 8 September 2016).

Centers for Disease Control and Prevention (CDC). 2014a. Foodborne Outbreak Online Database (FOOD). Available at: https://wwwn.cdc.gov/foodborneoutbreaks/(Accessed 20 September 2016).

Centers for Disease Control and Prevention (CDC). 2014b. National Surveillance of Bacterial Foodborne Illnesses (Enteric Diseases). http://www.cdc.gov/nationalsurveillance/botulism-surveillance.html (Accessed 16 September 2016).

Chironna, M., Germinario, C., De Medici, D., Fiore, A., Di Pasquale, S., Quartoa, M. and Barbuti, S. 2002. Detection of hepatitis A virus in mussels from different sources marketed in Puglia region (South Italy). Int. J. Food Microbiol. 75: 11–18.

Clark, M.A. and Barret, E.L. 1987. The phs gene and hydrogen sulfide production by Salmonella typhimurium. J. Bacteriol. 169(6): 2391–2397.

Consumer Reports. 2015. Consumer Reports finds potentially harmful bacteria and illegal antibiotic residues in tests of frozen shrimp. Available at: http://pressroom.consumerreports.org/pressroom/2015/04/consumer-reports-findspotentially-harmful-bacteria-and-illegal-antibiotic-residues-in-tests-of-frozenshrimp.html (Accessed 25 July 2016).

Cook, N., Knight, A. and Richards, G.A. 2016. Persistence and elimination of human norovirus in food and on food contact surfaces: a critical review. J. Food Prot. 79(7): 1273–1294.

Croci, L., Suffredini, E., Di Pasquale, S. and Cozzi, L. 2012. Detection of Norovirus and Feline calicivirus in spiked molluscs subjected to heat treatments. Food Control 25(1): 17–22.

Dan, L.I., Stals, A., Tang, Q. and Uyttendaele, M. 2014. Detection of noroviruses in shellfish and semi-processed fishery products from a Belgian seafood company. J. Food Prot. 77(8): 1342–1347.

David, O.M., Wandili, S., Kakai, R. and Waindi, E.N. 2009. Isolation of *Salmonella* and *Shigella* from fish harvested from the Winam Gulf of lake Victoria, Kenya. J. Infect. Dev. Ctries 3(2): 99–104.

Dennis, D.T. and Gage, K.L. 2004. Plague. pp. 1641–1648. *In*: Cohen, J. and Powderly, W.G. (eds.). Infectious Diseases, Second Ed., Vol. II. Mosby, Edinburgh.

Doménech-Sánchez, A., Laso, E., Pérez, M.J. and Berrocal, C.I. 2011. Emetic disease caused by *Bacillus cereus* after consumption of tuna fish in a beach club. Foodborne Pathog. Dis. 8(7): 835–837.

EFSA. 2010. Scientific opinion on risk assessment of parasites in fishery products. EFSA J. 8(4): 1543.

EFSA. 2012b. Scientific Opinion Scientific Opinion on Norovirus (NoV) in oysters: methods, limits and control options. EFSA J. 10(1): 2500.

Elbashir, S., Parveen, s., Schwarz, J., Rippen, T., Jahncke, M. and Depaola, A. 2018. Seafood pathogen and information on antimicrobial resistance: A review. Food Micro. 70: 85–93.

EU. 2004. Corrigendum to Regulation (EC) No. 853/2004 of the European Parliament and of the Council of 29 April 2004 laying down specific hygiene rules for food of animal origin. Off. J. Eur. Union L 226: 22–82.

Fallah, A.A., Saei-Dehkordia, S.S. and Mahzouniehb, M. 2013. Occurrence and antibiotic resistance profiles of *Listeria monocytogenes* isolated from seafood products and market and processing environments in Iran. Food Control 34(2): 630–636.

FAO/WHO. 2005. Risk assessment of *Vibrio vulnificus* in raw oysters: interpretative summary and technical report. Microbiological Risk Assessment Series No. 8. Rome, Italy: FAO.

FDA. 2011. Fish and Fishery Products Hazards and Controls Guidance, Fourth Ed. US Department of Health and Human Services Food and Drug Administration.

Ghasemian, R., Babamahmoodi, F. and Ahangarkani, F. 2016. Hepatitis A is a health hazard for Iranian pilgrims who go to Holly Karbala: a preliminary report. Hepat. Mon. 16(6): 38138. http://dx.doi.org/10.5812/hepatmon.38138.

Grabow, W.O.K. (1996). Waterborne diseases: Update on water quality assessment and control. Water SA 22: 193–202.

Hadler, S.C. 1991. Global impact of hepatitis A virus infection changing patterns. pp. 14–20. *In*: Hollinger, F.B., Lemon, S.M. and Margolis, H. (eds.). Viral Hepatitis and Liver Disease: Proceedings of the 1990 International Symposium on Viral Hepatitis and Liver Disease: Contemporary Issues and Future Prospects. Williams and Wilkins, Baltimore.

Hall, A.J., Eisenbart, V.G., Etingüe, A.L., Gould, L.H., Lopman, B.A. and Parashar, U.D. 2012. Epidemiology of foodborne norovirus outbreaks, United States, 2001–2008. Emerg. Infect. Dis. 18(10): 1566–1573.

Hall, A.J., Wikswo, M.E., Pringle, K., Gould, L.H. and Parashar, U.D. 2014. Vital signs: foodborne norovirus outbreaks-United States, 2009–2012. Morb. Mortal. Wkly. Rep. June 2014 75(22): 491–495.

Halliday, M.L., Kang, L.Y., Zhou, T.K., Hu, M.D., Pan, Q.C., Fu, T.Y., Huang, Y.S. and Hu, S.L. 1991. An epidemic of hepatitis A attributable to the ingestion of raw clams in Shanghai, China. J. Infect. Dis. 64(5): 852–859.

Han, F., Walker, R.D., Janes, M.E., Prinyawiwatkul, W. and Ge, B. 2007. Antimicrobial susceptibilities of *Vibrio parahaemolyticus* and *Vibrio vulnificus* isolates from Louisiana Gulf and retail raw oysters. Appl. Environ. Microbiol. 73: 7096–7098.

Heinitz, M.L., Ruble, R.D., Wagner, D.E. and Tatini, S.R. 2000. Incidence of *Salmonella* in fish and seafood. J. Food Prot. 63: 579–592.

Hirai, Y., Asahata-Tago, S., Ainoda, Y., Fujita, T. and Kikuchi, K. 2015. *Edwardsiella tarda* bacteremia. A rare but fatal water- and foodborne infection: Review of the literature and clinical cases from a single centre. The Canadian Journal of Infectious Diseases & Medical Microbiology 26(6): 313–318.

Huss, H.H., Ababouch, L. and Gram, L. 2003. Assessment and Management of Seafood Safety and Quality. FAO Fisheries Technical Paper 444, Rome, 230 pp.

Jancovich, J.K., Chinchar, V.G., Hyatt, A., Miyazaki, T., Williams, T. and Zhang, Q.Y. 2012. Family Iridoviridae. pp. 193–210. *In*: King, A.M.Q., Adams, M.J., Carstens, E.B. and Lefkowitz, E.J. (eds.). Virus Taxonomy: Ninth Report of the International Committee on Taxonomy of Viruses. Elsevier Academic Press, San Diego, CA.

Iwamoto, M., Ayers, T., Mahon, B.E. and Swerdlow, D.I. 2010. Epidemiology of seafood associated infections in the United States. Clin. Microbiol. Rev. 23: 399–411.

Japan Food Poisoning Statistics. 2009. Inspection and Safety Division, Department of Food Safety, Pharmaceutical and Food Safety Bureau, Ministry of Health Labour and Welfare Japan. Available at: http://www.mhlw.go.jp/english/topics/foodsafety/poisoning/dl/Food_Poisoning_Statistics_2009.pdf.

Jay, M.J., Loessner, M.J. and Golden, D.A. 2005. Modern Food Microbiology, Seventh Ed., ISBN 0387231803.

Joh, S.J., KIM, M.J., Kwon, H.M., Ahn, E.H., Jang, H. and Kwon, J.H. 2011. Characterization of *Edwardsiella tarda* isolated from farm-cultured eels, *Anguilla japonica*, in the Republic of Korea. J. Vet. Med. Sci. 73: 7–11.

Kim, J.G., Yousef, A.E. and Dave, S. 1999. Application of ozone for enhancing the microbiological safety and quality of foods: A review. J. Food Prot. 62: 1071–1087.

Köck, M., Farré, M., Martínez, E., Gajda-Schrantz, K., Ginebreda, A., Navarro, A. et al. 2010. Integrated ecotoxicological and chemical approach for the assessment of pesticide pollution in the Ebro River delta (Spain) J. Hydrol. 383: 73–82.

Kournay, M., Vasquez, M.A. and Saenz, R. 1977. Edwardsiellosis in man and animals in Panama: Clinical and epidemiological characteristics. Am. J. Trop. Med. Hyg. 26: 1183–90.

Kumar, R. 2009. Distribution trends of *Salmonella* serovars in India (2001–2005). Transactions of the Royal Society of Tropical Medicine and Hygiene 103: 390–394.

La Rosa, G., Fratini, M., Vennarucci, V.S., Guercio, A., Purpari, G. and Muscillo, M. 2012. GIV noroviruses and other enteric viruses in bivalves: a preliminary study. New Microbiol. 35: 27–34.

Lampel, K.A. 2005. *Shigella* species. pp. 341–356. *In*: Fratamico, P.M., Bhunia, A.K. and Smith, J.L. (eds.). Foodborne Pathogens Microbiology and Molecular Biology. Caister Academic Press, Norfolk, UK.

Leistner, L. 1995. Principles and applications of hurdle technology. pp. 1–21. *In*: Gould, G.W. (ed.). New Methods of Food Preservation, Springer.

Leistner, L. 2000. Basic aspects of food preservation by hurdle technology. Int. J. Food Microbiol. 55(1-3):181–6.

Lima dos Santos, C.A.M. and Howgate, P. 2011. Fishborne zoonotic parasites and aquaculture: a review. Aquaculture 318: 253–261.

Lindström, M., Kiviniemi, K. and Korkeala, H. 2006. Hazard and control of group II (non-proteolytic) *Clostridium botulinum* in modern food processing. Int. J. Food Microbiol. 108(1): 92–104.

Lopalco, P.L., Malfait, P., Salmaso, S., Germinario, C., Quarto, M. and Barbuti, S. 1997. A persisting outbreak of hepatitis A in Puglia, Italy, 1996: epidemiological follow up. Euro Surveill. 4: 31–32.

Mahmoud, B.S.M., Yamasaki, K., Miyashita, K., Kawai, Y., Shin, I.S. and Suzuki, T. 2006. Preservative effect of combined treatment with electrolysed NaCl solutions and essential oil compounds on carp fillets during convectional air-drying. Int. J. Food Microbiol, 106: 331–337.

Malfait, P., Lopalco, P.L., Salmaso, S., Germinario, C., Salamina, M., Quarto, M. and Barbuti, S. 1996. An outbreak of hepatitis A in Puglia, Italy, 1996. Euro Surveill. 1(5): 33–35.

Markkula, A., Autio, T., Lunde, J. and Korkeala, H. 2005. Raw and processed fish show identical *Listeria monocytogenes* genotypes with pulsed-field gel electrophoresis. J. Food Prot. 68(6): 1228–1231.

Martinez-Urtaza, J., Bowers, J.C., Trinanes, J. and DePaola, A. 2010. Climate anomalies and the increasing risk of *Vibrio parahaemolyticus* and *Vibrio vulnificus* illnesses. Food Res. Int. 43: 1780–1790.

McCoy, E., Morrison, J., Cook, V., Johnston, J., Eblen, D. and Guo, C. 2011. Foodborne agents associated with the consumption of aquaculture catfish. J. Food Prot. 74(3): 500–516.

Minami, A., Chaicumpa, W., Chongsa-Nguan, M., Samosornsuk, S., Monden, S., Takeshi, K. et al. 2010. Prevalence of foodborne pathogens in open markets and supermarkets in Thailand. Food Control 21: 221–226.

Miya, S., Takahashi, H., Ushikawa, T., Fujii, T. and Kimura, B. 2010. Risk of *Listeria monocytogenes* contamination of raw ready-to-eat seafood products available at retail outlets in Japan. Appl. Environ. Microbiol. 76(10): 3383–3386.

Miyoshi, S. 2006. *Vibrio vulnificus* infection and metalloprotease. J. Dermatol. 33: 589–595.

Muller, R.H., S. Jorks, S. Kleinsteuber and W. Babel. 1999. Comamonas acidovorans strain MC1: a new isolate capable of degrading the chiral herbicides dichlorprop and mecoprop and the herbicides 2,4-D and MCPA. Microbiol. Res. 154: 241–246.

Nakajima, H., Inoue, M. and Mori, T. (1991). Isolation of Yersinia, Campylobacter, Plesiomonas and Aeromonas from environmental water and fresh water fishes. Nippon Koshu Eisei Zasshi. 38: 815–820.

National Advisory Committee on Microbiological Criteria for Foods (NACMCF). 2008. Response to the questions posed by the Food and Drug Administration and the National Marine Fisheries Service regarding the determination of cooking parameters for safe seafood for consumers. J. Food Prot. 71(6): 1287–1308.

Nishibuchi, M. and DePaola, A. 2005. Vibrio species. pp. 251–271. *In*: Fratamico, P.M., Bhunia, A.K. and Smith, J.L. (eds.). Foodborne Pathogens Microbiology and Molecular Biology. Caister Academic Press, Norfolk, UK.

O'Connel, C.M.C., Sandlin, R.C. and Maurelli, A.T. 1995. Signal transduction and virulence gene regulation in *Shigella* spp.: temperature and (maybe) a whole lot more. pp. 111–127. *In*: Rappuoli, R. (ed.). Signal Transduction and Bacterial Virulence. R.G. Landes Company, Austen, Tex.

Oliver, J.D. 2005. Wound infections caused by *Vibrio vulnificus* and other marine bacteria. Epidemiol. Infect. 133: 383–391.

Oliver, J.D. 2006. *Vibrio vulnificus*. pp. 253–276. *In*: Belkin, S. and Colwell, R.R. (eds.). Oceans and Health: Pathogens in the Marine Environment. Springer Science, New York, NY.

Onogawa, T., Terayama, T. and Zenyoji, H. 1976. Distribution of *Edwardsiella tarda* and hydrogen sulfide-producing *Escherichia coli* in healthy persons. J. Jpn. Assoc. for Infect. Dis. (Kansenshogaku Zasshi) 50: 10–7.

Pagadala, S., Parveen, S., Schwarz, J.G., Rippen, T. and Luchansky, J.B. 2011. Comparison of automated BAX polymerase chain reaction and standard culture methods for detection of *Listeria monocytogenes* in blue crab meat (*Callinectus sapidus*) and blue crab processing plants. J. Food Prot. 74: 1930–1933.

Pagadala, S., Parveen, S., Schwarz, J.G., Rippen, T., Luchansky, J.B., Call, J.E., Tamplin, M.L. and Porrto-Fett, A.C.S. 2012. Prevalence, characterization and sources of *Listeria monocytogenes* in blue crab (*Callinectus sapidus*) meat and blue crab processing plants. J. Food Microbiol. 31: 263–270.

Parveen, S. and Tamplin, M.L. 2013. *Vibrio vulnificus*, *Vibrio parahaemolyticus* and *Vibrio cholera*. pp. 148–176. *In*: Roland, Labbe', G. and Santos, Garcia (eds.). Guide to Foodborne Pathogens. Second Ed. John Wiley & Sons.

Pavoni, E., Consoli, M., Suffredini, E., Arcangeli, G., Serracca, L., Battistini, R., Rossini, I., Croci, L. and Losio, M.N. 2013. Noroviruses in seafood: a 9-year monitoring in Italy. Foodborne Pathogens Dis. 10(6): 533–539.

Pinchuk, I.V., Beswick, E.J. and Reyes, V.E. 2010. Staphylococcal enterotoxins. Toxins 2: 2177–2197.

RASFF, The Rapid Alert System for Food and Feed. 2011. Annual Report 2011, Health and Consumers Directorate General, European Commission. RASFF, The Rapid Alert System for Food and Feed, 2011. Annual Report 2011, Health and Consumers Directorate General, European Commission.

Reddy, N.R., Solomon, H.M., Yep, H., Roman, M.G. and Rhodehamel, E.J. 1997. Shelf life and toxin development by *Clostridium botulinum* during storage of modified atmosphere-packaged fresh aquaculture salmon fillets. J. Food Prot. 9: 1055–1063.

Richards, G.P. 2014. Bacteriophage remediation of bacterial pathogens in aquaculture: a review of the technology. Bacteriophage 4(4): e975540. http:// dx.doi.org/10.4161/215 97061.2014.975540.

Rivera, I.N., Chun, J., Hug, A., Sack, R.B. and Colwell, A. 2001. Genotypes associated with virulence in environmental isolates of *Vibrio cholerae*. Appl. Environ. Microbiol. 67: 2421–2429.

Rodas-Suarez, O.R., Flores-Pedroche, J.F., Betancourt-Rule, J.M., Quinones Ramírez, E.I. and Vazquez-Salinas, C. 2006. Occurrence and antibiotic sensitivity of *Listeria monocytogenes* strains isolated from oysters, fish, and estuarine water. Appl. Environ. Microbiol. 72(11): 7410–7412.

Rose, Joan B., Paul R. Epstein, Erin K. Lipp, Benjamin H. Sherman, Susan M. Bernard and Jonathan A. Patz. 2001. Climate variability and change in the United States: Potential impacts on water and foodborne diseases caused by microbiologic agents. Environ. Health Perspect. 109 (supp. 2): 211–221.

Sala, M.R., Arias, C., Nguez, A.D., Bartolome, R. and Muntada, J.M. 2008. Short report: foodborne outbreak of gastroenteritis due to Norovirus and *Vibrio parahaemolyticus*. Epidemiol. Infect. 137(5): 1–4.

Scallan, E., Hoekstra, R.M., Angulo, F.J., Tauxe, R.V., Widdowson, M., Roy, S.L., Jones, J.L. and Griffin, P.M. 2011. Foodborne illness acquired in the United States major pathogens. Emerg. Infect. Dis. 17(1): 7–15.

Scharer, K., Savioz, S., Cerenela, N., Saegesser, G. and Stephan, R. 2011. Occurrence of *Vibrio* spp. in fish and shellfish collected from the Swiss market. J. Food Prot. 74(8): 1345–1347.

Shaw, K.S., Goldstein, R.E.R., He, X., Jacobs, J.M., Crump, B.C. and Sapkota, A.R. 2014. Antimicrobial susceptibility of *Vibrio vulnificus* and *Vibrio parahaemolyticus* recovered from recreational and commercial areas of Chesapeake Bay and Maryland coastal Bays. PLoS One 9: 1–11.

Smith, M.D. and Gradon, J.D. 2003. Bacteremia due to Comamonas species possibly associated with exposure to tropical fish. Southern Medical Journal 96(8): 815–817.

Sujatha, K., Senthilkumaar, P., Sangeetha, S. and Gopalakrishnan, M.D. 2011. Isolation of human pathogenic bacteria in two edible fishes, Priacanthus hamrur and Megalaspis cordyla at Royapuram waters of Chennai, India. Indian J. Sci. Technol. 4(5): 539–542.

Suzuki, A., Ichinose, M., Matsue, T., Amano, Y., Terayama, T., Izumiyama, S. and Endo, T. 2002. Occurrence of Legionella bacteria in a variety of environmental waters—From April, 1996 to November, 2000. Kansenshogaku Zasshi 76: 703–710.

Terio, V., Martella, V., Moschidou, P., Pinto, P.D., Tantillo, G. and Buonavoglia, C. 2010. Norovirus in retail shellfish. Food Microbiol. 27: 29–32.

Thompson, F.L., Gevers, D., Thompson, C.C., Dawyndt, P., Naser, S., Hoste, B., Munn, C.B. and Swings, J. 2005. Phylogeny and molecular identification of Vibrios on the basis of multilocus sequence analysis. J. Appl. Environ. Microbiol. 71(9): 5107–5115.

U.S. Food and Drug Administration - FDA. 2011. Fish and Fisheries Products Hazards and Control Guide, Fourth Ed. http://www.fda.gov/food/guidance compliance regulatory information/guidance documents/seafood/fish and fisheries products hazards and controls guide/default.htm (Accessed 22 May 2016).

U.S. Food and Drug Administration (FDA). 2012a. Investigation of Multistate Outbreak of Salmonella Bareilly Infections Associated with Nakaochi Scrape AA or AAA Frozen Raw Tuna from India. http://www.fda.gov/Food/FoodSafety/ CORENetwork/ucm298741. htm (Accessed 6 May 2016).

U.S. Food and Drug Administration (FDA). 2012b. *Vibrio cholerae* Serogroup NonO1. http:// www.fda.gov/Food/FoodborneIllnessContaminants/ Causes of Illness Bad Bug Book/ ucm070419.htm (Accessed 12 May 2016).

Vázquez-Sánchez, D., López-Cabo, M., Saá-Ibusquiza, P. and Rodríguez-Herrera, J.J. 2012. Incidence and characterization of *Staphylococcus aureus* in fishery products marketed in Galicia (Northwest Spain). Int. J. Food Microbiol. 157(2): 286–296.

Vidaček, S. 2014. Seafood. Risk and controls in the food supply chain. Chapter 8: 195–197.

Vidal, R., Solari, V., Mamani, N., Jiang, X., Vollaire, J., Roessler, P., Prado, V., Matson, D.O. and O'Ryan, M.L. 2005. Calici viruses and foodborne gastroenteritis. Chile. Emerg. Infect. Dis. 11(7): 1134–1137.

Virunhakul, P., Seemanopas, K., Wrongjinda, N., Maharkankul, W., Suwanlungsee, S., Kanjanakan, S., Chareanrian, S. and Treewanich, S. 2004. Risk analysis of problem in food chain for consumers in aquatic product. National Bureau of Agricultural Commodity and Food Standards. Bangkok 577.

Wachsmuth, K. and Morris, G.K. 1989. Shigella. pp. 448–660. *In*: Doyle, M.P. (ed). Foodborne Bacterial Pathogens. New York, NY: Marcel Dekker Inc.

Wang, F., Jiang, L., Yang, Q., Han, F., Chen, S., Pu, S., Vance, A. and Ge, B. 2011. Prevalence and antimicrobial susceptibility of major foodborne pathogens in imported seafood. J. Food Prot. 74(9): 1451–1461.

Webby, R.J., Carville, K.S., Kirk, M.D., Greening, G., Ratcliff, R.M., Crerar, S.K., Dempsey, K., Sarna, M., Stafford, R., Patel, M. and Hall, G. 2007. Internationally distributed frozen oyster meat causing multiple outbreaks of norovirus infection in Australia. Clin. Infect. Dis. Oxf. J. 44(15): 1026–1031.

Weber, D.E. (2000). Kinetics of microbial inactivation for alternative food processing technologies. J. Food Sci. Special supplement 65: 1–107.

Wong, T.Y., Tsui, H.Y., So, M.K., Lai, J.Y., Lai, S.T., Tse, C.W. and Ng, T.K. (2000). Plesiomonas shigelloides infection in Hong Kong: retrospective study of 167 laboratory-confirmed cases. Hong Kong Med. J. 6: 375–380.

World Health Organization (WHO). 1995. Control of foodborne trematode infections. WHO Tech. Rep. Ser. 849.

World Health Organization (WHO). 1999. Food safety issues associated with products from aquaculture. WHO Technol. Rep. Ser. 883.

World Health Organization (WHO). 2004. Report of Joint WHO/FAO workshop on food-borne trematode infections in Asia. Ha Noi, Vietnam, 26–28 November, WHO, WPRO 1–58.

World Health Organization (WHO). 2008. Guidance for Identifying Populations at Risk from Mercury Exposure. WHO, Geneva.

World Health Organization (WHO). 2012. Weekly epidemiological record 31-32(2): 289–304.

World Health Organization (WHO). 2016. Hepatitis A. Available at: http://www. who.int/ mediacentre/factsheets/fs328/en/ (Accessed 17 October 2016).

Woods, J.W., Calci, K.R., Marchant-Tambone, J.G. and Burkhardt 3rd, W. 2016. Detection and molecular characterization of norovirus from oysters implicated in outbreaks in the U.S. Food Microbiol. 59: 76–84.

Wozniak, R.A., Fouts, D.E., Spagnoletti, M., Colombo, M.M., Ceccarelli, D., Garriss, G. et al. 2009. Comparative ICE genomics: insights into the evolution of the SXT/R391 family of ICEs. PLoS Genet. 5: e1000786. http://dx.doi.org/10.1371/ journal.pgen.1000786.

Wyatt, L.E., Nickelson, R. and Vanderzant, C. 1979. *Edwardsiella tarda* in freshwater catfish and their environment. Appl. Environ. Microbiol. 38: 710–4.

Yan, H., Li, L., Alam, M.J., Shinoda, S., Miyoshi, S. and Shi, L. 2010. Prevalence and antimicrobial resistance of *Salmonella* in retail foods in northern China. International Journal of Food Microbiology 143: 230–234.

Yunzhang, Sun, Hongling, Yang, Zechun, Ling, Jianbo, Chang and Jidan, Ye. 2009. Gut microbiota of fast and slow growing grouper Epinephelus coioides. African Journal of Microbiology Research 3(11): 713–720.

Zhao, S., McDermott, P.F., Friedman, S., Qaiyumi, S., Abbott, J., Kiessling, C., Ayers, S., Singh, R., Hubert, S., Sofos, J. and White, D.J. 2006. Characterization of antimicrobial-resistant Salmonella isolated from imported foods. J. Food Prot. 69: 500–507.

3

Food-borne Pathogens Related to Seafood Products

Shogo Yamaki and *Koji Yamazaki**

1. Introduction

Seafood products are important sources of nutrients worldwide. However, raw seafood products, such as fish meat, shrimp, shellfish, and crab, can readily decompose and become naturally contaminated by bacteria, including pathogens. The contamination can increase the risk of illness from consumption of seafood products that are typically not heated prior to ingestion.

In the past, seafood products were consumed near their production location. However, the development of distribution and food-processing technology has made it possible to deliver food from the production site to distant sites. This may compound the risk of a more widely disseminated outbreak if inadequate products are distributed, which has increased the need for food manufacturers to ensure food safety. To prevent food-borne diseases, it is important for producers to know the possible contaminants and their characteristics.

The major microbial pathogens of seafood are listed in Table 3.1. They include *Vibrio* spp., *Salmonella* spp., *Listeria monocytogenes*, *Clostridium*

Faculty of Fisheries Sciences, Hokkaido University, 3-1-1, Minato, Hakodate, Hokkaido, 041-8611, Japan.
* Corresponding author: yamasaki@fish.hokudai.ac.jp

Table 3.1 Selected and common causative agents of foodborne diseases related to seafood products.

Organism or chemical	Symptom	Related seafood	Note
Gram-negative bacteria			
Vibrio spp.			
Vibrio parahaemolyticus	Watery diarrhea Adbominal cramps Nausea Bloody diarrhea	Fish Clam Oyster Shrimp	Most of pathogenic strains produce hemolysin, but environmental isolates cannot produce
Vibrio vulnificus	Septicemia	Raw oyster Undercooked shrimp	Preexisiting medical conditions of liver disease lead serve symptoms
Vibrio cholerae O1/O139	Cholera	Contaminated drinking water or seafood	Pay attention to overseas infection at endemic area by food and drinking water
Vibrio cholerae non-O1/non-O139	Gastroenteritis	Shrimp Mussel Oyster Fish	Most of *V. cholerae* infections are non-O1/non-139
Salmonella spp.	Diarrhea Abdominal pain Muscle aches Fever	Shrimp Shellfish Dried fish Squid	Many serovars cause food poisoning. Some serovars can survive in harsh environment during long storage
Gram-positive bacteria			
Listeria monocytogenes	Gastroenteritis Meningitis Encephalitis Mother-to-fetus infection	Ready-to-eat food Smoked salmon	High mortality rate. Capacity to grow under refrigerated temperatures, and salt and pH stress
Clostridium botulinum	Weakness Visual defects Respiratory paralysis Cardiac arrest	Vaccum-packed or canned food Fermented food	Spore-forming. Neurotoxins are heat-labile. Growth in anaerobic conditions
Viruses			
Norovirus	Abdominal pain Diarrhea Vomitting Fever	Raw or processed, but non-heated shellfish	Low estimated infectious dose. Transmitted from person-to-person
Chemicals			
Histamine	Hives Rash Flushing	Scombroid fish	Produced by histamine-producing bacteria. Stable for heating and freezing

botulinum, Norovirus, and histamine-producing bacteria. These pathogens can cause serious outbreaks, and differ in their distributions, characters, and symptoms. In this chapter, we describe the risks of food-borne diseases in seafood products.

2. *Vibrio* spp.

Vibrio spp. are Gram-negative, rod-shaped, and halophilic bacteria. The members of this genus are widespread and prominent members of natural aquatic environments (West 1989). The genus comprises many species, with at least 14 species being pathogenic in humans (Iwamoto et al. 2010). The Cholera and Other Vibrio Illness Surveillance (COVIS) system of the United States Centers for Disease Control and Prevention (CDC) have reported domestic *Vibrio* infection annually. *Vibrio parahaemolyticus*, *V. vulnificus*, *V. alginolyticus*, *V. fluvialis*, *V. mimicus*, and toxigenic or non-toxigenic *V. cholerae* have been reported as the cause of major *Vibrio* infection outbreaks. These cases have involved bacteria other than serogroups O1 and O139 (CDC 2012b, CDC 2013, CDC 2014). *V. parahaemolyticus*, *V. vulnificus*, and *V. cholerae* are the most common.

   V. parahaemolyticus is a leading pathogen in seafood-borne diseases worldwide (Su and Liu 2007). The affected seafood includes shellfish (clam, oyster), shrimp, and octopus (Daniels et al. 2000, Cabanillas-Beltrán et al. 2006). The major symptoms of *V. parahaemolyticus* infection are watery diarrhea, abdominal cramps, vomiting, nausea, and bloody diarrhea. *V. parahaemolyticus* pathogenicity is strongly related to the presence of the gene encoding the thermostable direct hemolysin (TDH). Presence of the TDH protein is evident as hemolytic activity on Wagatsuma blood agar, which is termed the Kanagawa phenomenon. A study from nearly 50 years ago reported that 96% of clinical isolates, but only 1% of environmental isolates from fish and seawater, were Kanagawa phenomenon-positive (Sakazaki et al. 1968). However, Kanagawa phenomenon-negative strains lacking TDH have been reported as the cause of gastroenteritis (Honda et al. 1987). Honda et al. (1988) reported a Kanagawa phenomenon-negative strain that produced a TDH-related hemolysin (TRH) as a virulence factor. Subsequent research revealed *tdh* and/or *trh* genes in most clinical isolates, with most environmental isolates lacking these genes (Nishibuchi and Kaper 1995). Thus, many environmental strains of *V. parahaemolyticus* in the environment do not possess pathogenicity.

   V. parahaemolyticus was one of the leading causes of food-borne diseases in Japan in the 1990s. The prevalence has markedly decreased since 2000 (Fig. 3.1). One of the main reasons for this decline has been the establishment and implementation of standards designed to curb *V. parahaemolyticus* infections. The standards have included storage (≤ 10°C),

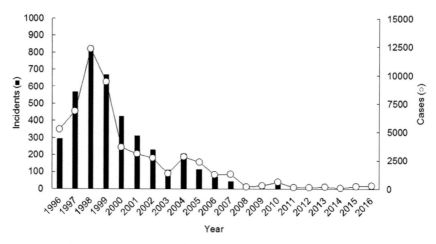

Figure 3.1 Changes in *V. parahaemolyticus* infection in Japan. Data was obtained from the list of food poisoning occurrences published by the Ministry of Health, Labour and Welfare in Japan (http://www.mhlw.go.jp/stf/seisakunitsuite/bunya/kenkou_iryou/shokuhin/syokuchu/04.html).

processing (using disinfected, artificial seawater, or water), and viable count (most probable number (MPN) per g < 100 for seafood eaten raw, and absence of viable bacteria in 25 g of boiled octopus and crab). Hara-Kudo et al. (2012) demonstrated that changes in the natural contamination of *V. parahaemolyticus* in seafood materials were not related to the decrease. Thus, the standards are considered to decrease *V. parahaemolyticus* population from processing to consumers during distribution.

V. *vulnificus* is the cause of septicemia, wound infection, and, more rarely, gastroenteritis (Butt et al. 2004). The fatality rate of seafood-borne septicemia is very high. Susceptibility to infection and chronic diseases influences the virulence of *V. vulnificus* infection. A study conducted in the U.S. reported that 80% of septicemia patients had pre-existing liver diseases (Shapiro et al. 1998). Major transmission routes of seafood-borne septicemia are ingestion of raw or undercooked shellfish, especially raw oysters. DePaola et al. (1994) reported densities of *V. vulnificus* in oyster and crab of about 5 log MPN/100 g during the spring and summer, with higher numbers in the intestines of inshore fish that eat mollusks and crustaceans. The presence of *V. vulnificus* in seawater is strongly related to seawater temperature, and isolation of *V. vulnificus* is often difficult at a seawater temperature below 13°C (DePaola et al. 1994, Pfeffer et al. 2003). Therefore, *V. vulnificus* counts of oyster can vary depending on season and seawater temperature (DePaola et al. 1994). Cases of oyster-associated *V. vulnificus* infections are more common during warmer months (Shapiro et al. 1998).

V. cholerae is another cause of seafood-borne poisoning. *V. cholerae* is commonly categorized into two serotypes: O1 and O139, and non-O1/non-O139 *V. cholerae* (Ottaviani et al. 2009). O1 and O139 have been traditionally associated with cholera. Although non-O1/non-O139 *V. cholerae* have been considered as part of the normal bacterial flora of aquatic ecosystems, incidents of diarrhea have been described. Major virulence factors of *V. cholerae* O1 and O139 are cholera toxin (*ctx*) and toxin-co-regulated pilus (*tcp*). Generally, *V. cholerae* non-O1/non-O139 do not possess cholera toxins (Chatterjee et al. 2009). However, putative virulence factors of non-O1/non-O139 include hemolysin (*hlyA*), heat-stable enterotoxin (*stn/sto*), outer membrane protein (*ompU*), repeats-in-toxin A toxin (*rtxA*), ToxR regulatory protein (*toxR*), and the type III secretion system (Rivera et al. 2001, Singh et al. 2001, Luo et al. 2013, Siriphap et al. 2017). Similar to other *Vibrio* species, *V. cholerae* infection results from ingestion of contaminated seafood or water. In countries that have implemented food hygiene regulations, infections due to O1 and O139 are mainly associated with overseas travel and consumption of local seafood. Non-O1/non-O139 *V. cholerae* contaminated various fresh or frozen seafood include mussel, prawn, cod, and anchovy (Ottaviani et al. 2009), with many infections reported following consumption of seafood domestically in the US (CDC 2011, CDC 2012b, CDC 2013).

3. *Salmonella* spp.

Salmonella spp. are Gram-negative, rod-shaped bacteria that are facultatively anaerobic. The two species are *S. enterica* and *S. bongori*. There are six subspecies of *S. enterica*: *S. enterica* subsp. *enterica*, *S. enterica* subsp. *salamae*, *S. enterica* subsp. *arizonae*, *S. enterica* subsp. *diarizonae*, *S. enterica* subsp. *houtenae*, and *S. enterica* subsp. *indica*. *Salmonella* spp. are classified by their cell surface components of lipopolysaccharide (O-antigen) and flagella (H-antigen). There are over 2,500 serovars in the genus *Salmonella* (Grimont and Weill 2007).

Food-borne diseases by *Salmonella* spp. are caused by two strains: typhoid and non-typhoid *Salmonella*. In general, typhoid *Salmonella* includes *S. enterica* serovar Typhi and serovar Paratyphi, which cause severe typhoid fever and paratyphoid fever, respectively. Typhoid fever begins several days to a few weeks after ingestion of bacteria, and features high fever, abdominal cramps, and severe diarrhea (Butt et al. 2004). Symptoms of paratyphoid fever are milder with a shorter incubation period (Bhan et al. 2005). Although both serovars are important causes of severe food-borne disease, most food-borne diseases associated with *Salmonella* spp. are due to infection by non-typhoid *Salmonella*. The latter salmonellosis features gastroenteritis, diarrhea, abdominal pain, muscle aches, and fever (Butt et al. 2004). Patients with non-typhoid salmonellosis usually recover within a week.

Salmonella spp. are widely present in the intestinal tract of animals (mammals, birds, and reptiles), which creates the potential for contamination of various foods. Examples of *Salmonella* outbreak related to seafood products are shown in Table 3.2. Most of the food-borne diseases due to non-typhoid *Salmonella* spp. involve consumption of contaminated foods. A report issued in 2000 by the U.S. Food and Drug Administration confirmed *Salmonella* spp. in 1.3% and 7.2% of domestic and imported seafood, respectively, in the U.S., with contamination of 2.6% of ready-to-eat imported seafood, such as cooked shrimp, shellfish, smoked fish, dried fish, and fish paste (Heinitz et al. 2000). Two of the most important *Salmonella* spp. are *S. enterica* serovar Enteritidis and serovar Typhimurium. *S. enterica* serovar Enteritidis is reportedly the most common serotype worldwide from human isolates, with *S. enterica* serovar Typhimurium found most frequently in non-human isolates. Other major serotypes from human or nonhuman isolates include *S. enterica* serovar Newport, serovar Heidelberg, and serovar Infantis (Galanis et al. 2006).

Prevention of contamination and ensuring bacterial killing during food processing are very important for inhibition of *Salmonella* infection, because *Salmonella* spp. can often survive harsh environments during food

Table 3.2 Example of *Salmonella* outbreaks related to seafood products.

Year	Country	Case	Death	Causative food	Serovar	Reference
1980–1994	United States[†]	49	N.D.	Whitefish	Enteritidis	Wallace et al. 1999
1980–1994	United States[†]	57 (3 outbreaks)	N.D.	Shrimp	Enteritidis, Typhimurium, and group D non-typhi	Wallace et al. 1999
1984	United Kingdom	87 (2 outbreaks)	0	Salmon	Montevideo	Cartwright and Evans 1988
1999	Japan	1505	0	Dried squid	Oranienburg	Miyakawa et al. 2006
2012	United States	425	0	Raw scraped ground tuna product	Bareilly and Nchanga	CDC 2012a
2015	United States	65	0	Frozen raw tuna	Paratyphi B and Weltevreden	CDC 2015

N.D.: No description.
[†]: Only outbreaks in New York were examined.

preservation. In Japan, more than 1,500 people fell sick during a diffuse outbreak of dried squid contaminated with *S. enterica* serovar Oranienburg in 1999 (Miyakawa et al. 2006). Contamination during the manufacturing process was the likely cause of the outbreak. The strain was able to survive incubation for 140 hours in 10% NaCl in peptone water, and 150–180 days in dried squid, although viable cell counts gradually decreased (Saito et al. 2000). Asakura et al. (2002) suggested that *S.* Oranienburg can adopt a viable but non-culturable (VNC) state in the presence of NaCl osmotic stress. This suggestion is important, given that food preservation often utilizes NaCl.

4. *Listeria monocytogenes*

Listeria monocytogenes is a Gram-positive bacterium, and is the causative agent of listeriosis. Listeriosis is a worldwide health concern. It is considered an emerging food-borne disease owing to changes in food production, preservation, and distribution, which are associated with extension of shelf life by cold chain because of the psychrotrophic growth capability (< 4°C) of *L. monocytogenes*. Symptoms of listeriosis include gastroenteritis, meningitis, encephalitis, mother-to-fetus infection, and septicemia. Listeriosis causes death in 25–30% of cases (Hamon et al. 2006). In the U.S. during 1998–2008, 24 outbreaks of listeriosis were reported, which resulted in 359 illnesses, 215 hospitalizations, and 38 deaths; serotype 4b was the most frequently involved (Cartwright et al. 2013). Deli meats, frankfurters, and cheese were investigated as causative foods. The U.S. Food and Drug Administration has set a "zero tolerance" for *L. monocytogenes*, especially for ready-to-eat foods that do not undergo further cooking (Klima and Montville 1995).

Food-borne listeriosis is relatively rare, but severe. Several recent outbreaks have occurred. In the U.S. during 2011, cantaloupe associated outbreak was occurred. In this outbreak, 147 people were infected and 33 died. *L. monocytogenes* serotype 1/2a and 1/2b were associated with the outbreak (McCollum et al. 2013). Other outbreaks were associated with ice cream and celery (Gaul et al. 2013, Pouillot et al. 2016, Buchanan et al. 2017). Seafood related listeriosis is rarer in comparison to illness associated with meats and dairy products. Reported seafood-borne outbreaks include shrimp in the U.S. (Riedo et al. 1994), "gravad" rainbow trout in Sweden (Ericsson et al. 1997), smoked mussels in New Zealand (Brett et al. 1998), vacuum-packed cold-smoked rainbow trout in Finland (Miettinen et al. 1999), and imitation crab meat in Canada (Farber et al. 2000; Table 3.3).

L. monocytogenes can live in diverse environments, such as soil, water, plants, animals, and humans (Weis and Seeliger 1975, Watkins and Sleath 1981, Okutani et al. 2004), suggesting that *L. monocytogenes* contaminates a variety of foods. Yamazaki et al. (2000) reported that in Japan, 4.5% of finfish (including salmon), 23.1% of smoked seafood (including smoked salmon),

Table 3.3 Example of *L. monocytogenes* outbreaks related to seafood products.

Year	Country	Case	Death	Causative food	Serotype	Reference
1980	New Zealand	22	6	Shellfish and raw fish	1b	Lenon et al. 1984
1989	United States	2	Unknown	Shrimp	4b	Riedo et al. 1994
1992	New Zealand	4	2	Smoked mussel	1/2a	Brett et al. 1998
1994–1995	Sweden	9	2	Graved or cold-smoked rainbow trout	4b	Ericsson et al. 1997
1996	Canada	2	0	Imitation crab meat	1/2b	Farber et al. 2000
1999	Finland	5	0	Cold-smoked rainbow trout	1/2b	Miettinen et al. 1999

and 22.2% of fermented seafood were contaminated with *L. monocytogenes*. Another report from Japan described *L. monocytogenes* contaminations in raw ready-to-eat seafood in 12.1% of minced tuna, 5.7% of salmon roe, 9.1% of cod roe, and 3.0% of smoked salmon (Miya et al. 2010). Testing of foods in the U.S. in 2000 and 2001 revealed 4.7% of seafood salads and 4.3% of smoked seafood were contaminated, with contamination exceeding 10^2 colony forming units (CFU)/g (Gombas et al. 2003).

Although the accurate minimum infective dose has not been determined, it is estimated to be 10^2–10^9 CFU depending on the immunological status of individuals (Jemmi and Stephan 2006). Therefore, it is important to control of the growth of *L. monocytogenes* in seafood. Storage under refrigeration is not an efficient control; growth of *L. monocytogenes* in smoked salmon at 4–5°C has been observed (Rørvik et al. 1991, Jørgensen and Huss 1998). Additional control methods using the "hurdle technology" are needed; these include pH, water activity, and food additives (Leistner and Gorris 1995). Proposed anti-listerial agents include bacteriocins and bacteriophages (Goodridge and Bisha 2011, Ghanbari et al. 2013).

Prevention of contamination is also important, and contamination of a smoked salmon processing plant has been shown (Rørvik et al. 1995). Contamination of food-processing plants is one of the most important issues in microbiological food hygiene, because food-manufacturing environments become persistent sources of contamination. *L. monocytogenes* can form biofilms on food-processing surfaces (Blackman and Frank 1996). In a food-processing plant, established clones of *L. monocytogenes* may persist for years (Møretrø and Langsrud 2004). Complete cleaning of a processing plant is essential to eradicate persistent contamination.

5. *Clostridium botulinum*

Clostridium botulinum is a Gram-positive rod. It is an obligate anaerobe. This bacterium can form a spore and survive in harsh environments. *C. botulinum* is divided into four phenotypic groups (I, II, III, and IV). Type I and IV are proteolytic, and type II and III are non-proteolytic. Each type produces botulinum neurotoxins. The neurotoxins also classified into seven groups (A to G) according to their antigenic properties. Human food-borne diseases typically involve type A or B (Peck et al. 2011), but type E neurotoxin is the most frequent in seafood-borne botulism, because bacteria harboring type E remain active at 3–5°C (Feldhusen 2000). Other types of *C. botulinum* cannot proliferate under refrigerated temperatures.

Botulinum neurotoxins are synthesized as 150-kDa single-chain proteins. Each type of neurotoxin is proteolytically digested to a 100-kDa heavy chain and 50-kDa light chain, both of which are cross-linked by a disulfide bond (Couesnon et al. 2006, Singh 2006). Gene expression of botulinum neurotoxin peaks at the late logarithmic or early stationary phase of growth (Bradshaw et al. 2004, Chen et al. 2008). Therefore, botulinum neurotoxin accumulates by growth of *C. botulinum* during food storage. The ingestion of the accumulated botulinum neurotoxin causes botulism. The symptoms of botulism are weakness, visual defects, and death by respiratory paralysis and cardiac arrest. The estimated lethal dose is 0.1–1 µg of neurotoxin (Feldhusen 2000).

C. botulinum is ubiquitous in the aquatic environment and seafood worldwide. *C. botulinum* type E has been reported in 90% of aquatic environment samples from Denmark, and 86% of samples from Greenland (Huss 1980). In a survey of Finnish freshwater and Baltic Sea sediment samples using a neurotoxin-specific polymerase chain reaction (PCR), 81% of seawater and 61% of freshwater samples were positive for *C. botulinum* type E (Hielm et al. 1998). Hyytiä et al. (1998) reported that 10–40% of raw fish, 4–14% of fish roe, and 2–10% of vacuum-packed seafood products in Finland were positive for the *C. botulinum* type E neurotoxin gene, and that *C. botulinum* type E might be a serious health risk associated with the consumption of seafood. In Japan, distribution of *C. botulinum* type E was demonstrated in river soil (Yamakawa and Nakamura 1992), and 7.7% of seafood from local supermarkets were contaminated by *C. botulinum* type C, D, or others (Haq and Sakaguchi 1980).

Seafood-borne botulism is a worldwide hazard (Peck 2006). The cause is usually inadequately processed, homemade foods. Examples of seafood-borne botulism are listed in Table 3.4. Many cases of seafood-borne botulism involve *C. botulinum* type E, and outbreaks of fermented foods such as "Izushi", "Faseikh", and "Rakfisk" were reported (Nakamura et al. 1956, Weber et al. 1993, Eriksen et al. 2004). Refinements in food processing and

Table 3.4 Example of *C. botulinum* outbreaks related to seafood products.

Year	Country	Case	Death	Causative food	Neurotoxin type	Reference
1951–1955	Japan	54	15	Izushi (suhshi made by fermenting fish)	E	Nakamura et al. 1956
1987	United States and Israel	8	1	Ribbetz or kapchunka (whitefish soaked in brine and air-dried)	E	Slater et al. 1989
1991	Egypt	91	18	Faseikh (fermented, salted mullet fish)	E	Weber et al. 1993
1997	Germany	2	0	Vacuum-packed hot-smoked whitefish	E	Korkeala et al. 1998
2002	United States	8	0	Muktuk from a beluga whale	E	McLaughlin et al. 2004
2003	Germany	3	0	Dried fish	E	Eriksen et al. 2004
2003	Norway	4	0	Rakfisk (salted and fermented fish)	N.D.	Eriksen et al. 2004
2005	United States	5	0	Salted whitefish	E	Sobel et al. 2007
2006	Finland	2	0	Vacuum-packed smoked whitefish	E	Lindström et al. 2006
2009	France	3	0	Vacuum-packed hot-smoked whitefish	E	King et al. 2009

N.D.: No description.

packaging technology have made it possible to distribute processed foods at low temperature without excessive processing, such as heating. However, it is difficult to kill the spores of *C. botulinum* without appropriate heating. In particular, vacuum packaging is an effective strategy for extension of shelf life. However, the anaerobic environment promotes proliferation of *C. botulinum* during storage. Botulism involving vacuum-packed fish has been reported (Table 3.4). *C. botulinum* is a ubiquitous bacterium. Therefore, food manufacturers should pay attention to *C. botulinum* as a natural contaminant. Certain killing of spores with appropriate heating or setting of proper storage conditions are important for inhibition of the outbreaks of botulism. Botulinum neurotoxin is heat-labile, and proper thermal processes during food processing or before consumption can inactivate neurotoxin and spores (Feldhusen 2000). Thorough cold storage ($< 3.3°C$) should be managed for unheated foods, and storage below pH 4.6 and a water activity of 0.94 can also prevent the growth of *C. botulinum*.

6. Histamine-producing Bacteria

Histamine-producing bacteria are related to histamine (scombroid) poisoning. Histamine poisoning is caused by consumption of foods that have accumulated a high level of histamine. Histamine-producing bacteria convert free histidine into histamine, using histidine decarboxylase. Therefore, histidine-rich foods are major causes of histamine poisoning. Temperature abused fish meats, including tuna, sardine, mackerel, and bonito, and fish sauce have the risk for accumulation of histamine. Symptoms of histamine poisoning are relatively mild and include allergy-like presentations, such as hives, rash, flushing, and facial swelling (Hungerford 2010), but several cases of life-threatening reaction have been reported (Sánchez-Guerrero et al. 1997, D'Aloia et al. 2011). Outbreaks of histamine poisoning related to the ingestion of raw or processed seafood have occurred worldwide, including the U.S., United Kingdom, South Africa, Japan, Italy, and Taiwan (Visciano et al. 2012).

Many bacteria are histamine-producers. Gram-negative bacteria include *Photobacterium damselae* subsp. *damselae*, *P. phosphoreum*, *P. kishitanii*, *P. angustum*, *Morganella morganii* subsp. *morganii*, *M. psychrotolerans*, *Proteus vulgaris*, and *Raoultella planticola* (Bermejo et al. 2003, Kanki et al. 2004, Rodtong et al. 2005, Emborg et al. 2006, Ferrario et al. 2012, Bjornsdottir-Butler et al. 2016). Histamine accumulation in fermented food (wine, cheese, and fish sauce) involves Gram-positive lactic acid bacteria, such as *Lactobacillus buchneri*, *L. hilgardii*, *Oenococcus oeni*, *Tetragenococcus halophilus*, and *T. muriaticus* (Lonvaud-Funel and Joyeux 1994, Kimura et al. 2001, Lucas et al. 2005, Martín et al. 2005, Satomi et al. 2008). Histidine decarboxylases include pyridoxal phosphate-dependent histidine decarboxylase in gram-negative bacteria (Kamath et al. 1991) and pyruvoyl-type histidine decarboxylase in gram-positive bacteria (van Poelje and Snell 1990).

Histamine-producing bacteria are widespread in fish-processing environments. Bjornsdottir-Butler et al. (2015) reported that 49% and 14% of 235 scombrotoxin-forming fish from the Gulf of Mexico were contaminated by *P. damselae* subsp. *damselae* and *M. morganii*, respectively. Another report described that contamination rates of *P. damselae* subsp. *damselae*, *P. phosphoreum*, *P. kishitanii*, and *M. morganii* were 42.7%, 13.6%, 14.4%, and 24.0%, respectively, from retail fish samples in Japan (Torido et al. 2014). Moreover, psychrotolerant histamine-producing *M. psychrotolerans* was detected from 44.3% of raw or processed seafood in Japan, and all isolates produced more than 4,000 mg/L of histamine during 48 hours of incubation at 25°C (Kato et al. 2017). These results indicate that histamine-producing bacteria generally contaminate fish meats. Therefore, storage under low temperature is the most important factor for inhibition of production of a large amount of histamine.

7. Noroviruses

Norovirus is a viral pathogen. The genus *Norovirus* is a member of the family *Caliciviridae*. Norovirus infection causes gastroenteritis that features abdominal pain, diarrhea, vomiting, and fever. The incubation period is 18–48 hours, and symptoms are self-limiting during 1–2 days (Butt et al. 2004). Norovirus infection is the most common seafood-borne disease worldwide. Noroviruses have been described as "small round structured viruses" and "Norwalk-like viruses" (Moore et al. 2015). Noroviruses are non-enveloped icosahedrons that range from 30–38 nm in diameter. The genome is positive-sense single-stranded RNA (ca. 7.5 kbp) with a polyadenylated 3'-terminus (Ushijima et al. 2014). There are five genogroups (GI to GV); groups GI, GII, and GIII are classified into nine, 22, and two groups, respectively (Ushijima et al. 2014). Among these genogroups, GI, GII, and GIV are virulent for humans (Moore et al. 2015). Groups GI and GII are the major causes of norovirus infection (Matthews et al. 2012, Vega et al. 2014); GII.4 is disseminated worldwide and is the cause of the majority of outbreaks (Siebenga et al. 2009).

Norovirus is typically transmitted person-to-person from an infected individual via the fecal-oral route or hand contact, or by eating contaminated foods (Butt et al. 2004). Raw or processed, but non-heated shellfish, especially oysters, are the leading causative food. Virus particles excreted from infected people contaminate seawater through sewage, and oysters concentrate virus particles into their body with contaminated seawater. Norovirus cannot increase in oysters, but persist in the environment. Therefore, in order to prevent Norovirus infection, sufficient heating of suspected food prior to ingestion is mandatory.

8. Other Seafood-borne Pathogens

Staphylococcus aureus and *Clostridium perfringens* also cause food poisoning associated with seafood. *S. aureus* is round, Gram-positive, and facultatively anaerobic. This bacterium is salt tolerant and can proliferate even in salted food, where other bacteria cannot proliferate. Outbreaks of *S. aureus* are caused by consumption of staphylococcal enterotoxins. Thus, latency is shorter than other infectious food poisonings. Symptoms of staphylococcus poisoning are vomiting, diarrhea, and abdominal pain. Staphylococcal enterotoxins can remain in food after heating because of their thermal stability. Therefore, inhibiting growth of *S. aureus* is important to prevent staphylococcal food poisoning. *S. aureus* is present on human skin. Simon and Sanjeev (2007) reported that 62% of factory workers were carriers of *S. aureus*, and among *S. aureus* isolated from factory workers, 28% of isolates was enterotoxigenic. Basti et al. (2006) demonstrated high rates of *S. aureus* contamination of smoked or salted Iranian fish, but almost no contamination of fresh fish. Contamination from food handlers is a particular concern.

C. perfringens is a Gram-positive, spore-forming, obligate anaerobic rod. This bacterium also produces enterotoxin, which accumulates in the intestine during the sporulation process. *C. perfringens* infection is caused by ingestion of foods contaminated with large numbers (> 10^6 CFU/g) of vegetative bacterial cells (García and Heredia 2011). *C. perfringens* infection is relatively uncommon in seafood, but *C. perfringens* that harbor the gene encoding enterotoxin can contaminate seafood (Wen and McClane 2004, Rahmati and Labbe 2008).

Aeromonas spp. are Gram-negative, facultative anaerobic bacteria. This genus is widely distributed in aquatic environments, including seafood. Some *Aeromonas* spp. are virulent in humans and fish. *Aeromonas* spp. contaminate about 70–90% of fish, shrimp, and ready-to-eat seafood products (Neyts et al. 2000, Pinto et al. 2012). Seafood-related *Aeromonas* infections are usually present as gastroenteritis (Butt et al. 2004).

References

Asakura, H., Makino, S., Takagi, T., Kuri, A., Kurazono, T., Watarai, M. and Shirahata, T. 2002. Passage in mice causes a change in the ability of *Salmonella enterica* serovar Oranienburg to survive NaCl osmotic stress: resuscitation from the viable but non-culturable state. FEMS Microbiol. Lett. 212: 87–93.

Basti, A.A., Misaghi, A., Salehi, T.Z. and Kamkar, A. 2006. Bacterial pathogens in fresh, smoked and salted Iranian fish. Food Cont. 17: 183–188.

Bermejo, A., Mondaca, M.A., Roeckel, M. and Marti, M.C. 2003. Growth and characterization of the histamine-forming bacteria of jack mackerel (*Trachurus symmetricus*). J. Food Process. Preserv. 26: 401–414.

Bhan, M.K., Bahl, R. and Bhatnagar, S. 2005. Typhoid and paratyphoid fever. Lancet 366: 749–762.

Bjornsdottir-Butler, K., Bowers, J.C. and Benner, R.A. Jr. 2015. Prevalence and characteriazation of high histamine-producing bacteria in Gulf of Mexico fish species. J. Food Prot. 78: 1335–1342.

Bjornsdottir-Butler, K., McCarthy, S.A., Dunlap, P.V. and Benner, R.A. Jr. 2016. *Photobacterium angustum* and *Photobacterium kishitanii*, psychrotrophic high-level histamine-producing bacteria indigenous to tuna. Appl. Environ. Microbiol. 82: 2167–2176.

Blackman, I.C. and Frank, J.F. 1996. Growth of *Listeria monocytogenes* as a biofilm on various food-processing surfaces. J. Food Prot. 8: 827–831.

Bradshaw, M., Dineen, S.S., Maks, N.D. and Johnson, E.A. 2004. Regulation of neurotoxin expression in *Clostridium botulinum* strains 62A, hall A-*hyper*, and NCTC 2916. Anaerobe 10: 321–333.

Brett, M.S.Y., Short, P. and McLauchlin, J. 1998. A small outbreak of listeriosis associated with smoked mussels. Int. J. Food Microbiol. 43: 223–229.

Buchanan, R.L., Gorris, L.G.M., Hayman, M.M., Jackson, T.C. and Whiting, R.C. 2017. A review of *Listeria monocytogenes*: an update on outbreaks, virulence, dose-response, ecology, and risk assessments. Food Cont. 75: 1–13.

Butt, A.A., Aldridge, K.E. and Sanders, C.V. 2004. Infections related to the ingestion of seafood. Part I: viral and bacterial infections. Lancet Infect. Dis. 4: 201–212.

Cabanillas-Beltrán, H., LLausás-Magaña, E., Romero, R., Espinoza, A., García-Gasca, A., Nishibuchi, M., Ishibashi, M. and Gomez-Gil, B. 2006. Outbreak of gastroenteritis caused by the pandemic *Vibrio parahaemolyticus* O3:K6 in Mexico. FEMS Microbiol. Lett. 265: 76–80.

Cartwright, E.J., Jackson, K.A., Johnson, S.D., Graves, L.M., Silk, B.J. and Mahon, B.E. 2013. Listeriosis outbreaks and associated food vehicles, United States, 1998–2008. Emerg. Infect. Dis. 19: 1–9.

Cartwright, K.A.V. and Evans, B.G. 1988. Salmon as a food-poisoning vehicle-two successive salmonella outbreaks. Epidem. Inf. 101: 249–257.

Centers for Disease Control and Prevention. 2011. National Enteric Disease Surveillance: Cholera and other *Vibrio* Illness Surveillance (COVIS) Annual Summary, 2011 [https://www.cdc.gov/vibrio/surveillance.html], accessed 2 May 2017.

Centers for Disease Control and Prevention. 2012a. Multistate outbreak of *Salmonella* Bareilly and *Salmonella* Nchanga infections associated with a raw scraped ground tuna product (final update, posted 26 July 2012) [https://www.cdc.gov/salmonella/bareilly-04-12/index.html], accessed 7 June 2017.

Centers for Disease Control and Prevention. 2012b. National Enteric Disease Surveillance: Cholera and other *Vibrio* Illness Surveillance (COVIS) Annual Summary, 2012 [https://www.cdc.gov/vibrio/surveillance.html], accessed 2 May 2017.

Centers for Disease Control and Prevention. 2013. National Enteric Disease Surveillance: Cholera and other *Vibrio* Illness Surveillance (COVIS) Annual Summary, 2013 [https://www.cdc.gov/vibrio/surveillance.html], accessed 2 May 2017.

Centers for Disease Control and Prevention. 2014. National Enteric Disease Surveillance: Cholera and other *Vibrio* Illness Surveillance (COVIS) Annual Summary, 2014 [https://www.cdc.gov/vibrio/surveillance.html], accessed 2 May 2017.

Centers for Disease Control and Prevention. 2015. Multistate outbreak of *Salmonella* Paratyphi B variant L(+) tartrate(+) and *Salmonella* Weltevreden infections linked to frozen raw tuna (final update, posted 19 August 2015) [https://www.cdc.gov/salmonella/paratyphi-b-05-15/index.html], accessed 7 June 2017.

Chatterjee, S., Ghosh, K., Raychoudhuri, A., Chowdhury, G., Bhattacharya, M.K., Mukhopadhyay, A.K., Ramamurthy, T., Bhattacharya, S.K., Klose, K.E. and Nandy, R.K. 2009. Incidence, virulence factors, and clonality among clinical strains of non-O1, non-O139 *Vibrio cholerae* isolates from hospitalized diarrheal patients in Kolkata, India. J. Clin. Microbiol. 47: 1087–1095.

Chen, Y., Korkeala, H., Lindén, J. and Lindström, M. 2008. Quantitative real-time reverse transciption-PCR analysis reveals stable and prolonged neurotoxin cluster gene activity in a *Clostridium botulinum* type E strains at refrigeration temperature. Appl. Environ. Microbiol. 74: 6132–6137.

Couesnon, A., Raffestin, S. and Popoff, M.R. 2006. Expression of botulinum neurotoxins A and E, and associated non-toxin genes, during the transition phase and stability at high temperature: analysis by quantitative reverse transcription-PCR. Microbiology 152: 759–770.

D'Aloia, A., Vizzardi, E., Pina, P.D., Bugatti, S., Magro, F.D., Raddino, R., Curnis, A. and Cas, L.D. 2011. A scombroid poisoning causing a life-threatening acute pulmonary edema and coronary syndrome in a young healthy patient. Cardiovasc. Toxicol. 11: 280–283.

Daniels, N.A., MacKinnon, L., Bishop, R., Altekruse, S., Ray, B., Hammond, R.M., Thompson, S., Wilson, S., Bean, N.H., Griffin, P.M. and Slutsker, L. 2000. *Vibrio parahaemolyticus* infections in United States, 1973–1998. J. Infect. Dis. 181: 1661–1666.

DePaola, A., Capers, G.M. and Alexander, D. 1994. Densities of *Vibrio vulnificus* in the intestines of fish from the U.S. Gulf coast. App. Envrion. Microbiol. 60: 984–988.

Emborg, J., Dalgaard, P. and Ahrens, P. 2006. *Morganella psychrotolerans* sp. nov., a histamine-producing bacterium isolated from various seafoods. Int. J. Syst. Evol. Microbiol. 56: 2473–2479.

Ericsson, H., Eklöw, A., Danielsson-Tham, M.-L., Loncarevic, S., Mentzing, L.-O., Persson, I., Unnerstad, H. and Tham, W. 1997. An outbreak of listeriosis suspected to have been caused by rainbow trout. J. Clin. Microbiol. 35: 2904–2907.

Eriksen, T., Brantsæter, A.B., Kiehl, W. and Steffens, I. 2004. Botulism infection after eating fish in Norway and Germany: two outbreak reports. Eurosurveillance 8: pii=2366.

Farber, J.M., Daley, E.M., Mackie, M.T. and Limerick, B. 2000. A small outbreak of listeriosis potentially linked to the consumption of imitation crab meat. Lett. Appl. Microbiol. 31: 100–194.

Feldhusen, F. 2000. The role of seafood in bacterial foodborne diseases. Microbes Infect. 2: 1651–1660.

Ferrario, C., Pegollo, C., Ricci, G., Borgo, F. and Fortina, G. 2012. PCR detection and identification of histamine-forming bacteria in filleted tuna fish samples. J. Food Sci. 77: 115–120.

Galanis, E., Wong, D.M.A.L.F., Patrick, M.E., Binsztein, N., Cieslik, A., Chalermchaikit, T., Aidara-Kane, A., Ellis, A., Angulo, F.J. and Wegener, H.C. 2006. Web-based surveillance and global *Salmonella* distribution, 2000–2002. Emerg. Infect. Dis. 12: 381–388.

García, S. and Heredia, N. 2011. *Clostridium perfringens*: a dynamic foodborne pathogen. Food Bioprocess Technol. 4: 624–630.

Gaul, L.K., Farag, N.H., Shim, T., Kingsley, M.A., Silk, B.J. and Hyytia-Trees, E. 2013. Hospital-acquired listeriosis outbreak caused by contaminated diced celery—Texas, 2010. Clin. Infect. Dis. 56: 20–26.

Ghanbari, M., Jami, M., Domig, K.J. and Kneifel, W. 2013. Seafood biopreservation by lactic acid bacteria—A review. LWT-Food Sci. Technol. 54: 315–324.

Gombas, D.E., Chen, Y., Clavero, R.S. and Scott, V.N. 2003. Survey of *Listeria monocytogenes* in ready-to-eat foods. J. Food Prot. 66: 559–569.

Goodridge, L.D. and Bisha, B. 2011. Phage-based biocontrol strategies to reduce foodborne pathogens in foods. Bacteriophages 1: 130–137.

Grimont, P.A.D. and Weill, F.X. 2007. Antigenic formulae of the *Salmonella* serovars. WHO Collaborating Centre for Reference and Research on *Salmonella* (Institut Pasteur, Paris, France).

Hamon, M., Bierne, H. and Cossart, P. 2006. *Listeria monocytogenes*: a multifaceted model. Nat. Rev. Microbiol. 4: 423–434.

Haq, I. and Sakaguchi, G. 1980. Prevalence of *Clostridium botulinum* in fishes from markets in Osaka. Japan. J. Med. Sci. Biol. 33: 1–6.

Hara-Kudo, Y., Saito, S., Ohtsuka, K., Yamasaki, S., Yahiro, S., Nishio, T., Iwade, Y., Otomo, Y., Konuma, H., Tanaka, H., Nakagawa, H., Sugiyama, K., Sugita-Konishi, Y. and Kumagai, S. 2012. Characteristics of a sharp decrease in *Vibrio parahaemolyticus* infections and seafood contamination in Japan. Int. J. Food Microbiol. 157: 95–101.

Heinitz, M.L., Ruble, R.D., Wagner, D.E. and Tatini, S.R. 2000. Incidence of *Salmonella* in fish and seafood. J. Food Prot. 5: 579–592.

Hielm, S., Hyytiä, E., Andersin, A.-B. and Korkeala, H. 1998. A high prevalence of *Clostridium botulinum* type E in Finnish freshwater and Baltic Sea sediment samples. J. Appl. Microbiol. 84: 133–137.

Honda, S., Goto, I., Minematsu, I., Ikeda, N., Asano, N., Ishibashi, M., Kinoshita, Y., Nishibuchi, M., Honda, T. and Miwatani, T. 1987. Gastroenteritis due to Kanagawa negative *Vibrio parahaemolyticus*. Lancet 329: 331–332.

Honda, T., Ni, Y. and Miwatani, T. 1988. Purification and characterization of a hemolysin produced by a clinical isolate of Kanagawa phenomenon-negative *Vibrio parahaemolyticus* and related to the thermostable direct hemolysin. Infect. Immun. 56: 961–965.

Hungerford, J.M. 2010. Scombroid poisoning: a review. Toxicon 56: 231–243.

Huss, H.H. 1980. Distribution of *Clostridium botulinum*. Appl. Environ. Microbiol. 39: 764–769.

Hyytiä, E., Hielm, S. and Korkeala, H. 1998. Prevalence of *Clostridium botulinum* type E in finnish fish and fishery products. Epidemiol. Infect. 120: 245–250.

Iwamoto, M., Ayers, T., Mahon, B.E. and Swerdlow, D.L. 2010. Epidemiology of seafood-associated infections in the United States. Clin. Microbiol. Rev. 23: 399–411.

Jemmi, T. and Stephan, R. 2006. *Listeria monocytogenes*: food-borne pathogen and hygiene indicator. Rev. Sci. Tech. 25: 571–580.

Jørgensen, L.V. and Huss, H.H. 1998. Prevalence and growth of *Listeria monocytogenes* in naturally contaminated seafood. Int. J. Food Microbiol. 42: 127–131.

Kamath, A.V., Vaaler, G.L. and Snell, E.E. 1991. Pyridoxal phosphate-dependent histidine decarboxylases. Cloning, sequencing and expression of genes from *Klebsiella planticola* and *Enterobacter aerogenes* and properties of the overexpressed enzymes. J. Biol. Chem. 266: 9432–9437.

Kanki, M., Yoda, T., Ishibashi, M. and Tsukamoto, T. 2004. *Photobacterium phosphoreum* caused a histamine fish poisoning incident. Int. J. Food Microbiol. 92: 79–87.

Kato, R., Wang, D., Yamaki, S., Kawai, Y. and Yamazaki, K. 2017. Distribution on fishery products and histamine producing ability of *Morganella psychrotolerans*, a psychrophilic histamine-producing bacterium. Jpn. J. Food Microbiol. 34: 158–165.

Kimura, B., Konagaya, Y. and Fujii, T. 2001. Histamine formation by *Tetragenococcus muriaticus*, a halophilic lactic acid bacterium isolated from fish sauce. Int. J. Food Microbiol. 70: 71–77.

King, L.A., Niskanen, T., Junnikkala, M., Moilanen, E., Lindström, M., Korkeala, H., Popoff, M., Mazuet, C., Callon, H., Pihier, N., Peloux, F., Ichai, C., Quintard, H., Dellamonica, P., Cua, E., Lasfargue, M., Pierre, F. and de Valk, H. 2009. Botulism and hot-smoked whitefish: a family cluster of type E botulism in France, September 2009. Eurosurveillance 14: pii=19394.

Klima, R.A. and Montville, T.J. 1995. The regulatory and industrial responses to listeriosis in the USA: a paradigm for dealing with emerging foodborne pathogens. Trends Food Sci. Tech. 6: 87–93.

Korkeala, H., Stengel, G., Hyytiä, E., Vogelsang, B., Bohl, A., Wihlman, H., Pakkala, P. and Hielm, S. 1998. Type E botulism associated with vacuum-packaged hot-smoked whitefish. Int. J. Food Microbiol. 43: 1–5.

Leistner, L. and Gorris, L.G.M. 1995. Food preservation by hurdle technology. Trends Food Sci. Tech. 6: 41–46.

Lenon, D.F., Lewis, B.R., Mantell, C.F., Becroft, D., Dove, B., Farmer, K., Tonkin, S., Yeates, N., Stamp, R. and Mickleson, K.F. 1984. Epidemic perinatal listeriosis. Pediatr. Infect. Dis. 3: 30–34.

Lindström, M., Vuorela, M., Hindering, K., Korkeala, H., Dahlsten, E., Raahenmaa, M. and Kuusi, M. 2006. Botulism associated with vacuum-packed smoked whitefish in Finland, June–July 2006. Eurosurveillance 11: pii=3004.

Lonvaud-Funel, A. and Joyeux, A. 1994. Histamine production by wine lactic acid bacteria: isolation of a histamine-producing strain of *Leuconostoc oenos*. J. Appl. Microbiol. 77: 401–407.

Lucas, P.M., Wolken, W.A.M., Claisse, O., Lolkema, J.S. and Lonvaud-Funel, A. 2005. Histamine-producing pathway encoded on an unstable plasmid in *Lactobacillus hilgardii* 0006. Appl. Environ. Microbiol. 71: 1417–1424.

Luo, Y., Ye, J., Jin, D., Ding, G., Zhang, Z., Mei, L., Octavia, S. and Lan, R. 2013. Molecular analysis of non-O1/non-O139 *Vibrio cholerae* isolated from hospitalised patients in China. BMC Microbiol. 13: 52.

Martín, M.C., Fernández, M., Linares, D.M. and Alvarez, M.A. 2005. Sequencing, characterization and transcriptional analysis of the histidine decarboxylase operon to *Lactobacillus buchneri*. Microbiology 151: 1219–1228.

Matthews, J.E., Dickey, B.W., Miller, R.D., Felzer, J.R., Dawson, B.P., Lee, A.S., Rocks, J.J., Kiel, J., Montes, J.S., Moe, C.L., Eisenberg, J.N.S. and Leon, J.S. 2012. The epidemiology of published norovirus outbreaks: a review of risk factors associated with attack rate and genogroup. Epidemil. Infect. 140: 1161–1172.

McCollum, J.T., Cronquist, A.B., Silk, B.J., Jackson, K.A., O'Connor, K.A., Cosgrove, S., Gossack, J.P., Parachini, S.S., Jain, N.S., Ettestad, P., Ibraheem, M., Cantu, V., Joshi, M., DuVernoy, T., Fogg Jr., N.W., Gorny, J.R., Mogen, K.M., Spires, C., Teitell, P., Joseph, L.A., Tarr, C.L., Imanishi, M., Neil, K.P., Tauxe, R.V. and Mahon, B.E. 2013. Multistate outbreak of listeriosis associated with cantaloupe. N. Engl. J. Med. 369: 944–953.

McLaughlin, J.B., Sobel, J., Lynn, T., Funk, E. and Middaugh, J.P. 2004. Botulism type E outbreak associated with eating a beached whale, Alaska. Emerg. Infect. Dis. 10: 1685–1687.

Miettinen, M.K., Siitonen, A., Heiskanen, P., Haajanen, H., Björkroth, K.J. and Korkeala, H.J. 1999. Molecular epidemiology of an outbreak of febrile gastroenteritis caused by *Listeria monocytogenes* in cold-smoked rainbow trout. J. Clin. Microbiol. 37: 2358–2360.

Miya, S., Takahashi, H., Ishikawa, T., Fujii, T. and Kimura, B. 2010. Risk of *Listeria monocytogenes* contamination of raw ready-to-eat seafood products available at retail outlets in Japan. Appl. Environ. Microbiol. 76: 3383–3386.

Miyakawa, S., Takahashi, K., Hattori, M., Itoh, K., Kurazono, T. and Amano, F. 2006. Outbreak of *Salmonella* Oranienburg infection in Japan. J. Environ. Biol. 27: 157–158.

Moore, M.D., Goulter, R.M. and Jaykus, L.-A. 2015. Human norovirus as a foodborne pathogen: challenges and developments. Annu. Rev. Food Sci. Technol. 6: 411–433.

Møretrø, T. and Langsrud, S. 2004. *Listeria monocytogenes*: biofilm formation and persistence in food-processing environments. Biofilms 1: 107–121.

Nakamura, Y., Iida, H., Saeki, K., Kanzawa, K. and Karashimada, T. 1956. Type E botulism in Hokkaido, Japan. Jap. J. M. Sc. & Biol. 9: 45–58.

Neyts, K., Huys, G., Uyttendaele, M., Swings, J. and Debevere, J. 2000. Incidence and identification of mesophilic *Aeromonas* spp. from retail foods. Lett. Appl. Microbiol. 31: 359–363.

Nishibuchi, M. and Kaper, J.B. 1995. Thermostable direct hemolysin gene of *Vibrio parahaemolyticus*: a virulence gene acquired by a marine bacterium. Infect. Immun. 63: 2093–2099.

Okutani, A., Okada, Y., Yamamoto, S. and Igimi, S. 2004. Overview of *Listeria monocytogenes* contamination in Japan. Int. J. Food Microbiol. 93: 131–140.

Ottaviani, D., Leoni, F., Rocchegiani, E., Santarelli, S., Masini, L., Trani, V.D., Canonico, C., Pianetti, A., Tega, L. and Carraturo, A. 2009. Prevalence and virulence properties of non-O1 non-O139 *Vibrio cholerae* strains from seafood and clinical samples collected in Italy. Int. J. Food Microbiol. 132: 47–53.

Peck, M.W. 2006. *Clostridium botulinum* and the safety of minimally heated chilled foods: and emerging issue? J. Appl. Microbiol. 101: 556–570.

Peck, M.W., Stringer, S.C. and Carter, A.T. 2011. *Clostridium botulinum* in the post-genomic era. Food Microbiol. 28: 183–191.

Pfeffer, C.S., Hite, M.F. and Oliver, J.D. 2003. Ecology of *Vibrio vulnificus* in estuarine waters of Eastern North Carolina. Appl. Environ. Microbiol. 69: 3526–3531.

Pinto, A.D., Terio, V., Pinto, P.D. and Tantillo, G. 2012. Detection of potentially pathogenic *Aeromonas* isolates from ready-to-eat seafood products by PCR analysis. Int. J. Food Sci. Tech. 47: 269–273.

Pouillot, R., Klontz, K.C., Chen, Y., Burall, L.S., Macarisin, D., Doyle, M., Bally, K.M., Strain, E., Datta, A.R., Hammack, T.S. and Doren, J.M.V. 2016. Infectious dose of *Listeria monocytogenes* in outbreak linked to ice cream, United States, 2015. Emerg. Infect. Dis. 22: 2113–2119.

Rahmati, T. and Labbe, R. 2008. Levels and toxigenicity of *Bacillus cereus* and *Clostridium perfringens* from retail seafood. J. Food Prot. 71: 1178–1185.

Riedo, F.X., Pinner, R.W., Tosca, M.d.L., Cartter, M.L., Graves, L.M., Reeves, M.W., Weaver, R.E., Plikaytis, B.D. and Broome, C.V. 1994. A point-source foodborne listeriosis outbreak: documented incubation period and possible mild illness. J. Infect. Dis. 170: 693–696.

Rivera, I.N.G., Chun, J., Huq, A., Sack, R.B. and Colwell, R.R. 2001. Genotypes associated with virulence in environmental isolates of *Vibrio cholerae*. Appl. Environ. Microbiol. 67: 2421–2429.

Rodtong, S., Nawong, S. and Yongsawatdigul, J. 2005. Histamine accumulation and histamine-forming bacteria in Indian anchovy (*Stolephorus indicus*). Food Microbiol. 22: 475–482.

Rørvik, L.M., Yndestad, M. and Skjerve, E. 1991. Growth of *Listeria monocytogenes* in vacuum-packed, smoked salmon during storage at 4°C. Int. J. Food Microbiol. 14: 111–118.

Rørvik, L.M., Caugant, D.A. and Yndestad, M. 1995. Contamination pattern of *Listeria monocytogenes* and other *Listeria* spp. in a salmon slaughterhouse and smoked salmon processing plant. Int. J. Food Microbiol. 25: 19–27.

Sakazaki, R., Tamura, K., Kato, T., Obara, Y., Yamai, S. and Hobo, K. 1968. Studies on the enteropathogenic, facultatively halophilic bacteria, *Vibrio parahaemolyticus*. III. Enteropathogenicity. Jpn. J. Med. Sci. Biol. 21: 325–331.

Saito, A., Otsuka, K., Hamada, Y., Ono, K. and Masaki, H. 2000. Effects of temperature, pH and sodium chloride concentration on the growth of *Salmonella* Oranienburg and *Salmonella* Chester isolated from dried squid and behaviors of those strains in the dried squid. Jpn. J. Food Microbiol. 17: 11–17.

Sánchez-Guerrero, I.M., Vidal, J.B. and Escudero, A.I. 1997. Scombroid poisoning: a potentially life-threatening allergic-like reaction. J. Allergy Clin. Immunol. 100: 433–434.

Satomi, M., Furushita, M., Oikawa, H., Yoshikawa-Takahashi, M. and Yano Y. 2008. Analysis of a 30 kbp plasmid encoding histidine decarboxylase gene in *Tetragenococcus halophilus* isolated from fish sauce. Int. J. Food Microbiol. 126: 202–209.

Shapiro, R.L., Altekruse, S., Hutwagner, L., Bishop, R., Hammond, R., Wilson, S., Ray, B., Thompson, S., Tauxe, R.V., Griffin, P.M. and the *Vibrio* Working Group. 1998. The role of Gulf coast oysters harvested in warmer months in *Vibrio vulnificus* infections in the United States, 1988–1996. J. Infect. Dis. 178: 752–759.

Siebenga, J.J., Vennema, H., Zheng, D.-P., Vinjé, J., Lee, B.E., Pang, X.-L., Ho, E.C.M., Lim, W., Choudekar, A., Broor, S., Halperin, T., Rasool, N.B.G., Hewitt, J., Greening, G.E., Jin, M., Duan, Z.-J., Lucero, Y., O'Ryan, M., Hoehne, M., Schreier, E., Ratcliff, R.M., White, P.A., Iritani, N., Reuter, G. and Koopmans, M. 2009. Norovirus illness is a global problem: emergence and spread of norovirus GII.4 variants, 2001–2007. J. Infect. Dis. 200: 802–812.

Simon, S.S. and Sanjeev, S. 2007. Prevalence of enterotoxigenic *Staphylococcus aureus* in fishery products and fish processing factory workers. Food Cont. 18: 1565–1568.

Singh, B.R. 2006. Botulinum neurotoxin structure, engineering, and novel cellular trafficking and targeting. Neurotox. Res. 9: 73–92.

Singh, D.V., Matte, M.H., Matte, G.R., Jiang, S., Sabeena, F., Shukla, B.N., Sanyal, S.C., Haq, A. and Colwell, R.R. 2001. Molecular analysis of *Vibrio cholerae* O1, O139, non-O1, and non-O139: clonal relationships between clinical and environmental isolates. Appl. Environ. Microbiol. 67: 910–921.

Siriphap, A., Leekitcharoenphon, P., Kaas, R.S., Theethankaew, C., Aarestrup, F.M., Sutheinkul, O. and Hendriksen, R.S. 2017. Characterization and genetic variation of *Vibrio cholerae* isolated from clinical and environmental sources in Thailand. PLoS ONE 12: e0169324.

Slater, P.E., Addiss, D.G., Cohen, A., Leventhal, A., Chassis, G., Zehavi, H., Bashari, A. and Costin, C. 1989. Foodborne botulism: an international outbreak. Int. J. Epidemiol. 18: 693–696.

Sobel. J., Malavet, M. and John, S. 2007. Outbreak of clinically mild botulism type E illness from home-salted fish in patients presenting with predominantly gastrointestinal symptoms. Clin. Infect. Dis. 45: e14–e16.

Su, Y.-C. and Liu, C. 2007. *Vibrio parahaemolyticus*: a concern of seafood safety. Food Microbiol. 24: 549–558.

Torido, Y., Ohshima, C., Takahashi, H., Miya, S., Iwakawa, A., Kuda, T. and Kimura, B. 2014. Distribution of psychrophilic and mesophilic histamine-producing bacteria in retailed fish in Japan. Food Cont. 46: 338–342.

Ushijima, H., Fujimoto, T., Müller, W.E.G. and Hayakawa, S. 2014. Norovirus and foodborne disease: a review. Food Safety 2: 37–54.

van Poelje, P.D. and Snell, E.E. 1990. Pyruvoyl-dependent enzymes. Annu. Rev. Biochem. 59: 29–59.

Vega, E., Barclay, L., Gregoricus, N., Shirley, S.H., Lee, D. and Vinjé, J. 2014. Genotypic and epidemiologic treands on norovirus outbreaks in the United States, 2009 to 2013. J. Clin. Microbiol. 52: 147–155.

Visciano, P., Schirone, M., Tofalo, R. and Suzzi, G. 2012. Biogenic amines in raw and processed seafood. Front. Microbiol. 3: 188.

Wallace, B.J., Guzewich, J.J., Cambridge, M., Altekruse, S. and Morse, D.L. 1999. Seafood-associated disease outbreaks in New York, 1980–1994. Am. J. Prev. Med. 17: 48–54.

Watkins, J. and Sleath, K.P. 1981. Isolation and enumeration of *Listeria monocytogenes* from sewage, sewage sludge and river water. J. Appl. Bacteriol. 50: 1–9.

Weber, J.T., Hibbs, R.G., Darwish, A., Mishu, B., Corwin, A.L., Rakha, M., Hatheway, C.L., El Sharkawy, S., El Rahim, S.A., Al Hamd, M.F.S., Sarn, J.E., Blake, P.A. and Tauxe, R.V. 1993. A massive outbreak of type E botulism associated with traditional salted fish in Cairo. J. Infect. Dis. 167: 451–454.

Weis, J. and Seeliger, H.P.R. 1975. Incidence of *Listeria monocytogenes* in nature. Appl. Microbiol. 30: 29–32.

Wen, Q. and McClane, B.A. 2004. Detection of enterotoxigenic *Clostridium perfringens* type A isolates in American retail foods. Appl. Environ. Microbiol. 70: 2685–2691.

West, P.A. 1989. The human pathogenic vibrios—A public health update with environmental perspectives. Epidem. Inf. 103: 1–34.

Yamakawa, K. and Nakamura, S. 1992. Prevalence of *Clostridium botulinum* type E and coexistence of *C. botulinum* nonproteolytic type B in the rever soil in Japan. Microbiol. Immunol. 36: 583–591.

Yamazaki, K., Tateyama, T., Kawai, Y. and Inoue, N. 2000. Occurrence of *Listeria monocytogenes* in retail fish and processed seafood products in Japan. Fish. Sci. 66: 1191–1193.

4

Detection Method of Food-borne Pathogens in Seafood

Chihiro Ohshima and *Hajime Takahashi**

1. Introduction

Food ingredients go through a complex distribution process before they are purchased by food companies and processed into products. Once there is an outbreak of mass food poisoning, its cause may lie in a raw ingredient produced in an unexpected place, and identifying the cause is extremely difficult. Food companies carry out daily tests for microorganisms in raw materials, manufacturing processes, and final products, in order to prevent harm due to microbes.

The Hazard Analysis Critical Control Point (HACCP) approach has been introduced by large food companies that produce specific food products, with hazard analysis for microbial harm carried out at every stage of production, and the frequency and importance of microbial testing have consequently increased. The detection and identification of microorganisms has conventionally been almost entirely based on the culture method. Food products contain a complex mixture of microorganisms, and detecting the specific microorganisms responsible for food poisoning in cultures from samples of this type requires the use of enrichment culture, selective

Laboratory of Food Microbiology, Tokyo University of Marine Science and Technology, Japan.
* Corresponding author: hajime@kaiyodai.ac.jp

enrichment culture, differential testing, and confirmation testing for various different food-borne pathogens. This is a process that requires at least a week. However, some products, such as seafood, have a short shelf life, and may already have been eaten by consumers by the time the test results have been received. The culture method alone is thus inadequate for this type of product.

Recent advances in molecular biology, particularly the development of the polymerase chain reaction (PCR) technique, are transforming not just the techniques used for microbe detection and identification, but even their taxonomic classification. Molecular biological procedures do not necessarily require the culture step that was formerly essential to the detection of microorganisms, and are also superior to the culture method in terms of speed. These techniques are therefore being introduced into food manufacturing plants, and hold great promise not only for detecting microorganisms, but also as methods of evaluating the safety of food products.

In this chapter, we describe the detection of food-borne microorganisms by the culture method, molecular detection methods, and rapid identification methods.

2. Basic Cultural Method for Seafood-related Bacteria

2.1 Viable bacterial count

The numbers of living bacteria in food products are measured using the test methods set out in the compositional standards for each country. In general, plate culture using a non-selective enrichment culture medium is used. In this method, food diluted in physiological saline is spread on the surface of agar culture medium that has been poured into petri dishes, or mixed with the dissolved culture medium in the petri dish, and the numbers of colonies growing on the plate are counted. In the method set out in the Japanese Food Sanitation Act, the culture conditions for viable bacterial count comprise culture on plate count agar (PCA) medium at 35°C for 48 hours, while the International Organization for Standardization (ISO) method stipulates culture at 30°C for 72 hours. However, food products contain bacteria with a wide variety of characteristics, and these conditions may not necessarily be favorable. They are only ideal for mesophiles, which account for many of the bacteria that cause food poisoning, and if the pathogens to be detected are psychrophilic bacteria found in products such as seafood, other culture conditions must be considered. The standard agar culture medium prescribed in these regulations is also unsuited to the growth of some auxotrophic bacteria with particular nutritional requirements (such as lactic acid bacteria), and this may lead to underestimation of the bacterial count in food products containing large numbers of such bacteria.

Figure 4.1 Spiral plater.

As one specific example, some lactic acid bacteria that adhere to raw meat ingredients and spoil processed meat products cannot be measured by this method, potentially resulting in the contradictory situation that a sample's bacterial count may not reveal any problem even though it has obviously begun to rot.

In these tests for enumerating bacteria, dilute solutions of food must be further diluted to create a series, which is a task that requires labor. The Spiral Plater which, as shown in the Figure 4.1, inoculates the culture medium while diluting the sample automatically in a non-serial manner without the use of serial dilutions, has recently come into widespread use, and this labor-saving device has greatly reduced the amount of culture medium used.

2.2 Coliform bacteria test

The most reliable way of confirming the safety of food products is to test for the various food-borne pathogens that may be found in the product concerned, but carrying out all the possible tests during everyday business operations is impracticable, given the time and effort required. In practice, many food manufacturers do not test for individual food-borne pathogens in their factories, but instead carry out tests for coliform bacteria or *Escherichia coli*. In Japan, such testing is obligatory for many food products.

E. coli lives in the gut of mammals and other animals, and most of its strains are harmless. *E. coli* strains with a specific serotype that cause diseases are known as pathogenic *E. coli*. If the presence of *E. coli* is detected in a food product, the fact that this bacterium is found mainly in the gut means that fecal contamination of the product is suspected. For this reason, the presence of *E. coli* is regarded as an indicator of fecal contamination.

Coliform bacteria are defined as "aerobic or facultative anaerobic gram-negative bacteria that ferment lactose and produce gas", and include both *E. coli* and other genera with similar characteristics. As coliform bacteria are widely present in the natural environment with no direct connection to human or mammalian feces, the detection of coliform bacteria is not in itself suggestive of fecal contamination. As these bacteria are killed by heating to around 70°C, however, coliform bacteria are widely used in food manufacturing plants as an indicator of secondary contamination (fecal contamination) of food products that have been processed by heating or a similar method.

The usual method of testing is first to emulsify the food product in a liquid such as physiological saline, then transfer it to a liquid culture medium known as a lactase fermentation tube for culture. The test is positive if lactose is used and gas is generated during this culture process, as in this case coliform bacteria must have been present in the food product. The Most Probable Number (MPN) method is used to quantify the coliform bacteria count in a food product from coliform bacteria measurements in lactose fermentation tubes. If the presence of coliform bacteria is detected, tests are performed to identify whether or not these include *E. coli*. This involves the use of differential culture using a culture medium such as *E. coli* (EC) or eosin methylene blue (EMB) medium from the lactose fermentation tubes found to be positive for coliform bacteria, with the presence of *E. coli* confirmed if growth is observed at 44.5°C.

2.3 Vibrio parahaemolyticus test

The food-borne pathogens with the closest link to seafood are *Vibrio parahaemolyticus*. As these bacteria are found in seawater, their complete elimination from seafood is impossible. However, as they grow at comparatively high temperatures, their growth can be suppressed by freezing or otherwise storing seafood at low temperature.

As *Vibrio parahaemolyticus* are marine bacteria, their optimum pH for growth is comparatively alkaline compared to those of other bacteria. A selective culture medium with a somewhat higher pH, which suppresses the growth of other bacteria while encouraging *Vibrio parahaemolyticus* to grow, is therefore used for their detection. The method set out in ISO/TS 21872-1 comprises culturing samples in alkaline peptone water (APW) at 37°C or 41.5°C for 6 hours, followed by inoculation onto thiosulfate–citrate–bile salts–sucrose (TCBS) medium or Chrom ID Vibrio agar, with typical colonies picked up and their properties tested.

2.4 Other food-borne pathogen tests

In addition to *Vibrio parahaemolyticus*, other pathogens associated with seafood products include *Listeria monocytogenes*, *Salmonella*, *Norovirus*, and hepatitis viruses. *L. monocytogenes* and *Salmonella* are bacterial pathogens that can be detected by similar method of enrichment culture and differential culture used to test for *Vibrio parahaemolyticus*, and test methods have been set out by the ISO and the Food and Drug Agency Bacteriological Analytical Manual (FDA-BAM) in the United States. As culture is either extremely difficult or takes a long time for both *Norovirus* and hepatitis viruses, rapid test methods using real-time PCR (RT-PCR) are recommended. These are described in detail in the section on PCR below.

3. Rapid Detection Method Based on Molecular Techniques

3.1 DNA extraction of bacteria in food

When investigating the presence or absence of pathogenic bacteria in food, it is essential to extract the target microbial DNA from a homogenate of that food. Separating bacteria from food and extracting their DNA from the culture medium is relatively simple and feasible with elementary DNA extraction methods, such as the boiling method. However, extracting DNA from food homogenates requires a more refined methodology. Extraction kits, such as those commercialized by Quiagen and Macherey-Nagel, which include steps in which the DNA is adsorbed to a silica membrane and washed with a wash buffer, have good yields that allow real-time PCR (described below) to correlate the amount of the extracted DNA with the bacterial number. Both these kits involve methods to purify the DNA by lysing the bacteria and other cells present in the food using a culture medium containing guanidine and a surfactant to denature the protein, capturing the DNA using a silica membrane, washing the membrane with wash buffer and alcohol, and then collecting the purified DNA in distilled water. Kits that use silica beads rather than membranes offer a similar performance. Although extraction methods using phenyl-chloroform offer the highest degree of purification because of DNA recovery via ethanol precipitation, they are not suitable for real-time PCR-based quantification of bacterial numbers because purified DNA levels do not correlate with pre-extraction bacterial numbers. Additionally, since bacteria are divided into gram-positive and gram-negative, which have vastly different cell surface structures, the extraction efficiency of their DNA is different. Therefore, it is necessary to carefully select DNA purification methods depending on the objective of the study, such as using phenyl-chloroform when the bacterial DNA is needed in experiments such as in DNA recombination, or using an extraction kit employing either a silica membrane or silica beads and culturing the purified bacteria when quantifying the bacterial number in food.

3.2 PCR and realtime-PCR

PCR is the abbreviation for "polymerase chain reaction", a reaction in which part of the DNA is artificially amplified. The principles of PCR were discovered in 1987 (Mullis and Faloona 1987, Saiki et al. 1988, 1985), and this technique is now used for the detection of microorganisms in both clinical and food hygiene settings (Law et al. 2015).

PCR is a reaction in which DNA polymerase, a buffer formulated to optimize the conditions for the enzyme, primers (oligonucleotides consisting of around 20 bases comprising sequences complementary to both ends of the region to be amplified), deoxyribonucleotide triphosphate, and the sample nucleotide genome are mixed and reacted together in repeated cycles comprising three steps at different temperatures. The reactions that occur at each of these three temperature steps are as follows. First, the reaction solution is heated to 95°C, to denature the double helix of the sample DNA into single strands. The temperature is then reduced to around 60°C, at which the primers bind to both ends of the region to be amplified. At this point, the sequences that have formed partial double strands by hybridization between the sample DNA and the primers are recognized and bound by the DNA polymerase. Finally, the temperature of the reaction solution is shifted to the optimum temperature for the enzyme, and the DNA polymerase synthesizes sequences complementary to the sample DNA, forming double-stranded DNA. From the second cycle on, the double-stranded DNA synthesized during the previous cycles also act as a template for synthesis in the same way as the sample genome, so that the amount of DNA doubles after each cycle of these three temperature steps. At the end of the reaction, a technique such as electrophoresis is generally used to confirm whether or not amplification products are present.

The primers hybridize only with their complementary sequences, and the DNA polymerase synthesizes double-stranded DNA with these as the points of origin. By designing primers that target genes specific to the target microbe, only the DNA of this target microbe is amplified, enabling the detection of specific harmful microorganisms in food. Detection tests using PCR have been developed for the various *Vibrio* species that cause food poisoning in seafood (Kim et al. 2015), *E. coli* (Molina et al. 2015, Oh et al. 2014, Toma et al. 2003) *Listeria* spp. and *L. monocytogenes* (Janabi et al. 2016, Phraephaisarn et al. 2016), *Salmonella* (Shabarinath et al. 2007). *E. coli* is further subdivided according to differences in pathogenicity and infection mechanism, and PCR techniques for detecting the various strains have also been developed (Iguchi et al. 2015). Seafood may also be contaminated by viruses that can cause food poisoning, including *Norovirus* and hepatitis A virus, and PCR techniques for these viruses have also been developed (Ando et al. 1995, Coelho et al. 2003, Vennema et al. 2002). Unlike bacteria, virus culture tests require the use of cells to act as viral hosts. PCR has therefore become the most common and most widely used test method. Recently, a method known as Multiplex PCR has also been developed, in which multiple amplification reactions can be carried out in a single PCR reaction solution through the design of primers for multiple target microorganisms (Chen et al. 2012, Espiñeira et al. 2010, Kawasaki et al. 2005).

In the culture method, it takes several days to a week before the results are available, but the development of PCR test techniques has reduced the time required to only around 4–5 hours. The culture method also requires culturing in several selective media or the performance of differential culture and property testing, whereas the only tasks required in PCR testing are the isolation of DNA and PCR itself, making the procedure much simpler. PCR is also highly specific, meaning that confirmatory tests such as property testing, which may be difficult to assess, are not required.

Real-time PCR is a PCR method in which the amplification of the target DNA can be monitored in real time. In regular PCR, a technique such as electrophoresis is used to confirm whether or not amplification products are present, but in Real-time PCR, fluorescence generated by the amplification of DNA is measured in each cycle, allowing this amplification to be confirmed. The more amplification products are produced, the more fluorescence is emitted, enabling the amount of DNA to be quantified on the basis of the fluorescence value in an improvement on regular PCR. By producing calibration curves based on known bacterial counts of food-borne pathogens and the fluorescence values produced by these counts, not only can the contamination of food with the pathogen under investigation be confirmed or ruled out, but the bacterial count in samples can also be investigated.

There are two main methods of generating fluorescence. The first is the use in reactions of fluorescent dyes known as intercalators, such as SYBR Green. Intercalators emit fluorescence when they are intercalated between the two strands that compose DNA, and thus fluoresce when they enter the double-stranded DNA replicated in PCR. Techniques utilizing intercalators to detect and quantify harmful microorganisms in seafood have been developed for pathogens including *V. parahaemolyticus* (Tyagi et al. 2009), *V. colerae* (Gubala 2006), *L. monocytogenes* (Traunsek et al. 2011), *E. coli* (Jothikumar and Griffiths 2002), *Norovirus* (Pang et al. 2004), and hepatitis A virus (Casas et al. 2007). Real-time PCR techniques are also being developed to quantify the viable bacteria count by using genes common to ordinary bacteria, and to quantify the Enterobacteriaceae count by using genes common to bacteria in that family.

The other method is the use of fluorescence-labeled probes. Probes are designed to include sequences complementary to those of the target bacteria species so that they bind to the inside of the binding sites on the two primers, with the addition of a fluorescent label to the 5′ terminal, and a substance known as a "quencher" that absorbs the light emitted by the fluorescent substance to the 3′ terminal. During the reaction, the primers and probe hybridize with the sample DNA, and as the DNA polymerase causes the extension reaction to progress, when it encounters the site to which the probe is bound, the enzyme's exonuclease activity causes the probe to dissociate. When this happens, the fluorescent substance is cleaved from the 5′ terminal, and the greater physiological distance between the quencher and the fluorescent substances allows the latter to emit fluorescence. This is known as the TaqMan probe method. In another method, probe sequences are made containing both DNA and RNA, so that hybridization between the sample DNA and the probe generates complementary DNA/RNA chains; when these are cleaved with RNase H, the physiological distance between the fluorescent substance and the quencher increases, resulting in the emission of fluorescence. This is called the cycling probe method. As in both the TaqMan and cycling probe methods, fluorescence is not emitted unless both the primer and the probe sequence are complementary to the sample DNA, they have better specificity for the target substance compared to intercalators. Tests to detect and quantify harmful microorganisms in seafood using probes have been developed for *V. parahaemolyticus* (Takahashi et al. 2005), *V. colerae* (Lyon 2001), *L. monocytogenes* (Traunsek et al. 2011), *E. coli* (Takahashi et al. 2009), *Norovirus,* and the hepatitis A virus (International organization of standalization 2017).

The levels of contamination of food products with *V. parahaemolyticus* and *L. monocytogenes* are limited by the government, and quantitative testing is therefore required. Real-time PCR is more useful than regular PCR in such cases.

3.3 Loop-Mediated isothermal Amplification (LAMP) method

In the loop-mediated isothermal amplification (LAMP) method, also known as isothermal PCR, the three-step temperature changes necessary in regular PCR are not required for the DNA amplification reaction.

The LAMP method selects a total of six sites in the target gene for amplification, three at each end, and uses four types of primer. The sample DNA, the different primers, deoxynucleotide triphopsphate, DNA polymerase with strand displacement activity (which synthesizes complementary strands while splitting any double-stranded DNA in the direction of extension during the process of synthesizing complementary DNA chains from template DNA), and a buffer are all mixed together, and the reaction proceeds when they are kept at a constant temperature (around 65°C). The detailed principles are as follows.

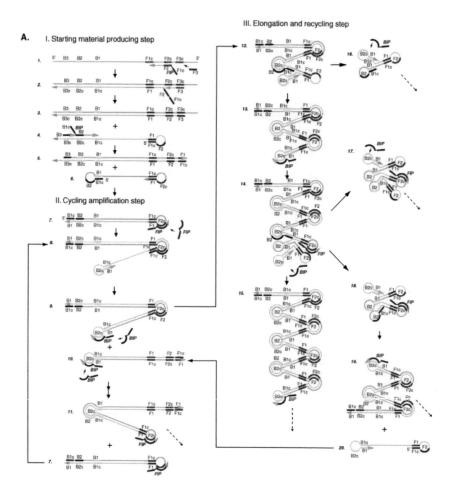

Figure 4-2. Schematic representation of mechanisms of Loop mediated Loop-Mediated isothermal Amplification (LAMP) method. Referred from Notomi et al., Nucleic Acids Res., 28;12(2000).

First, a primer hybridizes with its complementary sequence in the sample DNA. This is recognized by the DNA polymerase, causing the start of the dissociation of the two strands of the sample DNA and the new synthesis of complementary strands. At this point, the terminals of the hybridized primers have different sequences from those of the sample DNA, and different primers then hybridize with these dissociation sites. These newly hybridized sites are recognized by the DNA polymerase, and the dissociation of the two strands of the sample DNA and the new synthesis of complementary strands starts again. The terminals of the resulting dissociated single-strand DNA have primer-derived complementary sequences, and these hybridize with each other to form a loop structure. The occurrence of this reaction at both ends of the template DNA gives rise to dumbbell-shaped DNA with a loop structure at both ends. The two-stranded portion of this dumbbell-shaped DNA is recognized by the DNA polymerase and the synthesis and dissociation of complementary strands is repeated, amplifying the target gene.

The LAMP method has high amplification efficiency, being capable of creating 10^9–10^{10} copies of the template DNA within 15 minutes to 1 hour. It is also very highly specific. This phenomenon was first discovered by the Japanese company Eiken Chemical in 2000 (Notomi et al. 2000).

Numerous methods of detecting harmful microorganisms in food using this technique have been reported, and methods of detecting the pathogens *V. parahaemolyticus* (Zeng et al. 2014) and *V. vulnificus* (Han et al. 2011) in seafood have been developed. The addition of reverse transcriptase to the various elements of PCR in the LAMP method enables the simultaneous synthesis of cDNA from RNA on the basis of the same principle. This technique has therefore also been applied to the detection of viruses such as *Norovirus* (Fukuda et al. 2006) and the hepatitis A virus (Li et al. 2014, Song et al. 2012).

DNA amplification by the LAMP method does not require the sort of temperature changes needed in the PCR method, as the amplification reaction is carried out under constant temperature conditions. This has the advantage of eliminating the need for special equipments such as a thermal cycler, thereby removing restrictions on the test site. Like regular PCR, it also requires far less time for testing compared to the culture method.

3.4 DNA-DNA hybridization/DNA chip

The formation of complexes by nucleic acid molecules with complementary sequences is known as hybridization. DNA-DNA hybridization is a test method in which the gene to be detected and a probe with a complementary sequence are reacted with the target DNA (sample DNA), and the resulting double-stranded DNA is detected using radioisotopes or coloring generated by an enzyme-substrate reaction.

First, oligonucleotides for use as probes are synthesized by PCR or cloning, and these are denatured by heat or alkali to produce single-stranded probes. The gene regions to be detected are then amplified from the sample DNA extract solution using PCR or a similar method, and this reaction solution is reacted with the probes. For some types of sample, DNA extracted from the sample may be reacted directly with the probes. After this reaction, the unreacted probes are removed, and the double-stranded DNA formed by the target DNA and the probes is detected. The greater the homology of this double-stranded DNA, the stronger are its hydrogen bonds, and it is not removed by subsequent operations.

The terminals of the target DNA were formerly labeled with a radioisotope such as ^{32}P, but today non-radioactive labels are more generally used. The most commonly used non-radioactive form of labeling is biotin labeling. After hybridization, the biotin attached to the double-stranded DNA formed by the target DNA and the probes is bound to avidin, a protein that binds specifically and strongly to biotin. A biotinized coloring enzyme is then bound to the avidin, and finally reaction with the enzyme's substrate is carried out. As a result, color appears only at sites of hybridization with the biotin-labeled target DNA. Instead of a biotinized coloring enzyme, a fluorescent dye such as biotinized fluorescein isothiocyanate (FITC) may also be used.

A method based on this principle of detecting multiple types of microorganisms has been developed by setting up and attaching probes for several different microbial species on resin or glass, and making them react with labeled sample DNA to test for multiple genes in parallel. This is known as a DNA chip. Detection methods for food-borne pathogens in seafood using DNA chips or DNA hybridization techniques with different structures have been developed in *vibrio* spp. (Panicker et al. 2004), bacterial pathogen related to fishery products (Cao et al. 2011), and Norovirus (Mattison et al. 2011). Another method has also been developed for detecting multiple bacterial species by applying probes to detect multiple species of bacteria to a strip of filter paper, exposing this to labeled sample DNA solution, and then using the method described above to detect the formation of double-stranded DNA (Zhao et al. 2016). This method has the advantages of not requiring any special equipment for final detection and providing visible results. Many tests for food-borne pathogens based on the DNA hybridization technique are available in kit forms, and their rapid testing and simple operation make them easy to use for product quality control in food manufacturing plants.

3.5 High Resolution Melting Analysis (HRMA)

High-resolution melting analysis (HRMA) is a method of detecting differences between the base sequences of PCR products by utilizing the fact that the dissociation temperatures for double-stranded DNA vary depending on differences in their base sequences (Zhou 2005).

The dissociation temperature of the double-stranded DNA amplified by PCR differs because the state of dissociation on heating varies, depending on the GC content, strand length, and base sequence order of the PCR products. In HRMA, an intercalator is added to the PCR reaction solution and a melting step is carried out after the PCR reaction. The intercalator binds to the amplification products after PCR and emits fluorescence, after which the PCR products are heated from 60°C to around 95°C (the melting step). This heat causes the double-stranded DNA to dissociate, and the intercalator stops emitting fluorescence. Changes in temperature and fluorescence values during heating are recorded, and these differences enable variations between PCR products to be detected. In terms of which intercalator to use, saturated bound dyes such as Resolight dye are considered to enable the detection of smaller variations within amplification products compared to unsaturated bound dye, such as SYBR Green. As HRMA is capable of detecting these sorts of mutations and variations without the need for base sequencing, it is attracting attention as a low-cost option for gene analysis.

The use of this technique means that the presence of amplification products after PCR can be confirmed without the use of electrophoresis. By being aware of the melting temperatures (Tm) of the target products in advance, it is possible to judge whether or not the amplification products obtained from PCR have been amplified from the target microorganism.

HRMA can be carried out as a closed system from PCR amplification to analysis in a single tube, and its operations are also the same as those of PCR, without the need for any special analytical operations. It is thus attracting attention as a rapid and simple genetic detection method. If the melting temperatures of the target genes for analysis are different, these can be detected and distinguished by using multiplex PCR, enabling the detection of other species of microorganism at the same time. There have already been a number of reports of detection methods for harmful microorganisms in food using HRMA, with detection methods for pathogens related to seafood developed in pathogenic *E. coli* (Kagkli et al. 2012), Norovirus (Tajiri-Utagawa et al. 2009). Many studies have also described procedures for identifying species and genotypes by carrying out PCR and HRMA for gene regions for different species (Lüdeke et al. 2015, Ohshima et al. 2017, 2014).

4. Rapid Identification Method

4.1 Sequencing of 16S rRNA gene

When the culture method is used, bacterial species can only be identified by performing tests using several different culture media. The colonies that form on a differential culture medium after enrichment culture on a selective medium must be investigated for glucose fermentation and oxygen requirement, and their morphology must also be observed after gram staining. These tests thus take between several days and a week. The widespread use of molecular biological methods has greatly transformed microbial identification testing in recent years. The use of DNA sequencers has led to the revision of the taxonomy drawn up on the basis of the various properties obtained by the culture method into a phylogenetic tree that takes account of the speed of base substitution, and a number of groups of microorganisms have been reclassified into different genera or species.

The most widely used method of bacterial identification is base sequencing of the *16S rRNA* gene on the chromosomal genome. The *16S rRNA* gene codes for a ribosome protein, and among its entire length of around 1500 bases, it contains around ten regions that are highly conserved beyond the genus level (Weisburg et al. 1991). Universal primers have been designed for these highly conserved regions, and the variable domains within the gene are amplified by PCR, and their bases are sequenced. After base sequencing, the sequence information obtained is compared to an international database, which is searched for homologies. The species can then be identified based on its degree of resemblance to microbial species already registered in the database. The best-known database is run jointly by the DNA Data Bank of Japan, the National Center for Biotechnology Information, and the European Bioinformatics Institute of the European Molecular Biology Laboratory. Data on the *16S rRNA* gene for all the current internationally confirmed bacteria are registered in this database. The gyrase (*gyr*) gene and other housekeeping genes possessed by all bacteria can also be used for identification.

Base sequencing, including DNA extraction, requires 5–6 hours to complete, which is a far shorter time than is required for the identification of bacteria by the culture method. However, even though its cost has been coming down in recent years, base sequencing remains an expensive procedure.

The usual method of base sequencing is to determine the sequence of bases by the dye-terminator method using a capillary sequencer. This method entails extracting DNA from pure separate colonies and then performing PCR and base sequencing. In tests for food-borne pathogens in food products, cultures formed on selective media or differential media can be identified, and base sequencing is used as a test to confirm whether or not these colonies are actually bacteria that cause food poisoning.

The advent of next-generation sequencers has now enabled the base sequences of vast amounts of DNA to be identified at once. Methods have now been developed of identifying the presence of food-borne pathogens from data on all the bacteria present in food products, rather than testing pure separate colonies. In practical terms, this means extracting DNA from the diluted food solution, carrying out PCR using the *16S rRNA* universal primers, then using a next-generation sequencer to perform base sequencing, and carry out an exhaustive analysis of all the bacterial genomes, to determine whether or not they include any food-borne pathogens (Mayo et al. 2014).

4.2 MALDI-TOF MS

Matrix assisted laser desorption/ionization time of flight mass spectrometry (MALDI-TOF M) is a type of mass spectrometry in which molecules such as proteins are ionized and made to fly through a vacuum tube, utilizing the fact that flight times vary depending on differences in mass to measure the mass of different substances. Attempts to analyze microbial proteins directly by using this device began in around 2000. Microbial analysis can be used for molecules with a molecular weight in the range of around 2,000 Da to 20,000 Da. The levels of intensity of peak patterns obtained from different molecular weights are recorded, and the fact that these peak patterns vary between different types of microorganisms can be used for their identification. MALDI-TOF-MS for bacterial identification has been developed, and the peak patterns for important species of bacteria have been registered. Most of the proteins analyzed by MALDI-TOF-MS are constituents of bacterial cells. Proteins derived from bacterial ribosomes in particular account for 50–70% of all the proteins analyzed. The reason for this high proportion is that ribosomes exhibit a high level of expression, and the majority of their components are basic, meaning that they are easily ionized. The *16S rRNA* gene described above encodes one ribosome protein, and the fact that the sequence of bases in this gene varies in different species backs up the variation in the profiles of proteins expressed in the ribosomes of different types of bacteria.

The specific test procedure starts by mixing a colony formed on the culture medium with some of the matrix, spreading it on a plate, and allowing it to dry, after which its spectrum is acquired by MALDI-TOF-MS. The spectrum thus obtained is compared to the registered spectra for different bacteria, and the type of microorganism is identified from the pattern that is the closest match. The tasks involved before the sample is placed in the device take only around 5–10 minutes, and the time required for ionization and peak pattern acquisition is around 1 minute. This method is thus capable of extremely rapid bacterial identification.

The data provided with the devices that are currently available has basically been put together with a focus on medically important microbial species. In the field of food products, although food-borne pathogens can be identified with little difficulty, if the food-borne pathogens identified belong to a group other than one, such as coliform bacteria for which extensive data have been recorded, the possibility of misidentification as a closely related species cannot be excluded. Food spoilage microorganisms also vary widely depending on the type of food products. This gives the impression that there is still scope for the collection of more data. We are thus awaiting an upgrade for these data, and in some cases it may be quicker to construct a database for the desired groups of bacteria oneself.

As the MALDI-TOF MS works by displaying the commonest proteins as peaks, if the target microorganism could be made to grow preferentially to other microorganisms at the preculture stage, it could be identified directly from the culture solution. Actual examples with the aim of identifying a number of different microorganisms have already been reported, and there are hopes for the further development of this technique (Jadhav et al. 2014).

References

Ando, T., Monroe, S.S., Gentsch, J.R., Jin, Q., Lewis, D.C. and Glass, R.I. 1995. Detection and differentiation of antigenically distinct small round-structured viruses (Norwalk-like viruses) by reverse transcription-PCR and southern hybridization. J. Clin. Microbiol. 33: 64–71.

Cao, B., Li, R., Xiong, S., Yao, F., Liu, X., Wang, M., Feng, L. and Wang, L. 2011. Use of a DNA microarray for detection and identification of bacterial pathogens associated with fishery products. Appl. Environ. Microbiol. 77: 8219–8225. Doi: 10.1128/AEM.05914-11.

Casas, N., Amarita, F. and de Marañón, I.M. 2007. Evaluation of an extracting method for the detection of Hepatitis A virus in shellfish by SYBR-Green real-time RT-PCR. Int. J. Food Microbiol. 120: 179–185. Doi: 10.1016/j.ijfoodmicro.2007.01.017.

Chen, J., Tang, J., Liu, J., Cai, Z. and Bai, X. 2012. Development and evaluation of a multiplex PCR for simultaneous detection of five foodborne pathogens. J. Appl. Microbiol. 112: 823–830. Doi: 10.1111/j.1365-2672.2012.05240.x.

Coelho, C., Heinert, A.P., Simoes, C.M. and Barardi, C.R. 2003. Hepatitis A virus detection in oysters (Crassostrea gigas) in Santa Catarina State, Brazil, by reverse transcription-polymerase chain reaction. J. Food Prot. 66: 507–511.

Espiñeira, M., Atanassova, M., Vieites, J.M. and Santaclara, F.J. 2010. Validation of a method for the detection of five species, serogroups, biotypes and virulence factors of Vibrio by multiplex PCR in fish and seafood. Food Microbiol. 27: 122–131. Doi: 10.1016/j.fm.2009.09.004.

Fukuda, S., Takao, S., Kuwayama, M., Shimazu, Y. and Miyazaki, K. 2006. Rapid detection of norovirus from fecal specimens by real-time reverse transcription—Loop-mediated isothermal amplification assay. J. Clin. Microbiol. 44: 1376–81. Doi: 10.1128/JCM.44.4.1376.

Gubala, A.J. 2006. Multiplex real-time PCR detection of Vibrio cholerae. J. Microbiol. Methods 65: 278–293. Doi: 10.1016/j.mimet.2005.07.017.

Han, F., Wang, F. and Ge, B. 2011. Detecting potentially virulent Vibrio vulnificus strains in raw oysters by quantitative loop-mediated isothermal amplification. Appl. Environ. Microbiol. 77: 2589–2595. Doi: 10.1128/AEM.02992-10.

Iguchi, A., Iyoda, S., Seto, K., Morita-Ishihara, T., Scheutz, F. and Ohnishi, M. 2015. Escherichia coli O-genotyping PCR: A comprehensive and practical platform for molecular O serogrouping. J. Clin. Microbiol. 53: 2427–2432. Doi: 10.1128/JCM.00321-15.

International organization of standalization. 2017. ISO 15216-1:2017(en) Microbiology of the food chain—Horizontal method for determination of hepatitis A virus and norovirus using real-time RT-PCR—Part 1: Method for Quantification.

Jadhav, S., Sevior, D., Bhave, M. and Palombo, E.A. 2014. Detection of Listeria monocytogenes from selective enrichment broth using MALDI-TOF Mass Spectrometry. J. Proteomics 97: 100–106. Doi: 10.1016/j.jprot.2013.09.014.

Janabi, A.H.D., Kerkhof, L.J., McGuinness, L.R., Biddle, A.S. and McKeever, K.H. 2016. Comparison of a modified phenol/chloroform and commercial-kit methods for extracting DNA from horse fecal material. J. Microbiol. Methods 129: 14–19. Doi: 10.1016/j.mimet.2016.07.019.

Jothikumar, N. and Griffiths, M.W. 2002. Rapid detection of Escherichia coli O157:H7 with multiplex real-time PCR assays. Appl. Env. Microbiol. 68: 3169–3171. Doi: 10.1128/AEM.68.6.3169.

Kagkli, D.-M., Folloni, S., Barbau-Piednoir, E., Van den Eede, G. and Van den Bulcke, M. 2012. Towards a pathogenic Escherichia coli detection platform using multiplex SYBR®Green real-time PCR methods and high resolution melting analysis. PLoS One 7: e39287. Doi: 10.1371/journal.pone.0039287.

Kawasaki, S., Horikoshi, N., Okada, Y., Takeshita, K., Sameshima, T. and Kawamoto, S. 2005. Multiplex PCR for simultaneous detection of Salmonella spp., Listeria monocytogenes, and Escherichia coli O157:H7 in meat samples. J. Food Prot. 68: 551–556.

Kim, H.-J., Ryu, J.-O., Lee, S.-Y., Kim, E.-S. and Kim, H.-Y. 2015. Multiplex PCR for detection of the Vibrio genus and five pathogenic Vibrio species with primer sets designed using comparative genomics. BMC Microbiol. 15: 239. Doi: 10.1186/s12866-015-0577-3.

Law, J.W., Mutalib, N.A., Chan, K. and Lee, L. 2015. Rapid methods for the detection of foodborne bacterial pathogens: principles, applications, advantages and limitations 5: 1–19. Doi: 10.3389/fmicb.2014.00770.

Li, C., Chen, Z., Meng, C. and Liu, G. 2014. Rapid detection of duck hepatitis A virus genotype C using reverse transcription loop-mediated isothermal amplification. J. Virol. Methods 196: 193–198. Doi: 10.1016/j.jviromet.2013.11.009.

Lüdeke, C.H.M., Gonzalez-Escalona, N., Fischer, M. and Jones, J.L. 2015. Examination of clinical and environmental Vibrio parahaemolyticus isolates by multi-locus sequence typing (MLST) and multiple-locus variable-number tandem-repeat analysis (MLVA). Front. Microbiol. 6: 564. Doi: 10.3389/fmicb.2015.00564.

Lyon, W.J. 2001. TaqMan PCR for detection of Vibrio in pure cultures, raw oysters, and synthetic seawater TaqMan PCR for detection of Vibrio cholerae O1, O139, Non-O1, and Non-O139 in pure cultures, raw oysters, and synthetic seawater†. Society 67: 4685–4693. Doi: 10.1128/AEM.67.10.4685.

Mattison, K., Corneau, N., Berg, I., Bosch, A., Duizer, E., Gutiérrez-Aguirre, I., L'Homme, Y., Lucero, Y., Luo, Z., Martyres, A., Myrmel, M., O'Ryan, M., Pagotto, F., Sano, D., Svraka, S., Urzua, U. and Bidawid, S. 2011. Development and validation of a microarray for the confirmation and typing of norovirus RT-PCR products. J. Virol. Methods 173: 233–250. Doi: 10.1016/j.jviromet.2011.02.011.

Mayo, B., Rachid, C.T.C.C., Alegría, Á., Leite, A.M.O., Peixoto, R.S. and Delgado, S. 2014. Impact of next generation sequencing techniques in food microbiology. Curr. Genomics 15: 293–309. Doi: 10.2174/1389202915666140616233211.

Molina, F., López-Acedo, E., Tabla, R., Roa, I., Gómez, A. and Rebollo, J.E. 2015. Improved detection of Escherichia coli and coliform bacteria by multiplex PCR. BMC Biotechnol. 15: 48. Doi: 10.1186/s12896-015-0168-2.

Mullis, K.B. and Faloona, F.A. 1987. [21] Specific synthesis of DNA *in vitro* via a polymerase-catalyzed chain reaction. pp. 335–350. Doi: 10.1016/0076-6879(87)55023-6.

Notomi, T., Okayama, H., Masubuchi, H., Yonekawa, T., Watanabe, K., Amino, N. and Hase, T. 2000. Loop-mediated isothermal amplification of DNA. Nucleic Acids Res. 28: E63. Doi: 10.1093/nar/28.12.e63.

Oh, K.H., Kim, S.B., Park, M.S. and Cho, S.H. 2014. Development of a one-step PCR assay with nine primer pairs for the detection of five diarrheagenic Escherichia coli types. J. Microbiol. Biotechnol. 24: 862–868. Doi: 10.4014/jmb.1312.12031.

Ohshima, C., Takahashi, H., Phraephaisarn, C., Vesaratchavest, M., Keeratipibul, S., Kuda, T. and Kimura, B. 2014. Establishment of a simple and rapid identification method for Listeria spp. by using high-resolution melting analysis, and its application in food industry. PLoS One 9: 1–8. Doi: 10.1371/journal.pone.0099223.

Ohshima, C., Takahashi, H., Iwakawa, A., Kuda, T. and Kimura, B. 2017. A novel typing method for Listeria monocytogenes using high-resolution melting analysis (HRMA) of tandem repeat regions. Int. J. Food Microbiol. 253: 36–42. Doi: 10.1016/j.ijfoodmicro.2017.04.015.

Pang, X., Lee, B., Chui, L., Preiksaitis, J.K. and Monroe, S.S. 2004. Evaluation and validation of real-time reverse transcription-PCR assay using the light cycler system for detection and quantitation of norovirus. J. Clin. Microbiol. 42: 4679–4685. Doi: 10.1128/JCM.42.10.4679.

Panicker, G., Call, D.R., Krug, M.J., Bej, K. and Bej, A.K. 2004. Detection of pathogenic *Vibrio* spp. in Shellfish by using multiplex PCR and DNA microarrays. Appl. Environ. Microbiol. 70: 7436–7444. Doi: 10.1128/AEM.70.12.7436.

Phraephaisarn, C., Khumthong, R., Takahashi, H., Ohshima, C., Kodama, K., Techaruvichit, P., Vesaratchavest, M., Taharnklaew, R. and Keeratipibul, S. 2016. A novel biomarker for detection of Listeria species in food processing factory. Food Control 73: 1032–1038. Doi: 10.1016/j.foodcont.2016.10.001.

Saiki, R.K., Scharf, S., Faloona, F., Mullis, K.B., Horn, G.T., Erlich, H.A. and Arnheim, N. 1985. Enzymatic amplification of beta-globin genomic sequences and restriction site analysis for diagnosis of sickle cell anemia. Science 230: 1350–1354. Doi: DOI: 10.1126/science.2999980.

Saiki, R.K., Gelfand, D.H., Stoffel, S., Scharf, S.J., Higuchi, R., Horn, G.T., Mullis, K.B. and Erlich, H.A. 1988. Primer-directed enzymatic amplification of DNA with a thermostable DNA polymerase. Science 259: 487–491. Doi: 10.1126/science.239.4839.487.

Shabarinath, S., Sanath Kumar, H., Khushiramani, R., Karunasagar, I. and Karunasagar, I. 2007. Detection and characterization of Salmonella associated with tropical seafood. Int. J. Food Microbiol. 114: 227–233. Doi: 10.1016/j.ijfoodmicro.2006.09.012.

Song, C., Wan, H., Yu, S., Han, X., Qiu, X., Hu, Q., Tan, L. and Ding, C. 2012. Rapid detection of duck hepatitis virus type-1 by reverse transcription loop-mediated isothermal amplification. J. Virol. Methods 182: 76–81. Doi: 10.1016/j.jviromet.2012.03.013.

Tajiri-Utagawa, E., Hara, M., Takahashi, K., Watanabe, M. and Wakita, T. 2009. Development of a rapid high-throughput method for high-resolution melting analysis for routine detection and genotyping of noroviruses. J. Clin. Microbiol. 47: 435–440. Doi: 10.1128/JCM.01247-08.

Takahashi, H., Iwade, Y., Konuma, H. and Hara-Kudo, Y. 2005. Development of a quantitative real-time PCR method for estimation of the total number of Vibrio parahaemolyticus in contaminated shellfish and seawater. J. Food Prot. 68: 1083–1088.

Takahashi, H., Kimura, B., Tanaka, Y., Shinozaki, J., Suda, T. and Fujii, T. 2009. Real-time PCR and enrichment culture for sensitive detection and enumeration of Escherichia coli. J. Microbiol. Methods 79: 124–127. Doi: 10.1016/j.mimet.2009.08.002.

Toma, C., Lu, Y., Higa, N., Nakasone, N., Chinen, I., Baschkier, A., Rivas, M. and Iwanaga, M. 2003. Multiplex PCR assay for identification of human diarrheagenic Escherichia coli. J. Clin. Microbiol. 41: 2669–2671. Doi: 10.1128/JCM.41.6.2669.

Traunsek, U., Toplak, N., Jeršek, B., Lapanje, A., Majstorovic, T. and Kovac, M. 2011. Novel cost-efficient real-time PCR assays for detection and quantitation of Listeria monocytogenes. J. Microbiol. Methods 85: 40–46. Doi: 10.1016/j.mimet.2011.01.018.

Tyagi, A., Saravanan, V., Karunasagar, I. and Karunasagar, I. 2009. Detection of Vibrio parahaemolyticus in tropical shellfish by SYBR green real-time PCR and evaluation of three enrichment media. Int. J. Food Microbiol. 129: 124–130. Doi: 10.1016/j.ijfoodmicro.2008.11.006.

Vennema, H., De Bruin, E. and Koopmans, M. 2002. Rational optimization of generic primers used for Norwalk-like virus detection by reverse transcriptase polymerase chain reaction. J. Clin. Virol. 25: 233–235. Doi: 10.1016/S1386-6532(02)00126-9.

Weisburg, W.G., Barns, S.M., Pelletier, D.A. and Lane, D.J. 1991. 16S Ribosomal DNA amplification for phylogenetic study. J. Bacteriol. 173: 697–703. Doi: n.a.

Zeng, J., Wei, H., Zhang, L., Liu, X., Zhang, H., Cheng, J., Ma, D., Zhang, X., Fu, P. and Liu, L. 2014. Rapid detection of Vibrio parahaemolyticus in raw oysters using immunomagnetic separation combined with loop-mediated isothermal amplification. Int. J. Food Microbiol. 174: 123–128. Doi: 10.1016/j.ijfoodmicro.2014.01.004.

Zhao, Y., Wang, H., Zhang, P., Sun, C., Wang, X., Wang, X., Yang, R., Wang, C. and Zhou, L. 2016. Rapid multiplex detection of 10 foodborne pathogens with an up-converting phosphor technology-based 10-channel lateral flow assay. Sci. Rep. 6: 21342. Doi: 10.1038/srep21342.

Zhou, L., Wang, L., Palais, R., Pryor, R. and Wittwer, C. 2005. High-resolution DNA melting analysis for simultaneous mutation scanning and genotyping in solution. Clin. Chem. 51: 1770–1777. Doi: 10.1373/clinchem.2005.054924.

5

Control of Pathogens and Spoilage Bacteria on Seafood Products

Koji Yamazaki, * *Shogo Yamaki* and *Dominic Kasujja Bagenda*

1. Introduction

The methods used to control pathogens in seafood can be summarized into: (i) prevention of access of pathogens to the foods, (ii) inactivation of pathogens that access the food, and (iii) prevention or slowing down of growth of pathogens that may not been inactivated. Preventing access of pathogens has mainly been accomplished by aseptic packaging of heat-processed foods. On the other hand, inactivation of pathogens that have accessed the food can be done by heat pasteurization and sterilization, ionizing irradiation, addition of enzymes, use of high hydrostatic pressure, or electric shock treatments. Procedures to slow down or prevent growth of pathogens include: chilling and freezing, curing, drying, acidifying, fermenting, vacuum and modified atmosphere packing, addition of preservatives, or microstructure control in water-in-oil emulsions (Gould 1999).

For purposes of this chapter, the methods used in the control of pathogens have been divided into chemical, physical, and biological

Faculty of Fisheries Sciences, Hokkaido University, 3-1-1, Minato, Hakodate, Hokkaido, 041-8611, Japan.
* Corresponding author: yamasaki@fish.hokudai.ac.jp

methods. It must be emphasized that the methods outlined here are not substitutes for good sanitation, or agents to be used to improve the quality of partially spoiled seafood. Rather, these methods are best considered adjuncts to good sanitation and hygiene. Good sanitation and hygiene therefore take top priority in the control of pathogens in seafood.

2. Chemical Methods for Controlling Food-borne Pathogens in Seafood

2.1 Organic acids

Acid antimicrobials have found application in the control of pathogenic bacteria related to seafood. This group of antimicrobial agents includes sorbic acid, lactic acid, acetic acid, and citric acid. It is thought that in solution, most acids dissociate to release free protons. These free protons accumulate on the outer surfaces of microorganisms, resulting in functional destabilization of membranes due to denaturation of enzymes, and alteration of permeability. However, organic acids generally exhibit strong antimicrobial effects below their pKa. For example, the pKa of acetic acid, lactic acid, citric acid, and sorbic acid are about 4.8, 3.8, 3.1, and 4.8, respectively. When the pH of the solution is equal with pKa, the concentrations of undissociated and dissociated forms are equal. Undissociated molecules can penetrate the cell membrane and reach the cytoplasm. The cytoplasm is neutral, and release of protons from undissociated molecules acidifies the cytoplasm (Booth 1985, Cassio et al. 1987, Ray and Sandine 1992, Casal et al. 1998, Young and Foegeding 1993, Guldfeldt and Arneborg 1998). To maintain the homeostasis, the cell pumps protons out of the cell using adenosine triphosphate (ATP). Shortage of ATP results in cell death. Organic acids vary in their inhibitory strength. Based on experiments in broth systems, organic acids have been arranged from strongest to weakest in the following order: sorbic acid > propionic acid > acetic acid > formic acid > lactic acid > citric acid > adipic acid > malic acid > tartaric acid > gluconic acid (Matsuda et al. 1994).

Lactic acid is present in all life forms. Lactic acid and lactates have found application in the food industry as decontaminants and shelf life enhancers for fresh or semi-processed foods (Bogaert and Naidu 2000). Lactic acid and its salts are effective in preventing growth of pathogens of the genera *Clostridium*, *Listeria*, and *Vibrio* spp. (Notermans and Dufrenne 1981, Sun and Oliver 1994, Shelef and Yang 1991). However, reports on its application in seafood are not many. Lactic acid (300 ppm) has been reported to inhibit *V. vulnificus*, a pathogen associated with consumption of raw oyster (Sun and Oliver 1994).

Sorbic acid and its more water-soluble salts such as potassium sorbate are used all over the world as antimicrobial agents. Sorbic acid is a naturally occurring unsaturated fatty acid. Sorbic acid is effective against yeasts, molds, and many bacteria at concentrations of 0.02–0.3% (Sofos and Busta 1993). Sorbates can be applied by spraying, immersion into the sorbate solution, dusting with a sorbate powder, as well as coating or packaging material (Sofos 2000). Sorbates have been reported to be active against bacteria belonging to the genera *Clostridium*, *Vibrio*, and *Listeria* (Sofos et al. 1986, Beuchat 1980, El-Shenawy and Marth 1988). While the mechanisms of antimicrobial action against spore forming bacteria are not fully understood, inhibition of bacterial spore germination is thought to occur through action on spore membranes or enzymes involved in germination. Studies on the use of sorbate in seafood have shown that they improve preservation. However, one study indicated that toxin production by *Cl. botulinum* in vacuum packed shucked scallops was more rapid (27°C) when 0.1% sorbate was used (Fletcher et al. 1988). Nevertheless, in terms of acute toxicity, sorbic acid is considered to be one of the least harmful preservatives in use.

Acetic acid has been diluted to make vinegar and used in preservation of food because of its antimicrobial effect for a long time. It is reported that effectiveness of acetic acid rises with increasing concentration, increasing temperature, reducing pH, and reducing microbial load (Ray and Sandine 1992). Control of pathogens using acetic acid has been studied in combination with other antimicrobial treatments. Osmotic stress (by prewashing in 20% NaCl), monolaurin, sodium lauryl sulfate, hydrogen peroxide, sodium bicarbonate have all been reported to synergistically increase the effectiveness of acetic acid against *L. monocytogenes* and other pathogens (Dickson 1990, Oh and Marshall 1994, Tamblyn and Conner 1997). Acetic acid and salt also combine well to prevent outgrowth of *Cl. botulinum*, which is an important application in the fermentation of sushi in Northern Japan (Sasaki et al. 2004, Bagenda 2006). Acetic acid has a wide spectrum of activity, is cheap, widely available, and generally accepted well by consumers. It therefore has high potential for use in the control of food-borne pathogens in seafood.

Citric acid enhances the antimicrobial properties of many other substances by reducing pH or chelation of metal ions. On its own, citric acid has a much weaker antimicrobial effect than acetic acid and lactic acid (Sharma 2000, Branen and Keenan 1970). Consequently, derivatives of citric acid have weaker antimicrobial properties than derivatives of acetic acid and lactic acid. Nevertheless, Post et al. (1985) report that *Cl. botulinum* growth was arrested in shrimp puree acidified with citric acid to a pH of 4.2. No toxin was detected for a period of up to 8 weeks. Due to its pleasant sour taste, citric acid is used in several foods and beverages.

2.2 Reduced water activity

Sodium chloride, or salt as it is normally called, is a very important antimicrobial agent in the food industry because of its ability to reduce water activity. However the amount of salt that must be added to food in order to arrest microbial growth is large and would result in unpalatable food products. To achieve a safe product without sacrificing sensory attributes, small amounts of salt are used in combination with other preservation techniques. Salt inhibits microorganisms by lowering the water activity (Aw) of the substrate below 0.94–0.95, which is the minimum for growth of most pathogens. For example, the minimum Aw values for microbial growth at optimal temperatures are 0.97, 0.95, 0.94, and 0.92 for *Cl. botulinum* Type E, *E. coli*, *V. parahaemolyticus*, and *L. monocytogenes*, respectively. It is generally accepted that toxin production by *Cl. botulinum* is inhibited by 5% sodium chloride combined with pH values of less than 5.03 (Segner et al. 1966). The use of salt in the food industry is not without limitations. Yeasts and molds tolerate lower Aw conditions than bacteria. It has also been reported that *L. monocytogenes* can survive sodium chloride concentrations of up to 20% at −12°C for up to 30 days (Miller et al. 1997). Furthermore, the take of sodium chloride has been linked to hypertension, resulting in consumer' weariness of salted food. In recent times, lesser amounts of salt are being used in the control of pathogens in the food industry.

2.3 Plant-derived substances

Plants produce a wide array of metabolites that have antimicrobial properties. These metabolites are thought to play a role in defending the plant against pests and pathogens. These antimicrobial metabolites include alkaloids, flavonoids, isoflavonoids tannins, cummarins, glycosides, terpens, and phenolic compounds. Some of these metabolites have favorable flavors that have made plants valuable as spices and herbs for a long time (Souza et al. 2005). Plant metabolites are thought to cause their antimicrobial effect by cell membrane perturbation, leading to cell dysfunction. Some metabolites, such as phenols, may interfere with germination enzymes, and L-alanine utilization in a manner that affects germination of spores of pathogenic bacteria (Parker and Bradley 1968, Sierra 1970, Weinstein and Albersheim 1983). Also notable for their use in the control of pathogens in seafood are the glucosinolates. Glucosinolates occur in many species of plants of the family *Brassicaceae*. These include cabbages, mustards, horseradish, and Japanese horseradish (wasabi). Japanese horseradish is popular around the world for its allyl isothiocyanate (AITC) content. AITC is released from a naturally occurring glucosinolates, sinigrin, by the action of myrosinase (Farell 1985). AITC vapor from brown mustard was reported to show antimicrobial action against several bacteria, including *V. parahaemolyticus* (Isshiki et al. 1992, Hasegawa et al. 1999). A combination of gaseous AITC and modified atmosphere (CO_2:O_2:N_2 = 49:0.5:50.5%) packaging also prevented the growth of *Pseudomonas aeruginosa* on fresh catfish fillet (Pang et al. 2013). AITC and salicylaldehyde vapors inhibited histamine accumulation, as well as growth of *Morganella morganii* and *Raultella planticola* in bigeye tuna meat (Kamii et al. 2011). Furthermore, AITC treatment during anchovy marinating processes reportedly inactivates *Anisakis* larvae (Giarratana et al. 2015). AITC impregnated labels manufactured under the Wasaouro trademark by Mitsubishi Kagaku foods are now used to prevent growth of *Vibrio* spp. and other microorganisms in raw seafood lunch boxes all over Japan. Several other phytochemicals have been found to be effective against pathogens known to be problematic in seafood. Several examples are listed in Table 5.1. The major limitation of phytochemicals in the control of seafood-borne pathogens is their adverse effect on sensory aspects of food (Nychas 1999). For example, the effective antimicrobial dose for most spices is well above organoleptically acceptable levels. Nevertheless, in combination with other antimicrobial barriers, even moderate doses of spices can enhance microbial safety of food (Souza et al. 2005).

Table 5.1. Examples of plant metabolites effective against food borne pathogens related to seafood.

Sensitive pathogen	Phytochemical	Concentration	References
Cl. botulinum	Thymol	200μg/ml	Reddy et. al. (1982)
	Gallic acid	400μg/ml	Reddy et. al. (1982)
L. monocytogenes	Mint	42 mg/ml	Kumral and Sahin (2003)
	Thyme	17.5 mg/ml	Kumral and Sahin (2003)
	Garlic	81.3mg/ml	Kumral and Sahin (2003)
V. parahaemolyticus	Thyme	1,000 ppm	Aktug and Karapinar (1986)
	Bay leaf	5,000 ppm	Aktug and Karapinar (1986)
	Mint	6,000 ppm	Aktug and Karapinar (1986)
	Mustard	10% (in fish paste)	Al-Jedah et. al. (2000)

2.4 Surface sanitation with electrolyzed water

Electrolyzed water, also referred to as Electrolyzed Oxidizing (EO) water, is formed by adding a very small amount of NaCl (normally around 0.1%) to pure water, and conducting a current across an anode (acidic electrolyzed water, AEW) and cathode (basic electrolyzed water, BEW). Generally, AEW has a pH range of 2 to 3, an oxidation-reduction potential (ORP) of > 1,100 mV, and an active chlorine concentration of 20 to 60 ppm. On the other hand, BEW has a pH range of 11 to12, and an ORP of –800 to –900 mV. Since the mid 1980's, AEW has been certified for use in Japan as a medical product. The first form of AEW that was developed was the acidic type, and it was accepted quickly by the food industry in Japan. It was found to be useful at killing bacteria and parasites in raw fish without altering the sensory characteristics of the fish.

AEW exhibits strong antimicrobial action against *V. parahaemolyticus* and *V. vulnificus*. Holding oysters artificially contaminated with both pathogens in electrolyzed water (30 ppm chlorine) for 8 hours resulted in a significant decrease in counts of both *V. parahaemolyticus* and *V. vulnificus* (Ren and Su 2006). Ozer and Demirci (2006) reported that treatment of raw salmon surface with AEW achieved up to 1.07 logCFU/g (91.1%) and 1.12 logCFU/g (92.3%) reduction in *E. coli* O157:H7 and *L. monocytogenes* populations, respectively. Another study showed electrolyzed water ice (EW-ice) greatly inhibited the growth of aerobic and psychrotrophic bacteria on saury fillets during the refrigerated period, and increased the shelf life of the fish (Kim et al. 2006).

Although the antibacterial mechanism of AEW has not yet been fully understood, possibilities for its use instead of sodium hypochlorite solution are growing. AEW for disinfection of food contact surfaces or reduction of pathogens on fresh food products can be prepared on-site, reducing the need to store and handle chemicals.

3. Physical Methods for Controlling Pathogens in Seafood

3.1 High hydrostatic pressure

High hydrostatic pressure (HHP) is an alternative non-thermal food preservation method to avoid post-processing contamination, especially for foods whose nutritional, sensory, and functional characteristics are thermo-sensitive. HHP kills or sub-lethally injures cells by disruption of the cell wall and membrane, dissociation of protein and ribosomal subunit structures, and loss of activity of some enzymes (Hoover et al. 1989). Yuste et al. (2004) report that *V. parahaemolyticus* and *L. monocytogenes* are inhibited by 300 MPa for 5 minutes and 400 MPa for 5 minutes, respectively. Ma and Su (2011) showed that 293 Mpa for 2 min at 8°C reduced *V. parahaemolyticus* in oysters by more than 3.5 log MPN/g, and improved shelf-life from 6 to 16 days when stored at 5°C, and 8 to 18 days when stored on ice. Although HHP is well-accepted by European consumers (Baron et al. 1999), few reports have dealt with the effect of HHP on microorganisms in food matrices (Capellas et al. 2000, Yuste et al. 2000).

3.2 Modified atmosphere packaging

Gases can be used to control pathogens in seafood. Most notably, the use of Modified Atmosphere Packaging (MAP) has received a lot of research attention. The development of this method of control has been spurred by increased consumer demand for fresh and chilled convenience foods containing fewer preservatives. The three major gases used in MAP are oxygen, nitrogen, and carbon dioxide. Several other gases like ozone, sulphur dioxide, and chlorine have been studied, but safety concerns, legislation, cost, and effects on organoleptic quality limit their use.

Most studies on the bacteriostatic effects of MAP focus on carbon dioxide (CO_2). It is reported that CO_2 alters cell membrane function, including nutrient uptake and adsorption, inhibits enzymes or reduces the rates of their reactions, penetrates bacterial cells causing changes in intracellular pH, and/or changes the physiochemical properties of bacterial proteins (Farber 1991).

Examples include the use of CO_2 controlled atmosphere packaging to inhibit aerobic spoilage of fish (Dixon and Kell 1989). Also, packing shrimp in 100% CO_2 reportedly inhibits the growth of *L. monocytogenes* at 3°C (Rutherford et al. 2007). It has also been reported that the combined use of low temperature and CO_2 prevent *L. monocytogenes* from entering the viable but nonculturable (VBNC) state (Li et al. 2003).

The use of MAP has several limitations. In the United States, MAP is not recommended for fish products because of the high rate of contamination with psychrophilic strains of *Cl. botulinum* type E. Reduction of oxygen in packed seafood atmosphere can stimulate germination of spores and toxin production. One approach that may ensure safety of MAP seafood with respect to *Cl. botulinum* could be the use of pre-treatment in combination with MAP. The use of potassium sorbate, sodium chloride, and irradiation in combination with MAP is reportedly effective (Stammen et al. 1990). In situations where the holding time is short and refrigeration can be maintained, the risk of toxin production can be eliminated. In Japan, for example, there is a high demand for raw fish, especially red fleshed fish, such as common mackerel. Microbial deterioration of these types of fish is very high. Since transportation periods remain short (2–3 days) and refrigeration can be maintained, MAP may be used with minimal risk of toxin production by *Cl. botulinum* (Kimura et al. 1996). Another limitation of MAP is a phenomenon known as pack collapse, resulting from excessive absorption of CO_2 by fatty foods such as seafood (Parry 1993). Furthermore, dissolution of gas into the surface of fresh seafood can reduce their pH values sufficiently to weaken the water holding capacity of the proteins.

This causes a phenomenon known as in-pack drip that in turn reduces the shelf life of fish (Reddy et al. 1992).

Overall, the risks of occurrence of food-borne pathogens in MAP are usually less than those observed in aerobically stored foods (Church 1993).

3.3 Ozone

Ozone is a strong antimicrobial agent effective against food-borne pathogens associated with seafood, such as *L. monocytogenes* and *Vibrio* spp. (Korol et al. 1995). Ozone targets the cell wall and cell membrane, leading to oxidative damage and lysis of microorganisms (Horvath et al. 1985). Ozone, unlike chlorine, has a limited half-life, and the risk of accumulation of toxic residues is very low. Ozone also requires lesser doses and shorter contact time to exert its antimicrobial effect as compared to other sanitizers. For example, 0.35 mg/L of ozone is more effective against *V. cholerae* than 0.5 mg/L chlorine (Korol et al. 1995). Ozone has also been reported to inhibit spores of *Clostridium* spp. (Foegeding 1985). Ozonized slurry ice (a combination of slurry ice with ozone) reportedly improved the sensory and microbiological quality of sardine and improved shelf life (Campos et al. 2005). Unfortunately, the effect of ozone on physical, sensory, and nutritional aspects of food has not been studied in depth. Such information would be vital for the effective use of ozone in the control of food pathogens in seafood.

3.4 Irradiation

Food irradiation is the process of exposing food to ionizing radiation in order to achieve sterility. Food pathogens associated with seafood, such as *Vibrio* spp., have little resistance to irradiation. Rashid et al. (1992), Rodrick and Dixon (1993), and Loaharanu (1973), report that *V. cholerae*, *V. vulnificus*, and *V. parahaemolyticus* have D values in the range of 0.1–0.2 kGy. *Vibrio* spp. in seafood can therefore be inactivated sufficiently with irradiation doses as low as 1 kGy. Doses of 1.5 to 3.0 kGy are also reported to sufficiently control growth of *L. monocytogenes* in food (Loaharanu 1999). Other examples of irradiation application are listed in Table 5.2.

A major limitation to the use of irradiation is consumers' fear of radiation. To evade this problem, irradiation is sometimes referred to as cold pasteurization, emphasizing its similarities to the process of pasteurization. Even then, the use of irradiation to control pathogens is also strictly regulated. The World Health Organization states that irradiation of any food commodity up to an overall average dosage of 10 kiloGrays causes no toxicological hazard (WHO 1981). Likewise, many countries regulate the

Table 5.2. Examples of irradiation methods effective against food borne pathogens related to seafood.

Target pathogen	Technique	Seafood	Dose (kGy)	Bacterial reduction	Reference
V.parahaemolyticus	X-ray	RTE shrimp	3.0	7.6 log CFU/g	Mahmoud 2009
	X-ray	Whole shell oyster	3.0	4.3 log CFU/g	Mahmoud and Burrage 2009
	Gamma irradiation	Frozen shrimp	0.1 to 0.3	D_{10}	Bandeker et al. 1987
	Gamma irradiation	Oyster	1.0	6 log CFU/g	Jakabi et al. 2003
	Gamma irradiation	Smoked salmon	1.0	6.05 log CFU/g	Badr 2012
L. monocytogenes	X-ray	Catfish	2.0 to 3.0	4.8 log CFU/g	Gawborisut et al. 2012
	X ray	Smoked catfish	2.0	3.9 log CFU/g	Mahmoud et al. 2012
	X ray	Smoked salmon	1.0	3.7 log CFU/g	Mahmoud 2012
	X ray	Smoked mullet	2.0	4.0 log CFU/g	Robertson et al. 2006
	Gamma irradiation	Frozen seafood	0.43 to 0.66	D_{10}	Sommers and Rajkowski 2011
	Gamma irradiation	Catfish and tilapia	0.62	D_{10}	Rajkowski 2008
	Gamma irradiation	Smoked salmon	1.0	6.59 log CFU/g	Badr 2012

doses to which food matrices can be exposed (Steele 2000, Morehouse 2002). In the U.S., FDA has approved the use of ionizing radiation for inhibiting *V. parahaemolyticus* in fresh or frozen molluskan shellfish with an absorbed dose of less than 5.5 kGy (FDA 2011). Another limitation of irradiation is that there are threshold doses above which organoleptic changes in the food will occur. These must not be exceeded. Furthermore, irradiation will not eradicate spores or toxins of *Clostridium* spp., and so appropriate temperature controls need to be applied to avoid toxin production. Irradiation should therefore complement other methods in the control of pathogens in seafood.

4. Biological Methods for Controlling Food-borne Pathogens in Seafood

4.1 Lactic acid bacteria and their antibacterial substances

Bacteria have been used to control pathogens in food for a long time. Whole cultures as well as peptides and/or metabolites have been reported to be very useful in the inhibition of pathogenic organisms. In the food industry, probiotics, nisin, pediocin, reuterin, and sakacin have received a lot of research attention in the recent past. Probiotics have been used in various fermented foods since antiquity (Naidu and Clemens 2000). Lactic acid bacteria with probiotic roles are generally enteric flora believed to play a beneficial role in the intestinal tract. Probiotic cultures of *Lactobacillus* spp. are widely used in the dairy industry. Furthermore, through acidification and production of alcohols, *L. plantarum* and other lactic acid bacteria play an important role in control of pathogens in fermented seafood products in Japan (Matsushita 1937).

4.1.1 Nisins

Nisin, an antimicrobial peptide produced by *Lactococcus lactis* subsp. *lactic*, has been shown to be effective in the control of pathogens in seafood. Nisin is a small heat stable antimicrobial peptide described as a class I bacteriocin, a group that comprises lantibiotics (Klaenhammer 1993).

Nisin eliminates or inhibits the psychotrophic seafood pathogen *L. monocytogenes* in fresh and lightly preserved seafood products (Ben Embarek 1994, Benkeroum and Sandine 1988). Cold smoked salmon, for example, is usually vacuum packed with a salt content below 6% and a pH above 5.0. It is not heated before consumption, and so the risk of *L. monocytogenes* is high (Rørvik et al. 1995, Ben Embarek and Huss 1992, Eklund et al. 1995). A study by Nilsson et al. (1997) has shown that in a carbon dioxide atmosphere, nisin strongly inhibits *L. monocytogenes* on cold smoked salmon. Moreover, it had previously been reported that toxin production by *Cl. botulinum* type E spores in smoked mackerel fillets stored at 10°C and 26°C is delayed by spraying with nisin before packing in 100% carbon dioxide atmosphere (Taylor et al. 1990).

A number of variants of nisin have been reported over the years. In addition to the original nisin (A), other nisins (Z, Q, and U) have been reported (Mulders et al. 1991, Zendo et al. 2003, Wirawan et al. 2006). These are natural variants of the original nisin A, differing by a few amino acids (Table 5.3). Nisin A differs from nisin Z in one amino acid at position 27 (Mulders et al. 1991). The most recently isolated variant U is only 78% identical to nisin A or Z, and 85% identical to nisin Q. These differences in amino acid sequences result in differences in antimicrobial effect. For

Table 5.3. Amino acid sequences of nisin A and its variants Z, Q, and U.

Nisin variant	Amino acid position		
(Superfix indicates reference)	1–12	13–24	25–34
A[1]	ITSISLCTPGCK	TGALMGCNMKTA	TCHCSIHVSK-
Z[1]	ITSISLCTPGCK	TGALMGCNMKTA	TC<u>N</u>CSIHVSK-
Q[2]	ITSISLCTPGCK	TG<u>V</u>LMGCN<u>L</u>KTA	TC<u>N</u>CS<u>V</u>HVSK-
U[3]	ITS<u>K</u>SLCTPGCK	TG<u>I</u>LM<u>TC</u>P<u>L</u>KTA	TC<u>GC</u>H<u>FG</u>---

Positions at which variants differ from nisin A are bolded and underlined for emphasis.

References: [1]Mulders et al. (1991), [2]Zendo et al. (2003), [3]Wirawan et al. (2006).

example, Nisin Z has better diffusion properties, and is therefore more effective than nisin A (de Vos et al. 1993). Nisin A, on the other hand, is more effective than nisin U (Wirawan et al. 2006).

Nisin was the first bacteriocin to be permitted as a food preservative (Naidu and Clemens 2000). It is used in over 50 countries, mostly in dairy products and canned vegetables. Nisin has great potential for use in other foods as part of hurdle systems to prevent growth of pathogens such as *Cl. botulinum* and *L. monocytogenes*.

4.1.2 Pediocins

Pediocins are ribosomally synthesized antibacterial peptides produced by certain strains among the species that are currently included in the genus *Pediococcus* (Garvie 1986, Ray and Miller 2000). There are also species of *Lactococcus lactis* and *Lactobacillus plantarum* that have been reported to produce pediocins (Chikindas et al. 1995, Horn et al. 1998, Ennahar et al. 1996).

Pediocins are active against a wide range of gram positive bacteria, including genera of food-borne pathogens related to seafood such as *Listeria* and *Clostridium* spp. They inhibit bacteria by creating small pores in the cyltoplasmic membrane that cause loss of small molecules (less than 9400 Da), resulting in cell death (Chikindas et al. 1993). It must be noted that for each species that are sensitive to pediocins, there are resistant strains. Furthermore, bacterial spores are not sensitive to pediocins, but some studies have shown that pediocins have a bactericidal effect on germinated and outgrowing spores (Ray 1992, Kalchayanand et al. 1992, 1994, 1998a,b).

Pediocins can be effective in controlling growth of sub-lethally injured pathogens in food. Such sub-lethal injury can be achieved by pH values below 6, water activity below 0.9, low temperature storage, hydrostatic pressure treatment, or low heat treatment. In the production of fermented foods, pediocins have a specific application to control *L. monocytogenes*

(Naidu and Clemens 2000). By incorporating pediocin AcH during the formulation of the raw ingredients, pathogens in the final products can be reduced (Yang and Ray 1994, Bennik et al. 1997, Ennahar et al. 1998). A crude pediocin produced by a strain of *Pediococcus pentosaceus* isolated in our laboratory from retail fermented sushi in Japan inhibited *L. monocytogenes*. However, pediocin resistant strains occurred after 18 hours. Supplementing the pediocin with NaCl (1.5%), and injury of cells with 2% acetic acid for 10 minutes resulted in permanent inhibition *L. monocytogenes* for the entire incubation period (Bagenda 2006).

Apart from nisin and pediocins, several other bacteriocins, such as sakacin, piscicocin, and mundicin have been reported to be effective against gram-positive bacteria, including seafood pathogens, especially *L. monocytogenes*. Some bacteriocin-producing starter cultures exhibit effective protection against *L. monocytogenes* in fishery products. For example, *Carnobacterium piscicola* CS526, which is a class IIa bacteriocinogenic strain, inhibited the growth of *L. monocytogenes* in cold-smoked salmon stored at 12°C (Yamazaki et al. 2003). Another bacteriocinogenic strain, *C. piscicola* strain (A9b), showed strong anti-listerial activity and caused the growth suppression of *L. monocytogenes* in salmon juice (Nilsson et al. 2004). *Enterococcus munditii*, isolated from soil, inhibited the growth of *L. monocytogenes* on vacuum-packed cold smoked salmon during a four week shelf-life at 5°C (Bigwood et al. 2012).

The logistics of the production, storage, and use of these bacteriocins on a commercial scale remains a challenge for food scientists. Furthermore, safe use and effectiveness of these bacteriocins is still a point of research concern. Several factors, such as the type of seafood, number as well as type of microorganisms, temperature, and pH of the food have been reported to affect efficacy of bacteriocins in seafood.

4.2 Bacteriophages

Bacteriophages (phages) are bacterial viruses that infect and kill bacteria. They are considered to be the most ubiquitous organisms in nature. The therapeutic potential of phages has long been overshadowed by the discovery of antibiotics. However, with the emergence antibiotic resistant bacteria, phages and their lysins have recently been considered as an alternative antibacterial agent to control the pathogens. Phages are already in use in the food industry. The United States Department of Agriculture and FDA has approved the use bacteriophage-based additives, LISTEX™ P100 and SALMONELEX™ (Micreos Food Safety, Netherlands), to suppress the growth of *L. monocytogenes* and *Salmonella* spp. in some foods. LISTEX™ P100 exhibits a great anti-bacterial effect to *L .monocytogenes* in ready-to-eat foods including meat, poultry, and vegetables (Strydom and Witthuhn 2015). LISTEX™ P100 also exhibits listericidal effects in raw seafood, such as salmon fillets and catfish fillets surfaces (Soni and Nannapaneni 2010, Soni et al. 2010). It is also reported that *Vibrio* phages decrease *V. parahaemolyticus* counts in oyster and by approximately 2 log CFU/g in 5 to 30 min (Silva 2005). In addition, the specific phages FSP1 and Phda1 delayed the growth and histamine production for *Morganella morganii* and *Photobacterium damselae* subsp. *damselae*, in *in vitro* challenge tests (Yamaki et al. 2014, Yamaki et al. 2015). These pathogens are causative agents of opportunistic infections and histamine poisonings associated with consumption of seafood. Other examples of phage application in various foods are listed in Table 5.4.

5. Hurdle Technology

To ensure control of food-borne pathogens in seafood, a combination of several factors is always needed. These factors are referred to as hurdles, and this hurdle effect is crucial in ensuring safety of seafood products. Hurdle technology refers to the intelligent combination of hurdles to secure safety, stability, sensory, nutritive, and economic aspects of food products (Leistner 1999). Hurdles in a food system may be composed of factors such as high temperature during processing, low temperature during storage, water activity, acidity, redox potential of the product, as well as preservatives. The pathogens in the food cannot overcome all of these hurdles, and thus the food is microbiologically safe (Leistner 1999). Depending on the risk and the type of pathogen, the intensity of the various hurdles may be adjusted to suit consumer tastes and economic regimes without sacrificing safety aspects of the product. Fermented seafood products common in Japan provide a typical example hurdle technology. In the preparation of fermented *sushi*, pathogens are inhibited, while desired flora (lactic acid bacteria) is selected. Important hurdles in the early stages of fermentation are salt and vinegar. The fish used as raw material are cured in 20–30% of their weight of salt for one month before being desalted and pickled in vinegar. The main target of these hurdles is *Cl. botulinum*. Growth of lactic acid bacteria during fermentation results in acid production from metabolism of added sugars and rice. The result is a pH hurdle important in controlling growth of *Cl. botulinum*. It has been reported that the pH of fish drops from values of 6.4 to 4.9 during the ripening process (Sasaki et al. 2004).

Hurdle technology is invaluable in the control of food-borne pathogens in seafood. However, when selecting hurdles and their intensities, optimal ranges must be put into consideration. For example, the pH of fermented *sushi* must be low enough to inhibit *Cl. Botulinum*, but not so low as to make the final product unpalatable. Care must also be taken to ensure that the hurdles complement rather than counteract each other. For example, the

Table 5.4. Selected application of commertial bacteriophage products against the pathogens.

Target pathogen	Phage product	Food	Effect	Others	Reference
L. monocytogenes	Phage P100 (LISTEXTM)	Cheese	Suppression for 28 days at 4°C	Combination with Potassium lactae and sodium diacetate	Soni et al. 2012
	LISTEXTM	Melon, pear, apple	1–2 log CFU/g reduction in melon and pear	Apple and apple juice, no singificant reduction	Oliveira et al. 2014
		Each juices	8 and 2 log CFU/ml reduction in melon and pear juice		
	LISTEXTM	Turkey	Suppression for 14 at 4°C	Combination with Potassium lactae and sodium diacetate	Chibeu et al. 2013
		Roasted beef	Suppression for 28 days at 4°C		
	Phage P100 (LISTEXTM)	Hotdog	2 log CFU/g reduction	No effect in smoked salmon	Guenther et al. 2009
		Seafood mixture			
		Cabbege			
		Smoked salmon			
	LISTEXTM	Salmon fillet	Suppression for 10 days		Soni and Nannapaneni 2010
	LMP-102 (ListShieldTM)	Melon	5.7 and 2.3 log CFU/g reduction at 10°C	Combination with nisin	Leverentz et al. 2003
		Apple			

Table 5.4 contd. ...

...Table 5.4 contd.

Target pathogen	Phage product	Food	Effect	Others	Reference
E. coli	ECP-100 (EcoShieldTM)	Broccori	2 logCFU/g reduction at 10°C		Abuladze et al. 2008
		Tomato			
		Spinach			
		Beef			
	ECP-100 (EcoShieldTM)	Lettuce	1.6 log CFU/g reduction at 4°C		Sharma et al. 2009
		Melon	3.1 log CFU/g reduction at 4°C		
	EcoShieldTM	Beef	1 log CFU/g reduction		Carter et al. 2012
		Lettuce			
Salmonella enterica	SalmoFreshTM	Melon	1.5 log CFU/g reduction		Magnone et al. 2013
		Broccori	1 to 2.5 log CFU/g reduction		

use of MAP for fermented sushi may counteract other hurdles that have been used to prevent germination of *Cl. botulinum* spores.

A lot more research needs to be done to better understand the effects of each of the hurdles on the target pathogens, as well how best the hurdles can be combined to maximize sensory, nutritional, and economic value without compromising safety of seafood.

References

Abuladze, T., Li, M., Menetrez, M.Y., Dean, T., Senecal, A. and Sulakvelidze, A. 2008. Bacteriophages reduce experimental contamination of hard surfaces, tomato, spinach, broccoli, and ground beef by *Escherichia coli* O157:H7. Appl. Environ. Microbiol. 74: 6230–6238.

Aktug, S.E. and Karapinar, M. 1986. Sensitivity of some food-poisoning bacteria to thyme, mint and bay leaves. Int J. Food Microbiol. 3: 349–354.

Al-Jedah, J.H., Ali, M.Z. and Robinson, R.K. 2000. The inhibitory action of spices against pathogens that might be capable of growth in a fish sauce (mehiawah) from the Middle East. Int J. Food Microbiol. 57: 129–133.

Badr, H.M. 2012. Control of the potential health hazards of smoked fish by gamma irradiation. Int. J. Food Microbiol. 154: 177–186.

Bagenda, D.K. 2006. Bacteriocin producers and growth inhibition of *Clostridium botulinum* and *Listeria monocytogenes* in fermented Japanese seafood. Masters' thesis submitted to Hokkaido University, Graduate School of Fisheries Sciences, Hokkaido Japan.

Bandeker, J.R., Chandler, K. and Nerkan, D.P. 1987. Radiation control of *Vibrio parahaemolyticus* in shrimp. J. Food Safety 8: 83–88.

Baron, A., Bayer, O., Butz, P., Geisel, B., Gupta, B., Oltersdorf, U. and Tauscher, B. 1999. Consumer perception of high pressure processing: a three country survey. p. 18. In the Proceedings of the European Conference on Emerging Food Science and Technology, Tampere, Finland. The European Federation of Food Science and Technology.

Ben Embarek, P.K. and Huss, H.H. 1992. Growth of *Listeria monocytogenes* in lightly preserved fish products. pp. 203–303. *In*: Huss, H.H., Jakobsnen, M. and Liston, J. (eds.). Quality Assurance in the Fish Industry. Proc. Int. Conf., 26–30th August 1991, Copenhagen, Denmark. Amsterdam: Elsevier.

Ben Embarek, P.K. 1994. Presence detection and growth of *Listeria monocytogenes* in seafoods: A review. Int. J. Food Microbiol. 23: 17–34.

Benkeroum, N. and Sandine, W.E. 1988. Inhibitory action of nisin against *Listeria monocytogenes*. J. Diary Sci. 71: 3237–3245.

Bennik, M.H.J., Smid, E.J. and Gorris, L.G.M. 1997. Vegetable-associated *Pediococcus parvulus* produces pediocin PA-1. Appl. Environ. Microbiol. 63: 2074–2076.

Beuchat, L.R. 1980. Comparison of anti-vibrio activities of potassium sorbate, sodium benzoate, and glycerol and sucrose esters of fatty acids. Appl. Environ. Microbiol. 39: 1178–1182.

Bigwood, T., Hudson, J.A., Cooney, J., McIntyre, L., Billinhgton, C., Heinemann, J.A. and Wall, F. 2012. Inhibition of *Listeria monocytogenes* by *Enterococcus munditii* isolated from soil. Food Microbiol. 32: 354–360.

Bogaert, J.-C. and Naidu, A.S. 2000. Lactic acid. pp. 612–636. *In*: Naidu, A.S. (ed.). Natural Food Antimicrobial Systems. CRC Press New York.

Booth, I.R. 1985. Regulation of cytoplasmic pH in bacteria. Microbiol. Rev. 49: 359–378.

Branen, A.L. and Keenan, T.W. 1970. Growth stimulation of *Lactobacillus casei* by sodium citrate. J. Diary Sci. 53: 593–597.

Campos, C.A., Rodriguez, O., Losada, V., Aubourg, S.P. and Barros-Velazquez, J. 2005. Effects of storage in ozonized slurry ice on the sensory and microbial quality of sardine (*Sardina pilchardus*). Int. J. Food Microbiol. 103: 121–130.

Capellas, M., Mor-Mur, M., Gervilla, R., Yuste, J. and Guamis, B. 2000. Effect of high pressure combined with mild heat or nisin on inoculated bacteria and mesophiles of goat's milk fresh cheese. Food Microbiol. 17: 633–641.

Carter, C.D., Parks, A., Abuladze, T., Li, M., Woolston, J., Magnone, J., Senecal, A., Kropinski, A.M. and Sulakvelidze, A. 2012. Bacteriophage cocktail significantly reduces *Escherichia coli* O157:H7 contamination of lettuce and beef, but does not protect against recontamination. Bacteriophage 2: 178–185.

Casal, M., Cardoso, H. and Leão, C. 1998. Effects of ethanol and other alkanols on transport of acetic acid in *Saccharomyces cerevisiae*. Appl. Environ. Microbiol. 64: 665–668.

Cassio, F., Leao, C. and van Uden, N. 1987. Transport of lactate and other short chain monocarboxylates in the yeast *Saccharomyces cerevisiae*. Appl. Environ. Microbiol. 53: 509–513.

Chibeu, A., Agius, L., Gao, A., Sabour, P.M., Kropinski, A.M. and Balamurugan, S. 2013. Efficacy of bacteriophage LISTEX™ P100 combined with chemical antimicrobials in reducing *Listeria monocytogenes* in cooked turkey and roast beef. Int. J. Food Microbiol. 167: 208–214.

Chikindas, M.L., Garcia-Gacera, M.J., Driessen, A.J.M., Venema, K., Ledeboer, A.M., Nissen-Meyer, J., Nes, I.F., Abee, T., Konings, W.N. and Venema G. 1993. Pediocin PA-1, a bacteriocin of *Pediococcus acidilactici* PAC1.0, forms hydrophilic pores in the cytoplasmic membrane of target cells. Appl. Env. Microbiol. 59: 3577–3584.

Chikindas, M.L., Venema, K., Ledeboer, A.M., Venema, G. and Kok, J. 1995. Expression of lactococcin A and pediocin PA-1 in heterologous hosts. Lett. Appl. Microbiol. 21: 183–189.

Church, P.N. 1993. Meat products. pp. 229–268. *In*: Parry, R.T. (ed.). Principles and Applications of Modified Atmosphere Packaging of Food, Blackie Academic & Professional, Glasgow.

de Vos, W.M., Mulders, J.W., Siezen, R.J., Hugenholtz, J. and Kuipers, O.P. 1993. Properties of nisin Z and distribution of its gene, nisZ, in *Lactococcus lactis*. Appl. Environ. Microbiol. 59: 213–218.

Dickson, J.S. 1990. Surface moisture and osmotic stress as factors that affect the sanitizing of beef tissue surfaces. J. Food Prot. 53: 674–679.

Dixon, N.M. and Kell, D.B. 1989. The inhibition by CO_2 of the growth and metabolism of microorganisms. J. Appl Bacteriol. 67: 109–136.

Eklund, M.W., Poysky, F.T., Paranjpye, R.N., Lashbrook, L.C., Peterson, M.E. and Pelroy, G.A. 1995. Incidence and sources of *Listeria monocytogenes* in cold smoked fishery products and processing plants. J. Food Prot. 58: 502–508.

El-Shenawy, M.A. and Marth, E.H. 1988. Inhibition and inactivation of *Listeria monocytogenes* by sorbic acid. J. Food Prot. 51: 842–847.

Ennahar, S., Aoude-Werner, D., Sorokoine, O., Dorsselaer, A.V., Bringel, F., Hubert, J.-C. and Hasselmann, C. 1996. Production of pediocin AcH by Lactococcus plantarum WHE92 isolated from cheese. Appl. Environ. Microbiol. 62: 4381–4387.

Ennahar, S., Assobhel, D. and Hasslemann, C. 1998. Inhibition of *Listeria monocytogenes* in a smear surface soft cheese by *Lactobacillus plantarum* WHE92, a pediocin AcH producer. J. Food Prot. 61: 186–191.

Farber, J.M. 1991. Microbiological aspects of modified-atmosphere packaging technology—a review. J. Food Prot. 54: 58–70.

Farell, L.A. 1985. Spices Condiments. The AVI Publisher, Westport, Conn., pp. 128–146.

FDA. 2011. Code of Federal Regulations Title 21, CFR 179.26: ionizing rafiation for the treatment of food. Available from: https://www.accessdata.fda.gov/scripts/cdrh/cfdocs/cfcfr/cfrsearch.cfm?fr=179.26., Accessed 2016 April 1.

Fletcher, C.G., Murell, W.G., Statham, J.A., Stewart, B.J. and Bremner, H.A. 1988. Packaging of scallops with sorbate: An assessment of the hazard from *Clostridium botulinum*. J. Food Sci. 53: 349–352, 358.

Foegeding, P.M. 1985. Ozone inactivation of *Bacillus* and *Clostridium* spores and the importance of the spore coat to resistance. Food Microbiol. 2: 123–134.

Garvie, E.J. 1986. Genus pediococcus. pp. 1075–1079. *In*: Sneath, P.H.A. and Hold, J.G. (eds.). Bergeys' Manual of Systematic Bacteriology Baltimore. Williams and Wilkins Co.

Gawborisut, S., Kim, T.J., Sovann, S. and Silva, J.L. 2012. Microbial quality and safety of X-ray irradiated fresh catfish (*Ictalurua punctatus*) fillets stored under CO_2 atmosphere. J. Food Sci. 77: M533–538.

Giarratana, F., Panebianco, F., Muscolino, D., Beninati, C., Ziino, G. and Giuffrida, A. 2015. Effect of allyl isothiocyanate against *Anisakis larvae* during the anchovy marinating process. J. Food Prot. 78: 767–771.

Gould, G.W. 1999. Overview. *In*: Gould, G.W. (ed.). New Methods for Food Preservation. Aspen Publishers Maryland USA.

Guenther, S., Huwyler, D., Richard, S. and Loessner, M.J. 2009. Virulent bacteriophage for efficient biocontrol of *Listeria monocytogenes* in ready-to-eat-foods. Appl. Environ. Microbiol. 75: 93–100.

Guldefeldt, L.U. and Arneborg, N. 1998. Measurement of the effects of acetic acid and extracellular pH on intracellular pH of nonfermenting, individual *Saccharomyces cerevisiae* cells by fluorescence microscopy. Appl. Environ. Microbiol. 64: 530–534.

Hasegawa, N., Matsumoto, Y., Hoshino, A. and Iwashita, K. 1999. Comparison of effects of Wasabia japonica and allyl isothiocyanate on the growth of four strains of *Vibrio parahaemolyticus* in lean and fatty tuna meat suspensions. Int. J. Food Microbiol. 49: 27–34.

Hoover, D.G., Metrick, C., Papineau, A.M., Farkas, D.F. and Knorr, D. 1989. Biological effects of high hydrostatic pressure on food microorganisms. Food Technol. 43: 99–107.

Horn, N., Martinez, M.I., Martinez, J.M., Hernandez, P.E., Gasson, M.J., Rodriguez J.M. and Dodd, H.M. 1998. Production of Pediocin PA-1 by *Lactococcus lactis* using the lactococcin secretory apparatus. Appl. Environ. Microbiol. 64: 818–823.

Horvath, M., Bilitzky, L. and Huttner, J. 1985. Ozone. In Topics in Inorganic and General Chemistry-monograph 20 Amsterdam: Elsevier.

Isshiki, K., Tokuoka, K., Mori, R. and Chiba, S. 1992. Preliminary examination of allyl isothiocyanate vapor for food preservation. Biosci. Biotech. Biochem. 56: 1476–1477.

Jakabi, M., Gelli, D.S., Torre, J.C.M.D., Rodas, M.A.B., Franco, B.D.G.M., Destro, M.T. and Landgraf, M. 2003. Inactivation by ionizing radiation of *Salmonella Enteritidis, Salmonella Infantis*, and *Vibrio parahaemolyticus* in oyster (*Crassostrea brasiliana*). J. Food Prot. 66: 1025–1019.

Kalchayanand, N., Hanlin, M.B. and Ray, B. 1992. Sublethal injury makes Gram-negative and resistant Gram-positive bacteria sensitive to the bacteriocins, pediocin AcH and nisin. Lett. Appl. Microbiol. 15: 239–243.

Kalchayanand, N., Sikes, T., Dunne, C.P. and Ray, B. 1994. Hydrostatic pressure and electroporation have increased bactericidal efficiency in combination with bacteriocins. Appl. Environ. Microbiol. 60: 4174–4177.

Kalchayanand, N., Sikes, T., Dunne, C.P. and Ray, B. 1998a. Interaction of hydrostatic pressure, time and temperature of pressurization and pediocin AcH on inactivation of foodborne bacteria. J. Food Prot. 61: 425–431.

Kalchayanand, N., Sikes, T., Dunne, C.P. and Ray, B. 1998b. Factors influencing death and injury of foodborne pathogens by hydrostatic pressure pasteurization. Food Microbiol. 15: 207–214.

Kamii, E., Terada, G., Akiyama, J. and Isshiki, K. 2011. Antibacterial activity of essential oil vapor histamine-producing bacteria. Shokuhin Eiseigaku Zasshi 52: 276–280.

Kim, W.T., Lim, Y.S., Shin, I.S., Park, H., Chung, D. and Suzuki, T. 2006. Use of electrolyzed water ice for preventing freshness of pacific saury (*Cololabis saira*). J. Food Prot. 69: 2199–2204.

Kimura, B., Kuroda, S., Murakami, M. and Fujii, T. 1996. Growth of *Clostridium perfringens* in fish fillets with a controlled carbon dioxide atmosphere at abuse temperatures. J. Food Prot. 59: 704–710.

Klaenhammer, T.R. 1993. Genetics of bacteriocins produced by lactic acid bacteria. FEMS Microbiol. Revs. 12: 39–86.

Korol, S., Fortunato, M.S., Paz, M., Sanajuha, M.C., Lazaro, E., Santini, P. and D'Aquino, M. 1995. Water disinfection comparative activities of ozone and chlorine on a wide spectrum of bacteria. Rev. Argent. Microbiol. 27: 175–183.

Kumral, A. and Sahin, I. 2003. Effects of some spice extracts on *Escherichia coli, Salmonella typhimurium, Listeria monocytogenes, Yersinia enterolitica*, and *Enterobacter aerogenes*. Annals of Microbiol. 53: 427–435.

Leistner, L. 1999. Principles and applications of hurdle technology. *In*: Gould, G.W. (ed.). New Methods for Food Preservation. Aspen Publishers Maryland USA.

Leverentz, B., Conway, W.S., Camp, M.J., Janisiewicz, W.J., Abuladze, T., Yang, M., Saftner, R. and Sulakvelidze, A. 2003. Biocontrol of *Listeria monocytogenes* on fresh-cut produce by treatment with lytic bacteriophages and a bacteriocin. Appl. Environ. Microbiol. 69: 4519–4526.

Li, J., Kolling, G.L., Matthews, K.R. and Chikindas, M.L. 2003. Cold and carbon dioxide used as multi-hurdle preservation do not induce appearance of viable but non-culturable *Listeria monocytogenes*. J. Appl. Microbiol. 94: 48–53.

Loaharanu, P. 1973. Gamma irradiation of fishery products in Thailand with special reference to their microbiological and entomological aspects. *In*: Aspects of the Introduction of Food Irradiation in Developing Countries, STI/PUB/362. IAEA, Vienna.

Loaharanu, P. 1999. Food irradiation: current status and future prospects. *In*: Gould, G.W. (ed.). New Methods for Food Preservation. Aspen Publishers Maryland USA.

Ma, L. and Su. Y.C. 2011. Validation of high pressure processing for inactivating *Vibro parahaemolyticus* in Pacific oysters (*Crassostrea gigas*). Int. J. Food Microbiol. 144: 469–474.

Magnone, J.P., Marek, P.J., Sulakvelidze, A. and Senecal, A.G. 2013. Additive approach for inactivation of *Eshcerichia coli* O157:H7, *Salmonella*, and *Shigella* spp. on contaminated fresh fruits and vegetables using bacteriophage cocktail and produce wash. J. Food Prot. 76: 1336–1341.

Mahmoud, B.S.M. and Burrage, D.D. 2009. Inactivatin of *Vibrio parahaemolyticus* in pure culture, whole live and half shell oyster (*Crassostrea virginica*) by X-ray. Lett. Appl. Microbiol. 48: 572–578.

Mahmoud, B.S.M. 2009. Effect of X-ray treatments on inoculated *Escherichia coli* O157:H7, *Salmonella enterica, Shigella flexneri* and *Vibrio parahaemolyticus* in ready-to -eat shrimp. Food Microbiol. 26: 860–864.

Mahmoud, B.S. 2012. Control of *Listeria monocytogenes* and spoilage bacteria on smoked salmon during storage at 5°C after X-ray irradiation. Food Microbiol. 32: 317–320.

Mahmoud, B.S., Coker, R. and Su, Y.C. 2012. Reduction in *Listeria monocytogenes* and spoilage bacteria on smoked catfish using X-ray reatment. Lett. Appl. Microbiol. 54: 524–529.

Matsuda, T., Yano, T., Maruyama, A. and Kumagai, H. 1994. Antimicrobial activities of organic acids determined by minimum inhibitory concentrations at different pH ranged from 4.0 to 7.0. Jap. Soc. Food Sci. Technol. 41: 687–701.

Matsushita, K. 1937. Studies on the Prussian carp–'sushi'–II. Microorganisms isolated from Prussian carp 'sushi'. J. Agric. Chem. Soc. Japan 13: 635–638.

Miller, A.J., Call, J.E. and Eblen, B.S. 1997. Growth injury and survival potential of *Yersinia enterocolitica. Listeria monocytogenes*, and *Staphylococcus aureus* in brine chiller conditions. J. Food Prot. 60: 1334–1340.

Morehouse, K.M. 2002. Food irradiation—US regulatory considerations. Radiat. Pgys. Chem. 63: 281–284.

Mulders, J.W., Boerrigter, I.J., Rollema, H.S., Siezen, R.J. and de Vos, W.M. 1991. Identification and characterization of the lantibiotic nisin Z, a natural nisin variant. Eur. J. Biochem. 201: 581–584.

Naidu, A.S. and Clemens, R.A. 2000. Probiotics. pp. 431–462. *In*: Naidu, A.S. (ed.). Natural Food Antimicrobial Systems. CRC Press New York.

Nilsson, N., Huss, H.H. and Gram, L. 1997. Inhibition of *Listeria monocytogenes* in cold smoked salmon by nisin and carbon dioxide atmosphere. Int J. Food Microbiol. 38: 217–227.

Nilsson, L., Ng, Y.Y., Christiansen, J.N., Jorgensen, B.L., Grotinum, D. and Gram, L. 2004. The contribution of bacteriocins to inhibition of *Listeria monocytogenes* by *Carnobacterium piscicola* strains in cold-smoked salmon systems. J. Apple. Microbiol. 96: 133–143.

Notermans, S. and Dufrenne, J. 1981. Effect of glyceryl monolaurate on *Clostridium botulinum* in meat slurry. J. Food Safety 3: 83–88.

Nychas, G.J.E. 1999. Natural antimicrobials from plants. *In*: Gould, G.W. (ed.). New Methods for Food Preservation. Aspen Publishers Maryland USA.

Oh, D.H. and Marshall, D.L. 1994. Enhanced inhibition of *Listeria monocytogenes* by glycerol monolaurate with organic acids. J. Food Sci. 59: 249–252.

Oliveira, M., Viñas, I., Colàs, P., Anguera, M., Usall, J. and Abadias, M. 2014. Effectiveness of a bacteriophage in reducing *Listeria monocytogenes* on fresh-cut fruits and fruit juices. Food Microbiol. 38: 137–142.

Ozer, N.P. and Demirci, A. 2006. Electrolyzed oxidizing water treatment for decontamination of raw salmon inoculated with *Escherichia coli* O157:H7 and *Listeria monocytogenes* Scott A and response surface modeling. J. Food Enginer. 72: 234–241.

Pang, Y.H., Sheen, S., Zhou, S., Liu, L. and Yam, K.L. 2013. Antimicrobial effects of allyl isothiocyanate and modified atmosphere on *Pseudomonas aeruginosa* in fresh catfish fillet under abuse temperature. J. Food Sci. 78: M555–559.

Parker, M.S. and Bradley, T.J. 1968. A reversible inhibition of the germination of bacterial spores. Can. J. Microbiol. 14: 745–746.

Parry, R.T. 1993. Introduction. pp. 1–18. *In*: Parry, R.T. (ed.). Principles and Applications of Modified Atmosphere Packaging of Food. Blackie Academic & Professional, Glasgow.

Post, L.S., Amoroso, T.L. and Solberg, M. 1985. Inhibition of *Clostridium botulinum* type E in acidified food systems. J. Food Sci. 50: 966.

Rajkowski, K.T. 2008. Radiatioon D10-values on thawed and frozen catfish and tilapia for finfish isolates of *Listeria monocytogenes*. J. Food Prot. 71: 2278–2282.

Rashid, H.O., Ito, H. and Ishigaki, I. 1992. Distribution of pathogenic vibrios and other bacteria in imported frozen shrimps and their decontamination by gamma-irradiation. World J. Microbiol. Biotechnol. 8: 494–499.

Ray, B. 1992. Pediocin(s) of *Pediococcus acidilactici* as a food biopreservative. pp. 265–322. *In*: Ray, B. and Daeschel, M.A. (eds.). Food Biopreservatives of Microbial Origin. CRC Press Inc. Boca Raton, FL.

Ray, B. and Sandine, W.E. 1992. Acetic, proprionic, and lactic acids of starter culture bacteria as biopreservatives. pp. 103–136. *In*: Ray, B. (ed.). Food Biopreservatives of Microbial Origin. Boca Raton, Fl: CRC Press Inc.

Ray, B. and Miller, K.W. 2000. Pediocin. pp. 525–566. *In*: Naidu, A.S. (ed.). Natural Food Antimicrobial Systems. CRC Press New York.

Reddy, N.R., Pierson, M.D. and Lechowich, R.V. 1982. Inhibition of *Clostridium botulinum* by antioxidants, phenols, and related compounds. Appl. Environ. Microbiol. 43: 835–839.

Reddy, N.R., Armstrong, D.G., Rhodehamel, E.J. and Kautter, D.A. 1992. Shelf life extension and safety concerns about fresh fishery products packaged under modified atmospheres. A review. J. Food Safety 12: 87–118.

Ren, T. and Su, Y. 2006. Effects of electrolyzed oxidizing water treatment on reducing *Vibrio parahaemolyticus* and *Vibrio vulnificus* in raw oysters. J. Food Prot. 69: 1829–1834.

Robertson, C.B., Andrews, L.S., Marshall, D.L., Coggins, P., Schilling, M.W., Martin, R.E. and Collette, R. 2006. Effect of x-ray irradiation on reducing the risk of listeriosis in ready-to-eat vacuum-packaged smoked mullet. J. Food Prot. 69: 1561–1564.

Rodrick, G.E. and Dixon, D.W. 1993. The effects of gamma irradiation upon shell shock oysters with respect to shelf-life and *Vibrio vulnificus* levels. Paper presented at 2nd Research Co-ordination meeting on Irradiation in combination with other Process to Improve Food Quality, St. Hyacinthe, Canada, June 1993. IAEA, Vienna.

Rørvik, L.M., Caugant, D.A. and Yndestad, M. 1995. Contamination pattern of *Listeria monocytogenes* and other *Listeria* spp., in a salmon slaughter house and smoked salmon processing plant. Int. J. Food Microbiol. 25: 19–27.

Rutherford, T.J., Marshall, D.L., Andrews, L.S., Coggins, P.C., Wes Schilling, M. and Gerard, P. 2007. Combined effect of packaging atmosphere and storage temperature on growth of *Listeria monocytogenes* on ready-to-eat shrimp. Food Microbiol. 24: 703–710.

Sasaki, M., Kawai, Y., Yoshimizu, M. and Shinano, H. 2004. Changes in chemical composition and microbial flora of salmon Izushi during ripening process. Nippon Suisan Gakkaishi 70: 928–937.

Segner, W.P., Schmidt, C.F. and Boltz, J.K. 1966. Effect of sodium chloride and pH on outgrowth of spores of type E *Clostridium botulinum* at optimal and suboptimal temperatures. Appl. Microbiol. 14: 49–54.

Sharma, R.K. 2000. Citric acid. pp. 688–702. *In*: Naidu, A.S. (ed.). Natural Food Antimicrobial Systems. CRC Press New York.

Sharma, M., Patel, J.R., Conway, W.S., Ferguson, S. and Sulakvelidze, A. 2009. Effectiveness of bacteriophages in reducing *Escherichia coli* O157:H7 on fresh-cut cantaloupes and lettuce. J. Food Prot. 72: 1481–1485.

Shelef, L.A. and Yang, Q. 1991. Growth suppression of *Listeria monocytogenes* by lactates in broth, chicken and beef. J. Food Prot. 54: 283.

Sierra, G. 1970. Inhibition of the amino acid induced initiation of germination of bacterial spores by chlorocresol. Can J. Microbiol. 16: 51–52.

Silva, L.V.A.D. 2005. Control of *Vibriio vulnificus* and *Vibrio parahaemolyticus* in oyster (M.S. thesis), Louisiana State Univeristy.

Sofos, J.N., Pierson, M.D., Blocher, J.C. and Busta, F.F. 1986. Mode of action of sorbic acid on bacterial cells and spores. Int. J. Food Microbiol. 3: 1–7.

Sofos, J.N. and Busta, F.F. 1993. Sorbic acid and sorbates. pp. 49–94. *In*: Davidson, P.M. and Branen, A.L. (eds.). Antimicrobials in Food. New York: Marcel Dekker, Inc.

Sofos, J.N. 2000. Sorbic acid. pp. 637–659. *In*: Naidu, A.S. (ed.). Natural Food Antimicrobial Systems. CRC Press New York.

Sommers, C.H. and Rajkowski, K.T. 2011. Radiation inactivation of foodborne pathogens on frozen seafood products. J. Food Prot. 74: 641–644.

Soni, K.A. and Nannapaneni, R. 2010. Bacteriophage significantly reduces *Listeria monocytogenes* on raw salmon fillet tissue. J. Food Prot. 73: 32–38.

Soni, K.A., Nannapaneri, R. and Hagens, S. 2010. Reduction of *Listeria monocytogenes* on the surface of fresh channel catfish fillets by bacteriophage Listex P100. Foodborne Pathog. Dis. 7: 427–434.

Soni, K.A., Desai, M., Oladunjoye, A., Skrobot, F. and Nannapaneri, R. 2012. Reduction of *Listeria monocytogenes* in queso fresco cheese by a combination of listericidal and listeriostatic GRAS antimicrobials. Int. J. Food Microbiol. 155: 82–88.

Souza, E.L., Stamford, T.M.L., Lima, E.O., Trajano, V.N. and Filho, J.B. 2005. Antimicrobial effectiveness of spices: An approach for use in food conservation systems. Brazillian Archives of Biology and Technology 48: 549–558.

Stammen, K., Gerdes, D. and Caporaso, F. 1990. Modified atmosphere packing of seafood. Crit. Rev. Food. Sci. Ntr. 29: 301–331.

Steele, J.H. 2000. Food irradiation: a public health opportunity. Int. Infect Dis. 4: 62–66.

Strydom, A. and Witthuhn, C.R. 2015. *Listeria monocytogenes*: A target for bacteriophage biocontrol. Comp. Rev. Food Sci. Food Safety 14: 694–704.

Sun, Y. and Oliver, J.D. 1994. Antimicrobial action of some GRAS compounds against *Vibrio vulnificus*. Food Addit. Contam. 11: 549–558.

Tamblyn, K.C. and Conner, D.E. 1997. Bacterial activity of organic acids in combination with transdermal compounds against *Salmonella typhimurium* attached to the broiler skin. Food Microbiol. 14: 477–484.

Taylor, L.Y., Cann, D.D. and Welch, B.J. 1990. Antibotulinal properties of nisin in fresh fish packaged in an atmosphere of carbon dioxide. J. Food Prot. 53: 953–957.

Weinstein, L.I. and Albersheim, P. 1983. Host pathogen interactions. XXIII. The mechanism of antimicrobial action of glycinol, a plerocarpan phytoalexin synthesized by soybeans. Plant Physiol. 72: 557–560.

WHO. 1981. Wholesomeness of Irradiated Food. Report of a Joint FAO/IAEA/ WHO Expert Committee. Technical Report 651, WHO Geneva.

Wirawan, R.E., Klesse, N.A., Jack, R.W. and Tagg, J.R. 2006. Molecular and genetic characterization of a novel nisin variant produced by *Streptococcus uberis*. Appl. Env. Microbiol. 72: 1148–1156.

Yamaki, S., Omachi, T., Kawai, Y. and Yamazaki, K. 2014. Characterization of a novel *Morganella morganii* bacteriophage FSP1 isolated from river water. FEMS Microbiol. Lett. 359: 166–172.

Yamaki, S., Kawai, Y. and Yamazaki, K. 2015. Charcterization of a novel bacteriophage, Phda1, infecting the histamine-producing *Photobacterium damselae* subsp. *damselae*. J. Appl. Microbiol. 118: 1141–1550.

Yamazaki, K., Suzuki, U., Kawai, Y., Inoue, N. and Monteville, T.J. 2003. Inhibition of *Listeria monocytogenes* in cold smoked salmon by *Carnobacterium piscicola* CS526 isolated from frozen Surimi. J. Food Prot. 66: 1420–1425.

Yang, R. and Ray, B. 1994. Prevalence and biological control of bacteriocin producing psychrotrophic leuconostocs associated with spoilage of vacuum packaged processed meats. J. Food. Prot. 57: 209–217.

Young, K.M. and Foegeding, P.M. 1993. Acetic, lactic and citric acids and pH inhibition of *Listeria monocytogenes* Scott A and the effect on intracellular pH. J. Appl. Bacteriol. 74: 515–520.

Yuste, J., Pla, R. and Mor-Mur, M. 2000. *Salmonella* Enteritidis and aerobic mesophiles in inoculated poultry sausages manufactured with high-pressure processing. Lett. Appl. Microbiol. 31: 374–377.

Yuste, J., Capellas, M., Fung, D.Y.C. and Mor-Mur, M. 2004. Inactivation and sublethal injury of foodborne pathogens by high pressure processing: evaluation with conventional media and thin agar layer method. Food Research International 37: 861–866.

Zendo, T., Fukao, M., Ueda, K., Higuchi, T., Nakayama, J. and Sonomoto, K. 2003. Identification of the lantibiotic nisin Q, a new natural nisin variant produced by *Lactococcus lactis* 61-14 isolated from a river in Japan. Biosci. Biotechnol. Biochem. 67: 1616–1619.

6

Viral Diseases Affecting Aquaculture

Hisae Kasai,[1,]* Shotaro Nishikawa[1] and
Kenichi Watanabe[2,]*

1. Introduction

Global fish production has grown steadily in the last five decades, with food
fish supply increasing at an average annual rate of 3.2 percent, outpacing
world population growth at 1.6 percent (FAO 2014). World per capita
apparent fish consumption increased from an average of 9.9 kg in the
1960s to 19.2 kg in 2012 (preliminary estimate) (FAO 2014). This impressive
development has been driven by a combination of population growth, rising
incomes and urbanization, and facilitated by the strong expansion of fish
production and more efficient distribution channels (FAO 2014). In these
conditions, world aquaculture production continues to grow. According to
statistics collected globally by FAO, world aquaculture production attained
another all-time high of 90.4 million tons (live weight equivalent) in 2012
(US$ 144.4 billion), including 66.6 million tons of food fish (US$ 137.7 billion)
(FAO 2014). Table 6.1 shows aquaculture production by region. Aquaculture
industry is flourishing in Asian countries, especially China.

Unfortunately, disease outbreaks caused by pathogenic organisms
often cause a serious socio-economic damage for aquaculture field. In the

[1] Faculty of Fisheries Sciences, Hokkaido University, Japan.
[2] Faculty of Bioindustry, Tokyo University of Agriculture, Japan.
  Email: k4watana@bioindustry.nodai.ac.jp
* Corresponding authors

Table 6.1 Aquaculture production by region: quantity, and percentage of world total production (FAO 2014).

Selected groups and countries		1990	1995	2000	2005	2010	2012
Africa	**(tonnes)**	**81 015**	**110 292**	**399 688**	**646 182**	**1 286 591**	**1 485 367**
	(percentage)	**0.62**	**0.45**	**1.23**	**1.46**	**2.18**	**2.23**
North Africa	(tonnes)	63 831	75 316	343 986	545 217	928 530	1 030 675
	(percentage)	0.49	0.31	1.06	1.23	1.57	1.55
Sub-Saharan Africa	(tonnes)	17 184	34 976	55 702	100 965	358 062	454 691
	(percentage)	0.13	0.14	0.17	0.23	0.61	0.68
Americas	**(tonnes)**	**548 479**	**919 571**	**1 423 433**	**2 176 740**	**2 581 089**	**3 187 319**
	(percentage)	**4.19**	**3.77**	**4.39**	**4.91**	**4.37**	**4.78**
Caribbean	(tonnes)	12 169	28 260	39 704	29 790	37 301	28 736
	(percentage)	0.09	0.12	0.12	0.07	0.06	0.04
Latin America	(tonnes)	179 367	412 650	799 234	1 478 443	1 885 965	2 565 107
	(percentage)	1.37	1.69	2.47	3.34	3.19	3.85
North America	(tonnes)	356 943	478 661	584 495	668 507	657 823	593 476
	(percentage)	2.73	1.96	1.80	1.51	1.11	0.89
Asia	**(tonnes)**	**10 801 531**	**21 677 062**	**28 420 611**	**39 185 417**	**52 436 025**	**58 895 736**
	(percentage)	**82.61**	**88.90**	**87.67**	**88.46**	**88.82**	**88.39**
China	(tonnes)	6 482 402	15 855 653	21 522 095	28 120 690	36 734 215	41 108 306
	(percentage)	49.58	65.03	66.39	63.48	62.22	61.69
Central and Western Asia	(tonnes)	72 164	65 602	122 828	190 654	259 781	311 133
	(percentage)	0.55	0.27	0.38	0.43	0.44	0.47
Southern and Eastern Asia (excluding China)	(tonnes)	4 246 965	5 755 807	6 775 688	10 874 073	15 442 028	17 476 296
	(percentage)	32.48	23.61	20.90	24.55	26.16	26.23

Table 6.1 contd. ...

...Table 6.1 contd.

Selected groups and countries		1990	1995	2000	2005	2010	2012
Europe	(tonnes)	1 601 649	1 581 359	2 052 567	2 137 340	2 548 094	2 880 641
	(percentage)	12.25	6.49	6.33	4.83	4.32	4.32
European Union (Member Organization) (28)	(tonnes)	1 033 857	1 182 098	1 400 667	1 269 958	1 280 236	1 259 971
	(percentage)	7.91	4.85	4.32	2.87	2.17	1.89
Other European countries	(tonnes)	567 792	399 261	651 900	867 382	1 267 858	1 620 670
	(percentage)	4.34	1.64	2.01	1.96	2.15	2.43
Oceania	(tonnes)	42 005	94 238	121 482	151 466	185 617	184 191
	(percentage)	0.32	0.39	0.37	0.34	0.31	0.28
World	(tonnes)	13 074 679	24 382 522	32 417 781	44 297 145	59 037 416	66 633 253

case of Epizootic ulcerative syndrome, an infectious fungal (*Aphanomyces*) disease, in several Asian countries before 1990, damages exceeded US$ 10,000,000 (Lilley et al. 1998). A viral disease of shrimp caused by white spot syndrome virus affected most shrimp-producing countries in Asia and the Americas. Losses were in the range of more than US$ 400,000,000 in China in 1993, US$ 17,600,000 in India in 1994, and US$ 600,000,000 in 1997 in Thailand (Subasinghe et al. 2001). These disease problems are often associated with movement of live aquatic animals and animal products. This trend has been triggered by changing circumstances and perspectives, especially liberalization of world trade (Subasinghe et al. 2001). The World organization for animal health (OIE) has established a list of notifiable terrestrial and aquatic animal diseases (Table 6.2) in order to be line with the terminology of the Sanitary and Phytosanitary Agreement of the World Trade Organization. This classifies diseases as specific hazards, and gives

Table 6.2 OIE notifiable diseases (2017)*.

Disease of fish	Epizootic haematopoietic necrosis disease
	Infection with *Aphanomyces invadans* (epizootic ulcerative syndrome)
	Infection with *Gyrodactylus salaris*
	Infection with HPR-deleted or HPR0 infectious salmon anaemia virus
	Infection with salmonid alphavirus
	Infectious haematopoietic necrosis
	Koi herpesvirus disease
	Red sea bream iridoviral disease
	Spring viraemia of carp
	Viral haemorrhagic septicaemia
Disease of mollusks	Infection with abalone herpesvirus
	Infection with *Bonamia exitiosa*
	Infection with *Bonamia ostreae*
	Infection with *Marteilia refringens*
	Infection with *Perkinsus marinus*
	Infection with *Perkinsus olseni*
	Infection with *Xenohaliotis californiensis*
Disease of crustaceans	Acute hepatopancreatic necrosis disease
	Crayfish plague (*Aphanomyces astaci*)
	Infection with yellow head virus
	Infectious hypodermal and haematopoietic necrosis
	Infectious myonecrosis
	Necrotising hepatopancreatitis
	Taura syndrome
	White spot disease
	White tail disease

* http://www.oie.int/animal-health-in-the-world/oie-listed-diseases-2017/.

all listed diseases the same degree of importance in international trade. Characteristics of the listed diseases are high mortality, contagiousness, and restricted number of endemic countries. In this chapter, some viral diseases of aquatic animals are introduced.

2. Viral Diseases of Marine Fish

Due to improvements in aquacultural technologies and good economic value, many marine fish species are now being cultured. In addition, production of fish larvae for aquaculture has increased. Unfortunately, viruses such as Flounder herpesvirus (FHV), Pilchard herpesvirus (PHV), Aquatic birnaviruses, Lymphocystis disease virus (LCDV), Hirame rhabdovirus (HIRRV), Viral hemorrhagic septicemia virus (VHSV), Red sea bream iridovirus (RSIV), and Betanodaviruses seriously affect aquaculture operations. Here, we discuss the characteristics, genome size, serological classification, molecular classification, and pathogenesis of representative viruses, outlined in the OIE manual of diagnostic methods for aquatic animals (OIE Manual of diagnostic tests for aquatic animals 2016), AFS-FHS blue book (2014), and fish diseases and disorders (Woo and Bruno 2011).

2.1 Viral nervous necrosis

Viral nervous necrosis (VNN), otherwise known as viral encephalopathy and retinopathy (VER), is considered to be a serious disease of many marine fish species (more than 50 fish species), and also outbreak had been documented in freshwater farms (Bovo et al. 2011). Although Atlantic salmon (*Salmo salar*) has susceptibility to the disease (Korsnes et al. 2005), VNN has not been reported in salmonids. The disease was first reported in hatchery-reared larval and juvenile Japanese parrotfish (*Oplegnathus fasciatus*) in Japan (Yoshikoshi and Inoue 1990) and larval Asian sea bass (*Lates calcarifer*) in Australia (Glazebrook et al. 1990).

Disease Agent: The etiological agent is nervous necrosis virus (NNV). This virus was first identified as a new member of a family Nodaviridae from infected striped jack (*Pseudocaranx dentex*) larvae. Thus the name striped jack nervous necrosis virus (SJNNV) was suggested (Mori et al. 1992). Current taxonomical investigations classify the virus into a new genus *Betanodavirus* within the family Nodaviridae (Schneemann et al. 2005).

Virions are non-enveloped, spherical shape, with about 25–30 nm in diameter. Virions contain two positive-sense ssRNA molecules: RNA1 (3.1 kb) which encodes the RNA-dependent RNA replicase (110 kDa), and RNA2 (1.4 kb) which is the coat protein (42 kDa) gene. Based on a molecular phylogenetic analysis of the T4 region (427 bases) of the viral coat protein gene, betanodaviruses were classified into four major genotypes (Nishizawa et al. 1997), namely, Tiger puffer nervous necrosis virus (TPNNV), Striped jack nervous necrosis virus (SJNNV), Barfin flounder nervous necrosis virus (BFNNV), and Red spotted grouper nervous necrosis virus (RGNNV). The genetic differences are closely related to the serotypes as determined by virus neutralization tests with polyclonal antibodies, i.e., serotype A (SJNNV genotype), serotype B (TPNNV genotype), and serotype C (BFNNV and RGNNV genotypes) (Mori et al. 2003). Most isolated betanodaviruses are classified into serotype C. A close relationship between the genotypes and the host fish is also confirmed. The BFNNV genotype virus has been isolated from cold water fish species, while the other genotype virus has been isolated from warm water fish species. Figure 6.1 shows NNV virions in the retina of the infected barfin flounder (*Verasper moseri*) (Watanabe et al. 1999).

Although viral nervous necrosis virus most seriously infects hatchery-rearing larval and/or juvenile fish, several fish species, such as sevenband grouper *Epinephelus akaara* (Fukuda et al. 1996), European sea bass *Dicentrarchus labrax* (Le Breton et al. 1997), and Atlantic halibut *Hippoglossus hippoglossus* (Aspehaug et al. 1999) are susceptible to VNN in their young and/or adult stage fish (over 100 g of B.W.). Larval and juvenile fish often may be severely affected (more than 90% mortality), but as they become older the mortality is reduced (about 10–50%).

Diagnostic Methods: Diseased fish at all stages do not show definite clinical signs on the body surface and gills. However, affected juveniles or older fishes are characterized by abnormal swimming behaviors: spiral, whirling movement, lying down at rest, or belly-up floating with inflation of swim bladder. Similar swimming abnormalities are not usually observed

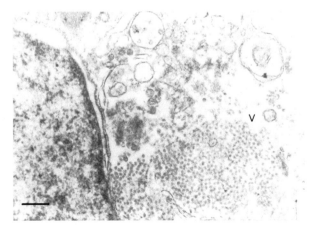

Figure 6.1 Nervous necrosis virus virions in the retina of the infected barfin flounder (Watanabe et al. 1999). V: virions, Bar: 200 nm.

in infected larvae. Microscopic observation reveals severe necrosis and vacuolation in the central nervous system (brain and spinal cord) and retina (Fig. 6.2) of infected fish.

To diagnose infected fish, processed samples are inoculated onto striped snakehead nerve (SSN-1; Frerichs et al. 1996) cell line or E-11 (cloning cell of SSN-1; Iwamoto et al. 2000) cell lines and incubated at 15–20°C (cold water fish) or 20–30°C (warm water fish). Inoculated cultures are incubated for 3–7 days, and observed for typical CPE (rounded, granular cells and intensive disintegration). The isolated virus must be confirmed by immunological procedures such as serum neutralization (Nishizawa et al. 1995), immunostaining (Arimoto et al. 1992). The most rapid and convenient method to diagnose clinically affected fish is RT-PCR (Nishizawa et al. 1994). However, for a confirmative diagnosis, either histopathology or virus isolation has to be performed.

Vertical transmission from broodstock to offspring frequently observed in the seed production of species, such as striped jack (Mushiake et al. 1994), barfin flounder (Watanabe et al. 2000), and Atlantic halibut (Grotmol and Totland 2000). Elimination of virus-carrying broodstocks, as well as disinfection of fertilized eggs and rearing water by disinfectants, are successful in preventing vertical transmission. (Mori et al. 1998, Watanabe et al. 1998, Grotmol and Totland 2000) Horizontal transmission (through virus contaminated water) is considered the most common transmission route (Le Breton et al. 1997, Mori et al. 1998, Watanabe et al. 1998). Chemical and physical treatments, such as iodine, UV irradiation, ozonized seawater, pH12, and sodium hypochlorite are effective in preventing water-borne transmission of betanodaviruses (Arimoto et al. 1996, Frerichs et al. 2000).

On the other hand, to reduce severe economic damage during seed production and aquaculture, the effective vaccination system is required.

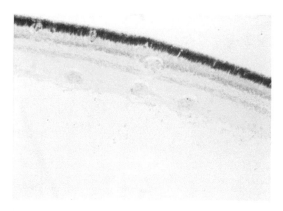

Figure 6.2 Light micrograph showing vacuolation in the retina of affected barfin flounder juvenile by VNN. Bar: 100 μm.

Recombinant viral coat protein expressed in *Escherichia coli* (Tanaka et al. 2001), virus-like particles expressed in a baculovirus expression system (Thiéry et al. 2006) or formalin-inactivated virus (Yamashita et al. 2005) may be effective in controlling VNN. In 2012, an inactivated RGNNV vaccine against sevenband grouper was commercialized in Japan.

2.2 Red seabream iridoviral disease

Red seabream iridoviral disease (RSIVD) is one of the most important viral diseases of farmed warmwater marine fish. The disease was first reported in red seabream (*Pagrus major*) cultured in Shikoku Island, Japan (Inouye et al. 1992). Since then, the disease has caused mass mortalities in cultured fish, such as the orders Perciformes and Pleuronectiformes species in the western part of Japan. Currently, the disease is found not only in Japan, but also widely in East and South-East Asian countries, and 41 fish species including two hybrid species are listed as host fish for RSIVD (Table 6.3; OIE 2012).

Disease Agent: RSIV is classified into the genus *Megalocytivirus* of the family *Iridoviridae* (Jancovich et al. 2011). RSIV is a non-enveloped icosahedral symmetry virus, 140–200 nm in diameter, with a thin lipid layer on the surface. The viral genome consists of a single, linear dsDNA molecule with the genome sizes being approximately 112–121 kbp (Kurita et al. 2002, He et al. 2001, Do et al. 2004, Lu et al. 2005, Ao and Chen 2006). RSIV has been divided into three groups by phylogenetic analysis of the major capsid protein gene, called infectious spleen and kidney necrosis virus (ISKNV) (He et al. 2000) group, RSIV group, and turbot reddish body iridovirus (TRBIV) (Shi et al. 2004) group (Wang et al. 2007).

Diagnostic Methods: In Japan, RSIVD occurs in the summer when water temperatures rise above 25°C (Kawakami and Nakajima 2002). Juveniles are more susceptible than adults, but observation in hatcheries has not been done (Muroga 2001). The affected red sea bream show severe anemia, petechiae of the gills, and enlargement of the spleen. RSIVD is mostly characterized by the presence of many enlarged cells (about 20 μm in diameter) in the spleen, heart, kidney, intestine, and gills. The enlarged cells are basophilic and Feulgen-positive, and may originate in infected leucocyte (Inouye et al. 1992). Virus isolation using cell culture method is not reliable due to lower permissibility of the GF cells. IFAT with a specific MAb (Nakajima and Sorimachi 1995) is frequently used to diagnose the disease, because of a simple and reliable method in overt infection cases of RSIV. PCR method to detect RSIV was first reported by Kurita et al. (1998). Currently, many PCR, nested PCR (Chao et al. 2002), real-time PCR assay (Caipang et al. 2003), and LAMP method (Caipang et al. 2004) have been

Table 6.3 Fish species affected by RSIV (OIE 2012).

Order	Family	Common name	Latin name
Perciformes	Carangidae	greater amberjack	*Seriola dumerili*
		hybrid of yellowtail amberjack and Japanese amberjack	*S. lalandi* × *S. quinqueradiata*
		Japanese amberjack	*S. quinqueradiata*
		Japanese jack mackerel	*Trachurus japonicus*
		snubnose pompano	*Trachinotus blochii*
		striped jack	*Pseudocaranx dentex*
		yellowtail amberjack	*Seriola lalandi*
	Centrarchidae	largemouth bass	*Micropterus salmoides*
	Centropomatidae	Asian sea bass	*Lates calcarifer*
		Japanese sea perch	*Lateolabrax japonicus*
		–	*Lateolabrax* sp.
	Haemulidae	chicken grunt	*Parapristipoma trilineatum*
		crescent sweetlips	*Plectorhinchus cinctus*
	Kyphosidae	largescale blackfish	*Girella punctata*
	Lethrinidae	Chinese emperor	*Lethrinus haematopterus*
		spangled emperor	*Lethrinus nebulosus*
	Moronidae	hybrid of striped sea bass and white bass	*Morone saxatilis* × *M. chrysops*
	Oplegnathidae	Japanese parrotfish	*Oplegnathus fasciatus*
		spotted knifejaw	*O. punctatus*
	Rachicentridae	Cobia	*Rachycentron canadum*
	Sciaenidae	croceine croaker	*Pseudosciaena crocea*
	Scombridae	chub mackerel	*Scomber japonicus*
		Japanese Spanish mackerel	*Scomberomorus niphonius*
		northern bluefin tuna	*Thunnus thynnus*
	Serranidae	brown-marbled grouper	*Epinephelus fuscoguttatus*
		convict grouper	*E. septemfasciatus*
		giant grouper	*E. lanceolatus*
		greasy grouper	*E. tauvina*
		Hong Kong grouper	*E. akaara*
		longtooth grouper	*E. bruneus*
		Malabar grouper	*E. malabaricus*
		orange-spotted grouper	*E. coioides*
		yellow grouper	*E. awoara*

Table 6.3 contd. ...

...Table 6.3 contd.

Order	Family	Common name	Latin name
	Sparidae	black porgy	*Acanthopagrus schlegeli*
		crimson sea bream	*Evynnis japonica*
		red sea bream	*Pagrus major*
		yellowfin sea bream	*Acanthopagrus latus*
Pleuronectiformes	Pleuronectidae	spotted halibut	*Verasper variegatus*
	Scophthalmidae	Japanese flounder	*Paralichthys olivaceus*
Scorpaeniformes	Sebastidae	Rockfish	*Sebastes schlegeli*
Tetraodontiformes	Tetraodontidae	Japanese puffer fish	*Takifugu rubripes*

developed for Megalocytiviruses detection. Since 1999, the formalin-killed vaccine by intraperitoneal injection method has been licensed in Japan, and to date is available for red sea bream, striped jack, Malabar (*E. malabaricus*), and orange-spotted grouper (*E. coioides*) and the genus *Seriola*.

3. Viral Diseases of Salmon and Trout

Salmonid fish including chum salmon (*Oncorhynchus keta*), pink salmon (*O. orbuscha*), sockeye (*O. nerka*), Chinook (*O. tshawytscha*), coho (*O. kisutch*), and masu salmon (*O. masou*) are important species with the unique characteristic of being hatchery reared and released into the North Pacific Ocean. Atlantic salmon (*Salmo salar*) and rainbow trout (*O. mysiss*) are important for aquaculture. Viral diseases, such as infectious pancreatic necrosis (IPN), infectious hematopoietic necrosis (IHN), viral hemorrhagic septicemia (VHS), Oncorhynchus masou virus disease (OMVD), infectious salmon anemia (ISA), erythrocytic inclusion body syndrome (EIBS), pancreas disease (PD), sleeping disease (SD), viral wiring disease (VWD), and viral erythrocytic necrosis (VEN) limit successful propagation and aquaculture of these species. Here, we discuss the characteristics, genome size, serological classification, molecular classification, and pathogenesis of representative viruses, outlined in the OIE manual of diagnostic methods for aquatic animals (OIE Manual of diagnostic tests for aquatic animals 2016), AFS-FHS blue book (2014), and fish diseases and disorders (Woo and Bruno 2011).

3.1 Infectious pancreatic necrosis

Infectious pancreatic necrosis (IPN) is an acute contagious systemic birnavirus disease of fry and fingerling trout. The disease most characteristically occurs in rainbow trout, brook trout (*Salvelinus fontinalis*), brown trout (*Salmo trutta*), and Atlantic salmon. However, IPNV and IPN-like birnaviruses have been isolated from a variety of non-salmonid fishes and invertebrates from freshwater, estuarine, and marine environments. IPNV (and IPN-like birnaviruses) is among geographically dispersed groups of viruses.

Disease Agent: IPNV is the type species of the genus Aquabirnavirus within the viral family designated *Birnaviridae* by the International Committee on Taxonomy of Viruses (ICTV) (King et al. 2012). The IPNV virion is non-enveloped, single shelled, and measures approximately 60 nm in diameter. The viral genome consists of a bi-segmented, double-stranded RNA. Aquatic birnavirus, including IPNV, is classified into two serogroups (A and B). Serogroup A isolates are divided into nine serotypes, termed A1–A9. Serogroup B contains only one serotype, but this is based on a limited number of isolates. Aquatic birnavirus, including IPNV, was classified into seven genogroups by sequence analysis of the VP2/NS junction region.

Diagnostic Methods: In salmonid fry, symptoms of infection include anorecia and a violent whirling pattern. Non-specific external symptoms include darkening of the skin, abdominal swelling, exophthalmia, pale gills, and petechial hemorrhages. IPN exhibits marked pancreatic necrosis and severe lesions in the intestinal mucosa. Pancreatic lesions can vary from small fogi to extensive acinar cell necrosis, with nuclear pyknosis, karyorrhexis, and basophilic cytoplasmic inclusions. Virological examination coupled with serological or molecular identification is required for confirmation of clinical or subclinical infection with IPNV. Cell cultures of blue gill fry (BF-2), Chinook salmon embryo (CHSE-214), or rainbow trout gonad (RTG-2) are commonly used for evaluation of clinical materials for detection of IPNV in salmonid fishes. Confirmatory identification of IPNV can be accomplished using immunologic-based or molecular-based assay formats.

3.2 Infectious hematopoietic necrosis

Infectious hematopoietic necrosis (IHN) is an acute, systemic, and usually virulent rhabdoviral disease that occurs in the wild, but is more typically seen in epizootic proportion among young trout and certain Pacific salmon under husbandry in coastal North America from California to Alaska, Europe, and the Far East. The disease most affects rainbow/steelhead trout, cutthroat trout (*Salmo clarki*), brown trout, Atlantic salmon, and Pacific salmon, including Chinook, sockeye/kokanee, chum, masu/yamame and amago salmon (*O. rhodurus*). Under experimental conditions, infections have been reported in non-salmonids such as pike fry, sea bream, and turbot.

Disease Agent: IHNV is the type species of the genus *Novirhabdovirus* within the viral family designated *Rhabdoviridae* by the ICTV (King et al. 2012). IHNV virion is bullet-shaped, and measures 170 × 70 nm. Its viral

Figure 6.3 Signs of IHN in masu salmon (*O. masou*).

genome consists of complementary, single-stranded RNA. IHNV isolates are not classified serologically, but can be grouped into three genetic types by sequence analysis of viral glycoprotein gene.

Diagnostic Methods: Moribund fish are lethargic, swim high in the water column, and are anorexic. They exhibit exophthalmia, darkening of body color, abdominal distension, pale gills, and hemorrhages at the bases of fins (Fig. 6.3). Fecal casts trailing from the vent have been reported, but are not always observed. Necrosis of splenic hematopoietic tissue and of the endocrine and exocrine tissue of the pancreas is diffused, but the liver can have areas of focal necrosis. Virological examination coupled with serological or molecular identification is required for confirmation of clinical or subclinical infection with IHNV. Cell cultures of BF-2, CHSE-214, RTG-2 or Epithelioma papulosum cyprini (EPC) are commonly used for evaluation of clinical materials for detection of IHNV in salmonid fishes. Confirmatory identification is the same as that for IPNV.

3.3 Viral hemorrhagic septicemia

Viral hemorrhagic septicemia (VHS) is an important source of mortality for cultured and wild fish in freshwater and marine environments in several regions of the northern hemisphere. Over 60 species of freshwater and marine fish are currently known to be natural hosts of VHS virus (VHSV), but rainbow trout are especially susceptible.

Disease Agent: VHSV is classified to the genus *Novirhabdovirus* within the viral family designated *Rhabdoviridae* by the ICTV (King et al. 2012). VHSV virion is bullet-shaped and measures 180–200 × 75 nm. The viral genome is complementary, single-stranded RNA. VHSV was classified into three serogroups; however there was considerable overlap of strains within and between these serotypes. VHSV can be grouped into four genotypes by sequence analysis. These genotypes show a geographic basis for their distribution rather than host specificity.

Diagnostic Methods: External clinical signs of disease include exophthalmia, abdominal distention, darkened coloration, anemia, lethargy, hyperactivity, and hemorrhages in the eyes, skin, gills, and at the base of fins. Internally, visceral mesenteries show diffuse hemorrhage, the kidneys and liver can be hyperemic, swollen, and discolored, and hemorrhaging may be seen in skeletal muscle. In the liver, kidney and spleen, focal to extensive necrotic changes may occur. Virological examination is same as IHNV. Confirmatory identification is same as IPNV.

3.4 Oncorhynchus masou virus disease

Oncorhynchus masou virus disease (OMVD) is an economically significant disease of farmed salmonid fish in Japan. This disease also occurs in wild fish. The disease most characteristically occurs in kokanee, coho, masu salmon, and rainbow trout. Surviving salmon often develop cutaneous tumors, particularly around the mouth (Fig. 6.4).

Diagnostic Methods: External clinical signs in infected fish include appetent exophthalmia, and petechiae on the body surface, especially beneath the lower jaw. Agonal or abnormal swimming behavior has not been observed. Internally, the liver shows white spot lesions, and in advanced cases the whole liver becomes pearly white. In some cases the spleen is found to

Figure 6.4 Tumor developing around the mouth of a chum salmon fingerling.

be swollen. Necrosis of epithelial cells and kidney have been observed in the young one-month old moribund specimens, while partial necrosis of the liver, spleen, and pancreas was seen in older moribund specimens. Virological examination coupled with serological or molecular identification is required for confirmation of clinical or subclinical infection with OMV. Cell cultures of CHSE-214 or RTG-2 are commonly used for evaluation of clinical materials for detection of OMV in salmonid fishes. Confirmatory identification is same as IPNV.

3.5 Infectious salmon anemia

Infectious salmon anemia (ISA) is one of the most important viral diseases of farmed Atlantic salmon. Sea-run brown trout, rainbow trout, and Atlantic herring (*Clupea harengus*) are potential asymptomatic carriers of the virus. Infectious salmon anemia has been found in Norway, the Faroe Islands, and the United Kingdom (Scotland and Shetland Islands). Infectious salmon anemia outbreaks occur periodically in Norway and Chile, as well as in limited regions of North America shared by the U.S. and Canada.

Disease Agent: ISA virus (ISAV) is classified as the type species of the genus *Isavirus* within the viral family designated *Orthomyxoviridae* by the ICTV (King et al. 2012). ISAV is a pleomorphic enveloped virus, 100–130 nm in diameter, with 10–12 nm surface projections. The viral genome consists of eight single-stranded RNA segments of negative polarity. ISAV has been divided into two major clusters or genotypes; North American and the European, and the third genotype may exist.

Diagnostic Methods: ISA should be a consideration if Atlantic salmon show increased mortality and signs of anemia, or lesions characteristic of this disease. It should always be investigated if the hematocrit is less than 10%. External clinical signs in infected fish are pale gills and muscle, petechial hemorrhage, dark liver and spleen, and ascites present. The presence of renal interstitial hemorrhaging, and tubular epithelial degeneration, necrosis, and casting within the posterior kidney have all been noted. Liver sections show multifocal to confluent hemorrhagic hepatic necrosis, focal congestion, and dilatation of hepatic sinusoids.

4. Conclusions and Suggestions

Viral, bacterial, fungal, and protozoan diseases often occur not only in finfish, but also in mollusks and crustaceans. The number of OIE listed diseases for mollusks and crustaceans is increasing due to the risks they pose for cultured and wild animals. Strengthening of global surveillance is important to decrease the risk of spreading such diseases to pathogens-free countries. Where the disease occurs, control strategies must be applied to reduce socio-economic damages for aquaculture field. The strategies are discussed in Chapter 7.

References

AFS-FHS (American Fisheries Society-Fish Health Section) (ed.). 2014. FHS blue book: suggested procedures for the detection and identification of certain finfish and shellfish pathogens, edition. AFS-FHS, Bethesda, Maryland.

Ao, J. and Chen, X. 2006. Identification and characterization of a novel gene encoding an RGD-containing protein in large yellow croaker iridovirus. Virology 355: 213–222.

Arimoto, M., Mushiake, K., Mizuta, Y., Nakai, T., Muroga, K. and Furusawa, I. 1992. Detection of striped jack nervous necrosis virus (SJNNV) by enzyme-linked immunosorbent assay (ELISA). Fish Pathol. 27: 191–195.

Arimoto, M., Sato, J., Maruyama, K., Mimura, G. and Furusawa, I. 1996. Effect of chemical and physical treatments on the inactivation of striped jack necrosis virus striped jack nervous necrosis virus (SJNNV). Aquaculture 143: 15–22.

Aspehaug, V., Devold, M. and Nylund, A. 1999. The phylogenetic relationship of nervous necrosis virus from halibut (*Hippoglossus hippoglossus*). Bulletin of the European Association of Fish Pathologists 19: 196–202.

Bovo, G., Gustinelli, A., Quaglio, F., Gobbo, F., Panzarin, V., Fusaro, A., Mutinelli, F., Caffara, M. and Fioravanti, M.L. 2011. Viral encephalopathy and retinopathy outbreak in freshwater fish farmed in Italy. Dis. Aquat. Org. 96: 45–54.

Caipang, C.M., Hirono, I. and Aoki, T. 2003. Development of a real-time PCR assay for the detection and quantification of red seabream iridovirus (RSIV). Fish Pathol. 38: 1–7.

Caipang, C.M., Haraguchi, I., Ohira, T., Hirono, I. and Aoki, T. 2004. Rapid detection of a fish iridovirus using loop-mediated isothermal amplification (LAMP). J. Virol. Met. 121: 155–161.

Chao, C.-B., Yang, S.-C., Tsai, H.-Y., Chen, C.-Y., Lin, C.-S. and Huang, H.-T. 2002. A Nested PCR for the detection of grouper iridovirus in Taiwan (TGIV) in cultured hybrid grouper, giant seaperch, and largemouth bass. J. Aquat. Anim. Health 14: 104–13.

Do, J.W., Moon, C.H., Kim, H.J., Ko, M.S., Kim, S.B., Son, J.H., Kim, J.S., An, E.J., Kim, M.K., Lee, S.K., Han, M.S., Cha, S.J., Park, M.S., Park, M.A., Kim, Y.C., Kim, J.W. and Park, J.W. 2004. Complete genomic DNA sequence of rock bream iridovirus. Virology 325: 351–363.

FAO. 2014. The State of World Fisheries and Aquaculture 2014. Rome, 243 p. (also available at www.fao.org/3/a-i3720e.pdf).

Frerichs, G.N., Rodger, H.D. and Peric, Z. 1996. Cell culture isolation of piscine neuropathy nodavirus from juvenile sea bass, *Dicentrarchus labrax*. J. Gen. Virol. 77: 2067–2071.

Frerichs, G.N., Tweedie, A., Starkey, W.G. and Richards, R.H. 2000. Temperature, pH and electrolyte sensitivity, and heat, UV and disinfectant inactivation of sea bass (*Dicentrarchus labrax*) neuropathy nodavirus. Aquaculture 185: 13–24.

Fukuda, Y., Nguyen, H.D., Furuhashi, M. and Nakai, T. 1996. Mass mortality of cultured sevenband grouper, *Epinephelus septemfasciatus*, associated with viral nervous necrosis. Fish Pathol. 31: 165–170.

Glazebrook, J.S., Heasman, M.P. and de Beer, S.W. 1990. Picorna-like viral particles associated with mass mortalities in larval barramundi, *Lates calcarifer* Bloch. J. Fish Dis. 13: 245–249.

Grotmol, S. and Totland, G.K. 2000. Surface disinfection of Atlantic halibut *Hippoglossus hippoglossus* eggs with ozonated sea-water inactivates nodavirus and increases survival of the larvae. Dis. Aquat. Org. 39: 89–96.

He, J.G., Wang, S.P., Zeng, K., Huang, Z.J. and Chan, S.M. 2000. Systemic disease caused by an iridovirus like agent in cultured mandarin fish, *Siniperca chuatsi* (Basilewsky), in China. J. Fish Dis. 23: 219–222.

He, J.G., Deng, M., Weng, S.P., Li, Z., Zhou, S.Y., Long, Q.X., Wang, X.Z. and Chan, S.M. 2001. Complete genome analysis of the mandarin fish infectious spleen and kidney necrosis iridovirus. Virology 291: 126–139.

Inouye, K., Yamano, Y., Maeno, Y., Nakajima, K., Matsuoka, M., Wada, Y. and Sorimachi, M. 1992. Iridovirus infection of cultured red sea bream, *Pagrus major*. Fish Pathol. 27: 19–27.

Iwamoto, T., Nakai, T., Mori, K., Arimoto, M. and Furusawa, I. 2000. Cloning of the fish cell line SSN-1 for piscine nodaviruses. Dis. Aquat. Org. 43: 81–89.

Jancovich, J.K., Chinchar, V.G., Hyatt, A., Miyazaki, T., Williams, T. and Zhang, Q.Y. 2011. Family iridoviridae. pp. 193–210. *In*: King, A.M.Q., Adams, M.J., Carstens, E.B. and Lefkowitz, E.J. (eds.). Virus Taxonomy Nineth Report of the International Committee on Taxonomy of Viruses. Academic Press, San Diego.

Kawakami, H. and Nakajima, K. 2002. Cultured fish species affected by red sea bream iridoviral disease from 1996 to 2000. Fish Pathol. 37: 45–47.

King, A.M.Q., Adams, M.J., Carstens, E.B. and Lefkowiz, E.J. (eds.). 2012. Virus Taxonomy: Ninth Report of the International Committee on Taxonomy of Viruses. Elsevier, USA, 1338 p.

Korsnes, K., Devold, M., Nerland, A.H. and Nylund, A. 2005. Viral encephalopathy and retinopathy (VER) in Atlantic salmon *Salmo salar* after intraperitoneal challenge with a nodavirus from Atlantic halibut *Hippoglossus hippoglossus*. Dis. Aquat. Org. 68: 7–15.

Kurita, J., Nakajima, K., Hirono, I. and Aoki, T. 1998. Polymerase chain reaction (PCR) amplification of DNA of red sea bream iridovirus (RSIV). Fish Pathol. 33: 17–23.

Kurita, J., Nakajima, K., Hirono, I. and Aoki, T. 2002. Polymerase chain reaction (PCR) amplification of DNA of red sea bream iridovirus (RSIV). Fish Sci. 68(Suppl. II): 1113–1115.

Le Breton, A., Grisez, L., Sweetman, J. and Olievier, F. 1997. Viral nervous necrosis (VNN) associated with mass mortalities in cage-reared sea bass, *Dicentrarchus labrax* (L.). J. Fish Dis. 20: 145–151.

Lilley, J.H., Callinan, R.B., Chinabut, S., Kanchanakhan, S., MacRae, I.H. and Phillips, M.J. 1998. Epizootic Ulcerative Syndrome (EUS) Technical Handbook.

Lu, L., Zhou, S.Y., Chen, C., Weng, S.P., Chan, S.M. and He, J.G. 2005. Complete genome sequence analysis of an iridovirus isolated from the orange-spotted grouper, Epinephelus coioides. Virology 339: 81–100.

Mori, K., Nakai, T., Muroga, K., Arimoto, M., Mushiake, K. and Furusawa, I. 1992. Properties of a new virus belonging to nodaviridae found in larval striped jack (*Pseudocaranx dentex*) with nervous necrosis. Virology 187: 368–371.

Mori, K., Mushiake, K. and Arimoto, M. 1998. Control measures for viral nervous necrosis in striped jack. Fish Pathol. 33: 443–444.

Mori, K., Mangyoku, T., Iwamoto, T., Arimoto, M., Tanaka, S. and Nakai, T. 2003. Serological relationships among genotypic variants of betanodavirus. Dis. Aquat. Org. 57: 19–26.

Muroga, K. 2001. Viral and bacterial diseases of marine fish and shellfish in Japanese hatcheries. Aquaculture 202: 23–44.

Mushiake, K., Nishizawa, T., Nakai, T., Furusawa, I. and Muroga, K. 1994. Control of VNN in striped jack: selection of spawners based on the detection of SJNNV gene by polymerase chain reaction (PCR). Fish Pathol. 29: 177–182.

Nakajima, K. and Sorimachi, M. 1995. Production of monoclonal antibodies against red sea bream iridovirus. Fish Pathol. 30: 47–52.

Nishizawa, T., Mori, K., Nakai, T., Furusawa, I. and Muroga, K. 1994. Polymerase chain reaction (PCR) amplification of RNA of striped jack nervous necrosis virus (SJNNV). Dis. Aquat. Org. 18: 103–107.

Nishizawa, T., Mori, K., Furuhashi, M., Nakai, T., Furusawa, I. and Muroga, K. 1995. Comparison of the coat protein genes of five fish nodaviruses, the causative agents of viral nervous necrosis in marine fish. J. Gen. Virol. 76: 1563–1569.

Nishizawa, T., Furuhashi, M., Nagai, T., Nakai, T. and Muroga, K. 1997. Genomic classification of fish nodaviruses by molecular phylogenetic analysis of the coat protein gene. Appl. Env. Microbiol. 63: 1633–1636.

OIE. 2012. Red sea bream iridoviral disease. pp. 345–356. *In*: Manual of Diagnostic Tests for Aquatic Animals 7th Ed., World Organization for Animal Health, Paris.

Schneemann, A., Ball, L.A., Delsert, C., Johnson, J.E. and Nishizawa, T. 2005. Family nodaviridae. pp. 865–872. *In*: Fauquet, C.M., Mayo, M.A., Maniloff, J., Desselberger, U. and Ball, L.A. (eds.). Virus Taxonomy, Eighth Report of the International Committee on Taxonomy of Viruses. Elsevier Academic Press, London, UK.

Shi, C.Y., Wang, Y.G., Yang, S.L., Huang, J. and Wang, Q.Y. 2004. The first report of an iridovirus-like agent infection in farmed turbot, *Scophthalmus maximus*, in China. Aquaculture 236: 11–25.

Subasinghe, R.P., Bondad-Reantaso, M.G. and McGladdery, S.E. 2001. Aquaculture development, health and wealth. pp. 167–191. *In*: Subasinghe, R.P., Bueno, P., Phillips, M.J., Hough, C., McGladdery, S.E. and Arthur, J.R. (eds.). Aquaculture in the Third Millennium. Technical Proceedings of the Conference on Aquaculture in the Third Millennium, Bangkok, Thailand, 20–25 February 2000. NACA, Bangkok and FAO, Rome.

Tanaka, S., Mori, K., Arimoto, M., Iwamoto, T. and Nakai, T. 2001. Protective immunity of sevenband grouper, *Epinephelus septemfasciatus* Thunberg, against experimental viral nervous necrosis. J. Fish Dis. 24: 15–22.

Thiéry, R., Cozien, J., Cabon, J., Lamour, F., Baud, M. and Schneemann, A. 2006. Induction of a protective immune response against viral nervous necrosis in the European sea bass *Dicentrarchus labrax* by using betanodavirus virus-like particles. J. Virol. 80: 10201–10207.

Wang, Y.Q., Lu, L., Weng, S.P., Huang, J.N., Chan, S.M. and He, J.G. 2007. Molecular epidemiology and phylogenetic analysis of a marine fish infectious spleen and kidney necrosis virus-like (ISKNV-like) virus. Archive Virology 152: 763–773.

Watanabe, K., Suzuki, S., Nishizawa, T., Suzuki, K., Yoshimizu, M. and Ezura, Y. 1998. Control strategy for viral nervous necrosis of barfin flounder. Fish Pathol. 33: 443–444.

Watanabe, K., Yoshimizu, M., Ishima, M., Kawamata, K. and Ezura, Y. 1999. Occurrence of viral nervous necrosis in hatchery-reared barfin flounder. Bull. Fac. Fish. Hokkaido Univ. 50: 101–113.

Watanabe, K., Nishizawa, T. and Yoshimizu, M. 2000. Selection of broodstock candidates of barfin flounder using an ELISA system with recombinant protein of barfin flounder nervous necrosis virus. Dis. Aquat. Org. 41: 219–223.

Woo, P.T.K. and Bruno, D.W. (eds.). 2011. Fish Diseases and Disorders. CABI, UK, 930 p.

World Organization for Animal Health. 2016. Manual of Diagnostic Tests for Aquatic Animals. OIE, Paris, 589 p.

Yamashita, Y., Fujita, Y., Kawakami, H. and Nakai, T. 2005. The efficacy of inactivated virus vaccine against viral nervous necrosis (VNN). Fish Pathol. 40: 15–21.

Yoshikoshi, K. and Inoue, K. 1990. Viral nervous necrosis in hatchery-reared larvae and juveniles of Japanese parrotfish, *Oplegnathus fasciatus* (Temminck & Schlegel). J. Fish Dis. 13: 69–77.

7

Prevention and Treatment of Diseases Caused by Fish Pathogens

*Hisae Kasai** and *Shotaro Nishikawa*

1. Introduction

Fish aquaculture is economically important worldwide. However, viral, bacterial, fungal, and parasitic diseases limit successful propagation in aquaculture. Viral diseases found in salmonid fish are infectious hematopoietic necrosis (IHN), *Oncorhynchus masou* virus (salmonid herpesvirus 2) disease (OMVD), erythrocytic inclusion body syndrome (EIBS), chum salmon virus disease, and viral erythrocytic necrosis (see Chapter 6). Diseases found affecting cultured marine fish include lymphocystis disease (LCD), viral ascites, rhabdovirus disease, viral epithelial hyperplasia, reovirus infection, viral nervous necrosis (VNN), and viral hemorrhagic septicemia (VHS).

In aquaculture, disease prevention is often preferred to treatment for two main reasons. First, effective medication may not be available. In the case of viral infections, there is usually no effective medication. Even if medication exists, it may not be available because its production cost

Faculty of Fisheries Sciences, Hokkaido University, Japan.
* Corresponding author: hisae@fish.hokudai.ac.jp

is not economically viable for farmers and pharmaceutical companies. Examples of such situations include anti-herpesvirus agents, such as phosphonoacetate, acyclovir (ACV; 9-(2-hydroxymethyl) guanine), (E)-5-(2-bromovinil)-2′-deoxyuridine (BVdU), and 9-β-D-arabinofuranosyladenine (Ara-A), which effectively inhibit replication of OMV (Kimura et al. 1983a, 1983b, 1988), but are not commercially available. The second reason why prevention is preferable to treatment is that there is a risk of development of pathogenic strains resistant to the antimicrobial agents in medication. This situation applies to antimicrobial agents and anthelmintic agents used to treat fish bacterial infections and parasitic infections. Potentially these resistant strains can have an impact on the therapy of fish diseases, the therapy of human diseases, or the environment of the fish farms (Smith et al. 1994). Users must follow the usage and dosage instructions for reducing that risk. In Japan, regulations controlling the use of standard chemotherapy for bacterial or parasitic infections for cultured fish were approved by Food Safety and Consumer Affairs Bureau, Ministry of Agriculture, Forestry and Fisheries in Japan (The Use of Aquatic Medicine 29th Report 2016, http://www.maff.go.jp/j/syouan/suisan/suisan_yobo/pdf/29.pdf). Approvals are needed for each fish species (fish classified to Order Perciformes and Clupeiformes are shown in Tables 7.1 and 7.2, respectively).

Methods currently used to control and prevent the diseases in hatcheries are (1) hygiene and sanitation, (2) disinfection of water supplies and waste water, (3) selection of pathogen-free broodstock, (4) health monitoring of hatched fry, (5) vaccination, (6) control of normal intestinal flora, and (7) temperature control.

2. Hygiene and Sanitation

General sanitation measures are standard practice in hatcheries and seed producing facilities. Special care must be taken to avoid the movement of equipment from one tank to another, and all articles should be disinfected after use. Methods used to sanitize a rearing unit should take into account chemical toxicity to fish, effects of temperature, and consequences of prolonged use. It should be remembered that workers themselves often act as vectors for pathogens, and therefore proper disinfection of hands and boots is required to prevent dissemination of pathogens. Although it may be difficult to sanitize a rearing unit during use, tanks and raceways should be disinfected with chlorine before and after use (Ahne et al. 1989, Kimura and Yoshimizu 1991). Equipments, nets, brushes may be disinfected with ozonated or electrolyzed sea water containing 0.5 mg/L of total residual oxidants (TROs) or chlorine for 30 minutes (Watanabe and Yoshimizu 1998, 2001). These tools should be used exclusively for each individual fish tank.

Table 7.1 List of drugs approved in Japan for fish belonging to Order Perciformes.

Treatable disease or condition	Classification	Active ingredient	Method used to administer the drug
Vibriosis	Antimicrobial agent	Thiamphenicol Sulfamonomethoxine Oxytetracycline hydrochloride Oxytetracycline quaternary salt Sulfisozole sodium (yellowtail only)	Oral administration
Pseudotuberculosis	Antimicrobial agent	Oxolinic acid Thiamphenicol Florfenicol Bicozamycin benzoate Fosfomycin calcium Amoxicillin Ampicillin Sulfisozole sodium (yellowtail only)	Oral administration
Edwardsiellosis	Antimicrobial agent	Fosfomycin calcium	Oral administration
Streptococcosis	Antimicrobial agent	Florfenicol Lincomycin hydrochloride Oxytetracycline quaternary salt Doxycycline hydrochloride Erythromycin Josamycin Spiramycin embonate	Oral administration
Nocardiosis	Antimicrobial agent	Sulfamonomethoxine Sulfisozole sodium (yellowtail only)	Oral administration
Benedenia seriolae infection	Anthelmintic agent	Praziquantel	Oral administration
Bivagina tai infection	Anthelmintic agent	Hydrogen peroxide	Immersion
Feed-associated liver damage	Amino acid and preparations	Glutathione	Oral administration
Cardicola opisthorchis infection	Anthelmintic agent	Praziquantel (bluefin tuna is also available)	Oral administration
White spot disease	Anthelmintic agent	Lysozyme chloride (red seabream only)	Oral administration

Since some viruses and bacteria are transmitted vertically from adult to progeny via infected eggs or sperms, disinfection of the surface of fertilized eggs has proven to be effective in breaking the infection cycle for several viruses, such as rhabdovirus, herpesvirus, and nodavirus (Yoshimizu et al.

Table 7.2 List of drugs approved in Japan for fish belonging to Order Clupeiformes.

Purpose or treatable disease	Classification	Active ingredient	Method used to administer the drug
Vibriosis (mariculture)	Antimicrobial agent	Oxolinic acid Oxytetracycline hydrochloride	Oral administration
Furunculosis (mariculture)	Antimicrobial agent	Oxolinic acid	Oral administration
Disinfection of eggs (mariculture)	Disinfectant	Bronopol	Immersion
Vibriosis (freshwater aquaculture)	Antimicrobial agent	Oxolinic acid Oxolinic acid (ayu only) Florfenicol Sulfamonomethoxine Sulfamonomethoxine sodium Compounding ingredients (sulfamonomethoxine sodium, ormetoprim) (ayu only) Oxytetracycline hydrochloride Sulfisozole sodium (rainbow trout and ayu only)	Oral administration Immersion Oral administration Oral administration Immersion Oral administration

Oral administration |
Furunculosis (freshwater aquaculture)	Antimicrobial agent	Oxolinic acid Florfenicol Sulfamonomethoxine Sulfamonomethoxine sodium Oxytetracycline hydrochloride	Oral administration Oral administration Oral administration Immersion Oral administration
Streptococcosis (freshwater aquaculture)	Antimicrobial agent	Oxytetracycline hydrochloride	Oral administration
Disinfection of eggs (freshwater aquaculture)	Disinfectant	Bronopol Povidone iodine (salmonid fish only)	Immersion
Cold water disease	Antimicrobial agent	Sulfisozole sodium (rainbow trout and ayu only)	Oral administration

1989, Watanabe et al. 2000). In the case of salmonid, eggs can be disinfected with iodine (50 ppm for 20 min) just after fertilize and eyed stage (Yoshimizu and Nomura 1989), and flounder eggs can be disinfected with ozonated seawater (0.5 mg/L of TROs for 10 min) at the morula stage (Watanabe and Yoshimizu 2000). Ideally, individual broodstock should be separated during fertilization (Fig. 7.1a), and reared in separate containers until the broodstock was certified as specific pathogen-free (SPF) (Fig. 7.1b). Working environment, including tools, should be disinfected after use (Fig. 7.1c,d).

Figure 7.1 Hygiene and sanitation at hatchery.

3. Disinfection of Water Supplies and Wastewater

Water supplies for seed production and aquaculture may also be pathways for the introduction and spread of infectious diseases. A pathogen-free water source is essential for success in aquaculture. Water commonly used in aquaculture comes from coastal waters or rivers and may contain fish pathogens. Such open water supplies should not be used without prior treatment. Disinfection of wastewater before it is discharged is necessary to avoid contamination of the environment with pathogens. Disinfection may be done using ultraviolet (UV), oxidants produced by ozonization of seawater, or hypochlorite produced by electrolyzation of seawater for disinfection of water. In addition to evaluating the disinfection efficacy of these three methods for a hatchery water supply and wastewater, their effects on survival of cultured fish have been assessed (Kasai et al. 2002).

3.1 U.V. susceptibility of fish pathogens and the effects of UV treatment on hatchery water

The disinfectant effects of UV irradiation on fish pathogenic bacteria, viruses, and fungi were determined using cell suspensions of bacteria, punched agar medium disk covered with aquatic fungi, and cell free suspensions of viruses. Of the viable bacterial cells of Gram-negative bacteria and Gram-positive bacteria, 99.9% or more were killed by UV irradiation at doses of 4.0×10^3 and 2.0×10^4 μW·sec/cm^2, respectively.

The hyphae of aquatic fungi showed relatively lower susceptibility to UV irradiation; levels that inhibited the growth of hyphae were 1.5×10^5 to 2.5×10^5 µW·sec/cm². Fish rhabdoviruses, herpesviruses, and iridovirus were found to be sensitive to UV irradiation. More than 99% infectivity decrease (ID_{99}) was observed at a dose of 1.0 to 3.0×10^3 µW·sec/cm². Susceptibility of birnaviruses, reovirus, and nodavirus was found to be lower with an observed ID_{99} of 1.5 to 2.5×10^5 µW·sec/cm² (Fig. 7.2) (Yoshimizu et al. 1986, Kasai et al. 2002, 2005).

In the studies on infectious hematopoietic necrosis virus (IHNV), infectivity in virus contaminated river water and pond water, was 0.56 and 5.6 $TCID_{50}$/L, respectively, when measured using the molecular filtration method (Yoshimizu et al. 1991). UV treatment of river water with 10^4 µW·sec/cm² prevented an IHN outbreak. Furthermore, UV treatment of the hatchery water supply also decreased the viable bacterial counts and fungal infection rates in salmonid eggs (Kimura et al. 1980). Figure 7.3 illustrates one example of equipment used for UV disinfection of hatchery water.

In the case of myxozoan parasites, *K. yasunagai* actinospores are relatively vulnerable to UV irradiation, and the minimum effective dose of preventing infection in Yellowtail amberjack (*Seriola lalandi*) lies between 5×10^3 and 1.5×10^4 µW·sec/cm² (Shin et al. 2016). Similarly *K. neurophila* infection to striped trumpeter (*Latris lineata*) was prevented with a dose of $> 4.0 \times 10^4$ µW·sec/cm² UV (Cobcroft and Battaglene 2013). Additionally, treatment with UV irradiation at the dose of 4.6×10^4 µW·sec/cm² was effective for the prevention of *K. septempunctata* infection of Japanese flounder (*Paralichthys olivaceus*) (Nishioka et al. 2016).

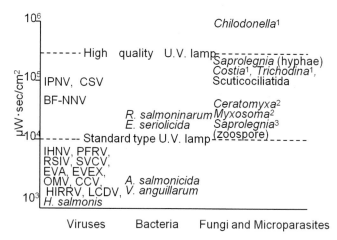

Figure 7.2 U.V. susceptibility of fish pathogens. [1]Vlasenko (1969), [2]Hoffman (1974), [3]Normandeau (1968).

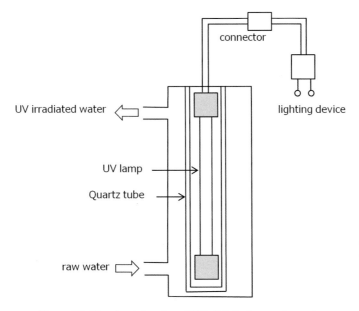

connector

UV irradiated water

lighting device

UV lamp

Quartz tube

raw water

Figure 7.3 Structural drawing of UV disinfection equipment.

3.2 Disinfectant effect of oxidant produced by ozonization of sea-water on fish pathogens

Treatment of natural seawater with ozone produced oxidants that showed a disinfectant effect. Total residual oxidants (TROs) produced in seawater were stable for 1 hour or more. Disinfectant effect of TROs against fish pathogenic organisms was observed at a dose of 0.5 mg/L for 15 to 30 s or 0.1 mg/L for 60 s, and killed more than 99.9% of bacterial cells of *Vibrio anguillarum, Lactococcus garvieae, Aeromonas salmonicida, A. hydrophila* and *E. coli,* and inactivated 99% or more of IHNV, hirame rhabdovirus (HIRRV), and OMV. To inactivate or kill more than 99% of yellowtail ascites virus (YAV), infectious pancreatic necrosis virus (IPNV), chum salmon virus (CSV), and a Scuticociliatida (ciliata), higher doses of 0.5 to 1.0 mg/L for 1 min were required (Table 7.3) (Yoshimizu et al. 1995, Ito et al. 1997).

However TROs were toxic to fish. Barfin flounder (*Verasper moseri*) and herring (*Clupea pallasii*) died after 16 and 2 hours of exposure to TROs of 0.1 and 0.5 mg/L, respectively (Ito et al. 1996). Nevertheless, Japanese flounder could be cultured in ozonized seawater after the TROs were removed using charcoal, resulting in survival rates similar to fish cultured in UV treated or non-treated seawater (Kasai et al. 2002). Figure 7.4 illustrates a method for ozonization of seawater.

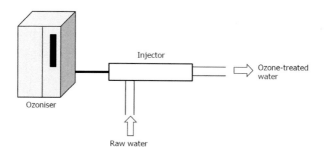

Figure 7.4 Diagrammatic illustration of ozonization.

3.3 Effects of electrolyzed salt water on fish pathogenic bacteria and viruses

The bactericidal and virucidal effects of hypochlorite produced by electrolysis of salt water were examined against pathogenic bacteria and viruses of fish. The principle of electrolysis of seawater is shown in Fig. 7.5. This disinfection system utilizes hypochlorite produced from electrical decomposition of seawater using an electrolyzer. Sodium chloride solutions, ranging from 0.5 to 3% were electrolyzed, and the concentration of chlorine produced was measured. Similar concentrations of chlorine were produced when 1.0% or higher NaCl solution and seawater were electrolyzed. A 3% solution of sodium chloride containing pathogenic bacteria or virus was electrolyzed, and the organisms were exposed to chlorine. More than 99.9% of *V. anguillarum* and *A. salmonicida* cells were killed when the bacteria were exposed to 0.1 mg/L chlorine for 1 min. In addition, 99.9% infectivity reduction of yellowtail ascites virus (YTAV) and HIRRV was achieved after treatment with 0.45 mg/L chlorine for 1 min (Table 7.4) (Kasai et al. 2000).

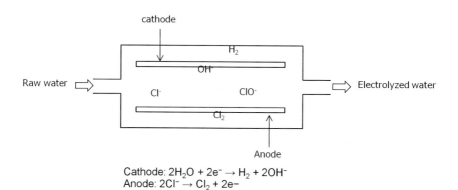

Cathode: $2H_2O + 2e^- \rightarrow H_2 + 2OH^-$
Anode: $2Cl^- \rightarrow Cl_2 + 2e-$

Figure 7.5 Principle of electrolysis of seawater.

Table 7.3 Effect of total residual oxidants (TROs) produced by ozonization of seawater on infectivities of fish pathogens.

Fish pathogens	TROs concentration (mg/l)	Treatment time (sec)	Reduction rate (%)	Initial number (log)
Yellow ascites virus (YAV)	0.5	60	> 99	4.3[1]
Hirame rhabdovirus (HIRRV)	0.5	15	> 99	5.6[1]
Infectious pancreatic necrosis virus (IPNV)	0.5	60	> 99	4.1[1]
Infectious haematopoietic virus (IHNV)	0.5	15	> 99	4.1[1]
Onchorhynchus masou virus (OMV)	0.5	15	> 99	3.1[1]
Chum salmon virus (CSV)	0.5	60	> 99	4.1[1]
Vibrio anguillarum NCMB6	0.5	15	> 99.9	5.6[2]
Lactococcus garvieae 538	0.5	15	> 99.9	5.8[2]
Aeromonas salmonicida ATTC14174	0.5	15	> 99.9	5.1[2]
Aeromonas hydrophila IAM1018	0.5	15	> 99.9	4.6[2]
Scuticociliatida BR9001	0.8	30	> 99.9	5.5[3]

[1]Initial viral infectivity ($TCID_{50}$/ml). [2]Initial viable bacterial number (CFU/ml). [3]Initial viable number.

Table 7.4 Concentration of chlorine produced by electrolysis of salt water and treatment time required to reduce the viability of bacteria and the infectivity of viruses by 99.9%.

Fish pathogens	Chlorine concentration (mg/l)	Treatment time (min)	Initial number (log)	Reduction rate (%)
Vibrio anguillarum NCMB6	0.07	1	6.7[1]	> 99.99
Aeromonas salmonicida ATTC14174	0.06	1	6.6[1]	99.96
Escherichia coli O-26	0.14	1	6.6[1]	99.98
Yellow ascites virus (YAV)	0.45	1	4.5[2]	99.92
Hirame rhabdovirus (HIRRV)	0.34	1	4.5[2]	99.97

[1]Initial viable bacterial number (CFU/ml). [2]Initial viral infectivity ($TCID_{50}$/ml).

The bactericidal and virucidal effects of hypochlorite produced by electrolysis were greater than that of the chemical reagent. The purity of the sodium chloride used for electrolysis influenced the efficacy of hypochlorite produced. Sodium chloride obtained as a super grade chemical reagent was more effective than food-grade sodium chloride. Nevertheless, a sufficient disinfectant effect was observed even in electrolyzed seawater, a method which may have wide applications in aquaculture. To use electrolyzed seawater for culture, the chlorine has to be removed with charcoal because of its toxicity (Kasai et al. 2000).

3.4 Disinfection of wastewater

In studies on the disinfection of hatchery wastewater, the bactericidal effect of hypochlorite produced using a continuous flow electrolyzer was investigated. The number of viable bacteria in the wastewater was reduced by more than 99% when the water was treated with chlorine at a concentration of 0.5 mg/L for 1 min, and over 99.9% of the bacteria cells were killed when treated with 1.28 mg/L for 1 min (Kasai et al. 2001). Viability of bacteria was reduced by more than 99% after treatment with 0.5 mg/L of chlorite for 1 min. The bactericidal effect of electrolysis was almost the same as that of ultraviolet irradiation (1.0×10^5 µW·sec/cm²) or ozonization (TROs 0.5 mg/L, 1 min) of seawater. Electrolization can be used to treat larger volumes of wastewater compared to ultraviolet irradiation or ozonization.

All three disinfection methods above eliminated 96.6 to 99.8% of bacteria in hatchery water supplies. Survival rate of Japanese flounder *Paralichthys olivaceus* and barfin flounder cultured in UV irradiated, ozonized, and electrolyzed seawater have been compared. No statistically significant differences in survival rates were found between the three groups of fish cultured with treated water. Ozonized and electrolyzed seawater have been shown to be effective for disinfecting equipment used in aquaculture, and ozonized seawater is effective for disinfecting fertilized barfin flounder eggs contaminated with nervous necrosis virus. Therefore, ozonization and electrolization of seawater seem to be effective methods for disinfection in fish culture (Kasai et al. 2002).

4. Pathogen-free Broodstock

Monitoring the health of broodstock is very important for seed production in aquaculture. Health inspections of broodstock are conducted to ensure that fish are free from certain important diseases. Specialized diagnostic techniques are required to make specific pathogen-free broodstock for routine inspections. The tests have been made easier and more rapid by the development of enzyme-linked immunosorbent assay (ELISA) (Watanabe et al. 2000).

For salmonid fish, Yoshimizu et al. (1985) recommended a method for collection of ovarian fluid for routine inspection. Fertilized eggs were disinfected with 100 ppm iodophor for 10 min (OIE 2016). It was also suggested that eyed eggs were an indication that the inside of the egg membrane is pathogen-free (Yoshimizu et al. 1989).

At a flounder hatchery, tagging was used for identification of individual fish. For example, to control the barfin flounder and Japanese flounder nervous necrosis virus (BF-, and JF-NNV), a standard sandwich ELISA and RT-PCR is used to select healthy broodstock. This technique targeted antibody against partial BF-NNV coat protein and virus specific gene sequences. ELISA was done 3 months before spawning and the negative fish by ELISA were reared for the broodstock (Watanabe et al. 2000). Eggs and sperms were tested by RT-PCR, and specimens inoculated to SSN-1 cells at the same time. The eggs or sperms that were positive were removed.

5. Monitoring Health of Hatched Fry

For monitoring purposes, it is advisable that fry from each spawner are cultured in separate tanks. Although this is difficult in a salmonid hatchery, it can be achieved for flounder. If fry show abnormal swimming or disease signs, they should be isolated for diagnosis as soon as possible. Moreover, health monitoring should be done using a variety of methods for viral detection, such as cell culture, fluorescent antibody techniques (FAT), immuno-peroxidase stain (IPT), antigen detecting ELISA, and PCR test. RT-PCR is suitable for detection of fish nodavirus and flounder ascites virus. FAT is commonly used to diagnose the viral epithelial hyperplasia and LCD, and HIRRV, and reovirus.

6. Vaccination

Vaccination is the most effective method to control the diseases for which avoidance is not possible. Several commercial vaccines are available to protect the fish against important pathogens. In Norway, combination vaccines containing five pathogens are available. In Canada, DNA vaccine against IHNV is available. In Japan, vaccines against vibriosis, *Lactococcus garvieae* infection, *Streptococcus iniae* infection, edwardsiellosis, pseudotuberculosis, *S. parauberis* infection, red seabream iridoviral disease, and VNN are available (Table 7.5).

7. Control of Normal Bacterial Flora

Generally, normal bacterial flora plays an important role in inhibiting the growth of pathogenic bacteria in the intestine or on the skin, and also to stimulate the immune response of the host animals. Sometimes, bacterial flora of larvae cultured in the disinfected water is not normal. It is important to establish the normal bacterial flora of the fish before they are released

Table 7.5 List of vaccines approved in Japan for fish diseases.

Method used to administer the vaccine	Disease	Fish in which vaccine can be used
Oral administration	*Lactococcus garviae* infection (αSt.)	Yellowtail Genus Seriola
Immersion	Vibriosis (Vi.)	Salmonid fish Ayu
Injection (monovalent)	αSt. *Streptococcus iniae* infection (βSt.) Edwardsiellosis Red seabream iridoviral disease (RSIVD) Viral nervous necrosis	Yellowtail Genus Seriola Japanese flounder Japanese flounder Yellowtail Genus Seriola Red seabream Striped jack Grouper (4 species) Sevenband grouper
Injection (combination)	Vi. + αSt. RSIVD + αSt. αSt. + Pseudotuberculosis (Ps.) βSt. + *S. parauberis* infection RSIVD + Vi. + αSt. Ps. + αSt. + Vi. αSt. + Vi. + *S. dysgalactiae* infection RSIVD + Vi. + αSt. + Ps.	Yellowtail Genus Seriola Amberjack Genus Seriola Yellowtail Amberjack Japanese flounder Yellowtail Genus Seriola Amberjack Yellowtail Amberjack Amberjack Yellowtail Genus Seriola

to the river or ocean. Many bacterial strains that produce the anti-viral substances against fish viruses have been reported. In one study, rainbow trout and masu salmon fed with bacteria isolated from normal intestinal flora and showed anti-IHNV activity, and higher resistance to artificial infection with IHNV (Yoshimizu et al. 1992). In another study, barfin flounder, disinfected at the egg stage and hatched in disinfected water fed with *Artemia* supplemented with *Vibrio* spp. isolated from the normal intestinal flora, showed anti-viral resistance against IHNV, OMV, and BF-NNV. Anti-IHNV, OMV, and BFNNV activity was observed in homogenates of intestines of fish fed with the *Artemia*. These barfin flounder fed with *Artemia* containing *Vibrio* sp. also showed higher resistance to natural infection by BFNNV (Yoshimizu and Ezura 1999).

8. Temperature Control

It is well known that many diseases of aquatic animals are temperature-dependent. In the case of koi herpesvirus (*Cyprinid herpesvirus* 3), mortality was observed in carp that had been experimentally exposed to KHV between 18°C and 28°C, but not at 13°C (Gilad et al. 2004). In the case of HIRRV, natural outbreaks of infections disappeared when the water temperature increased to 15°C. Cumulative mortality of artificially infected Japanese flounder (Intraperitoneal injection; $10^{5.3}$ TCID$_{50}$/fish) which were reared at 5, 10, 15, and 20°C, were 40%, 60%, 10%, and 0%, respectively. The highest virus infectivity was obtained from the fish that cultured at 5°C, followed by the 10°C (Oseko et al. 1988). We strongly recommended that Japanese flounder should be cultured at the water temperature up to 18°C, and outbreaks of HIRRV infection were not reported since 1988.

References

Ahne, W., Winton, J.R. and Kimura, T. 1989. Prevention of infectious diseases in aquaculture. J. Vet. Med. 36: 561–567.

Cobcroft, J.M. and Battaglene, S.C. 2013. Ultraviolet irradiation is an effective alternative to ozonation as a seawater treatment to prevent *Kudoa neurophila* (Myxozoa: Myxosporea) infection of striped trumpeter, *Latris lineata* (Forster). J. Fish Dis. 36: 57–65.

Gilad, O., Yun, S., Zagmutt-Vergara, F.J., Leutenegger, C.M., Bercovier, H. and Hedrick, R.P. 2004. Concentrations of a Koi herpesvirus (KHV) in tissues of experimentally infected *Cyprinus carpio* koi as assessed by real-time TaqMan PCR. Dis. Aquat. Org. 60: 179–187.

Hoffman, G.L. 1974. Disinfection of contaminated water by ultraviolet irradiation, with emphasis on whirling disease (*Myxosoma cerebralis*) and its effect on fish. Trans. Amer. Fish. Soc. 3: 541–550.

Ito, S., Yoshimizu, M., Oh, M.J., Hyuga, S., Watanabe, K., Hayakawa, Y. and Ezura, Y. 1996. Effect of ozonized seawater on bacterial population and survival of cultured flounders (*Paralichthys olivaceus* and *Verasper moseri*). Suisanzoshoku 44: 457–463.

Ito, S., Yoshimizu, M. and Ezura, Y. 1997. Disinfectant effects of low level of total residual oxidants in artificial seawater on fish pathogenic microorganisms. Nippon Suisan Gakkaishi 63: 97–102.

Kasai, H., Ishikawa, M., Hori, Y., Watanabe, K. and Yoshimizu, M. 2000. Disinfectant effects of electrolyzed salt water on fish pathogenic bacteria and viruses. Nippon Suisan Gakkaishi 66: 1020–1025.

Kasai, H., Watanabe, K. and Yoshimizu, M. 2001. Bactericidal effect of continuous flow electrolyzer on hatchery waste-seawater. Nippon Suisan Gakkaishi 67: 222–225.

Kasai, H., Yoshimizu, M. and Ezura, Y. 2002. Disinfection of water for aquaculture. Fish. Sci. 68(Suppl.I): 821–824.

Kasai, H., Muto, Y. and Yoshimizu, M. 2005. Virucidal effects of ultraviolet, heat treatment and disinfectants against Koi herpesvirus (KHV). Fish Pathol. 40: 137–138.

Kimura, T., Yoshimizu, M. and Atoda, M. 1980. Disinfection of hatchery water supply by ultraviolet (U.V.) irradiation-Ill. Effect of disinfection of hatchery water supply by ultraviolet irradiation on hatching rate of salmonid eggs. Fish pathol. 14: 139–142.

Kimura, T., Suzuki, S. and Yoshimizu, M. 1983a. *In vitro* antiviral effect of 9-(2-hydroxyethoxymethyl) guanine on the fish herpesvirus, *Oncorhynchus masou* virus (OMV). Antiviral Res. 3: 93–101.

Kimura, T., Suzuki, S. and Yoshimizu, M. 1983b. *In vivo* antiviral effect of 9-(2-hydroxyethoxymethyl) guanine on experimental infection of chum salmon (*Oncorhynchus keta*) fry with *Oncorhynchus masou* virus (OMV). Antiviral Res. 3: 103–108.

Kimura, T., Nishizawa, T., Yoshimizu, M. and De Clercq, E. 1988. Inhibitory activity of (E)-5-(2-Bromovinyl)-2'-Deoxyuridine on the salmonid herpesviruses, *Oncorhynchus masou* virus (OMV) and *herpesvirus salmonis*. Microbiol. Immunol. 32: 57–65.

Kimura, T. and Yoshimizu, M. 1991. Disinfection methods of aquaculture unit. pp. 220–226. *In*: Takano, M. and Yokoyama, R. (eds.). New Handbook of Sterilization Technology. Science-Forum, Tokyo, Japan.

Nishioka, T., Satoh, J., Mekata, T., Mori, K., Ohta, K., Morioka, T., Lu, M., Yokoyama, H. and Yoshinaga, T. 2016. Efficacy of sand filtration and ultraviolet irradiation as seawater treatment to prevent *Kudoa septempunctata* (Myxozoa: multivalvulida) infection in olive flounder *Paralichthys olivaceus*. Fish pathol. 51: 23–27.

Normandeau, D.A. 1968. Progress report, Project F-I4-R-3, State of New Hampshire.

OIE (the World Organization for Animal Health). 2016. Aquatic Animal Health Code, OIE, Paris, France, pp. 68.

Oseko, N., Yoshimizu, M. and Kimura, T. 1988. Effect of water temperature on artificial infection of *Rhabdovirus olivaceus* (hirame rhabdovirus: HRV) to hirame (Japanese flounder, *Paralichthys olivaceus*). Fish Pathol. 23: 125–132.

Shin, S.P., Nishimura, T., Ogawa, K. and Shirakashi, S. 2016. Determination of effective dose of ultraviolet irradiation to influent water for prevention of Kudoa yasunagai infection. Fish Pathol. 51: 128–131.

Smith, P., Hiney, M.P. and O.B. Samuelsen. 1994. Bacterial resistance to antimicrobial agents used in fish farming: a critical evaluation of method and meaning. Annu. Rev. Fish Dis. 4: 273–313.

Vlasenko, M.I. 1969. Ultraviolet rays as a method for the control of diseases of fish eggs and young fishes. Problems Ichthyol. 9: 697–705.

Watanabe, K. and Yoshimizu, M. 1998. Disinfection of equipment for aquaculture and fertilized eggs by ozonated seawater. Fish Pathol. 33: 145–146.

Watanabe, K. and Yoshimizu, M. 2000. Disinfection of viral nervous necrosis virus contaminated fertilized barfin flounder eggs by ozonated seawater. Nippon Suisan Gakkaishi 66: 1066–1067.

Watanabe, K., Nishizawa, T. and Yoshimizu, M. 2000. Selection of brood stock candidates of barfin flounder using an ELISA system with recombinant protein of barfin flounder nervous necrosis virus. Dis. Aquat. Org. 41: 219–223.

Watanabe, K. and Yoshimizu, M. 2001. Disinfection of equipment for aquaculture by electrolyzed seawater. Nippon Suisan Gakkaishi 67: 304–305.

Yoshimizu, M., Kimura, T. and Winton, J.R. 1985. An improved technique for collecting reproductive fluid samples from salmonid fishes. Prog. Fish. Cult. 47: 199–200.

Yoshimizu, M., Takizawa, H. and Kimura, T. 1986. U.V. susceptibility of some fish pathogenic virus. Fish Pathol. 21: 47–52.

Yoshimizu, M. and Nomura, T. 1989. An improved method for isolation of fish pathogenic bacteria and viruses from mature salmonids. Tech. Rep. Hokkaido Salmon Hatchery. Fish and Egg 158: 49–59.

Yoshimizu, M., Sami, M. and Kimura, T. 1989. Survivability of infectious hematopoietic necrosis virus in fertilized eggs of masu and chum salmon. J. Aquat. Animal Health 1: 13–20.

Yoshimizu, M., Sami, M., Kohara, M., Yamazaki, T. and Kimura, T. 1991. Detection of IHNV in hatchery water by molecular filtration method and effectiveness of U.V. irradiation on IHNV infectivity. Nippon Suisan Gakkaishi 57: 555–560.

Yoshimizu, M., Fushimi, Y., Kouno, K., Shinada, C., Ezura, Y. and Kimura, T. 1992. Biological control of infectious hematopoietic necrosis by antiviral substance producing bacteria. pp. 301–307. *In*: Kimura T. (ed.). Salmonid Diseases. Hokkaido University Press, Sapporo, Japan.

Yoshimizu, M., Hyuga, S., Oh, M.J., Ito, S., Ezura, Y. and Minura, G. 1995. Disinfectant effect of oxidant produced by ozonation of seawater on fish pathogenic viruses, bacteria, and ciliata. pp. 203–209. *In*: Shariff, M., Arthur, J.R. and Subasighe, R. (eds.). Diseases in Asian Aquacultur II. Fish Health Section, Asian Fisheries Society, Manila, Phillipin.

Yoshimizu, M. and Ezura, Y. 1999. Biological control of fish viral diseases by anti-viral substance producing bacteria. Microb. Environ. 14: 269–275.

8

Predictive Microbiology on Seafood Safety

Shigenobu Koseki

1. Overview of Predictive Microbiology Relating to Seafood Safety

1.1 Listeria monocytogenes

Seafood is one of the most perishable and also has a high risk of food-borne diseases due to consumption by raw and/or minimal processing. Many efforts have been conducted in ensuring microbiological safety of seafood. Among the related pathogenic bacteria to seafood, *Listeria monocytogenes* has long been focused on controlling their behavior. One of the approaches for controlling pathogenic bacteria is prediction of bacterial growth in/on seafood. Predictive models for bacterial growth in/on seafood have been mainly described as a function of time and temperature, because the most influential parameter related to bacterial growth is temperature. Other parameters relating with bacterial growth of minimally processed seafood, such as smoked salmon, are pH, sodium chloride concentration, water activity, and phenolic compounds occurring from smoking process.

A high incidence of *Listeria* spp. in ready-to-eat (RTE) seafood from retail stores has been documented, due to the ability of *L. monocytogenes* to

Research Faculty of Agricultural Science, Hokkaido University, Kita 9, Nishi 9, Kita-ku, Sapporo 060-8589, Japan.
E-mail: koseki@bpe.agr.hokudai.ac.jp

grow during the refrigerated storage of naturally contaminated products (Lianou and Sofos 2007). In predictive microbiology studies, the growth of *L. monocytogenes* in cold-smoked salmon has been extensively investigated (Augustin and Carlier 2000, Augustin et al. 2005, Baty et al. 2002, Delignette-Muller et al. 2005); in particular, Danish research groups have long been studying *Listeria monocytogenes* growth in cold-smoked salmon and lightly preserved ready-to-eat shrimp (Mejlholm et al. 2015, Mejlholm and Dalgaard 2015, 2007a, 2007b). Recently, some of the reported predictive models were validated by large numbers of experimental data sets derived from real foods, and the performance was assessed as an international collaborative study (Mejlholm et al. 2010). Among the models, the most complex growth model including nine environmental factors showed the highest accuracy of prediction. However, since the original data of the previous studies was derived from lightly preserved foods, such as cold-smoked salmon, the models in the previous studies were mainly based on the data range of lower water activity (a_w, 0.93–0.98). In addition, Dalgaard et al. (2002) in the same Danish research group developed predictive models for estimation of histamine production during distribution based on the growth prediction of *Morganella morganii* (Emborg and Dalgaard 2008, 2006, Emborg et al. 2005). The achievement of the research group are illustrated as a PC software called "Food Safety and Spoilage Predictor (FSSP)", which can be downloaded from the website and used free of charge (Dalgaard et al. 2002).

1.2 Vibrio spp.

Another important bacterium for seafood safety is *Vibrio* spp., such as *Vibrio parahaemolyticus* and *Vibrio vulnificus*. In particular, these bacteria are related to raw oyster. University of Tasmania's research group firstly developed growth models of *Vibrio* spp. in culture media condition as a function of temperature and water activity (Miles et al. 1997). The model of the square-root type includes a novel term to describe the effects of super-optimal water activity, and can be used to predict generation times for the temperature range (8–45°C) and water activity range (0.936–0.995) which permit growth of *V. parahaemolyticus*. The same research group has developed the predictive models for growth of *V. parahaemolyticus* in raw oyster. They reported that growth/inactivation rates for *V. parahaemolyticus* were 0.006, 0.004, 0.005, 0.003, 0.030, 0.075, 0.095, and 0.282 log10 CFU/h at 3.6, 6.2, 9.6, 12.6, 18.4, 20.0, 25.7, and 30.4°C, respectively (Fernandez-Piquer et al. 2011). Square root and Arrhenius-type secondary models were generated for *V. parahaemolyticus* growth and inactivation kinetic data, respectively. In addition, they developed a stochastic model to quantitatively assess the populations of *V. parahaemolyticus* and total viable bacteria in Pacific oysters for six different supply chain scenarios (Fernandez-Piquer et al. 2013). Probabilistic distributions and predictions for the percentage of Pacific oysters containing *V. parahaemolyticus* and high levels of viable bacteria at the point of consumption were generated for each simulated scenario. Yoon et al. (2008) also developed growth kinetics model of pathogenic and nonpathogenic *V. parahaemolyticus* isolated in Korea in broth and oyster slurry. Yang et al. (2009) examined growth and survival curves of a strain of pandemic *V. parahaemolyticus* TGqx01 (serotype O3:K6) on salmon meat at different storage temperatures (ranging from 0°C to 35°C), and developed kinetic models. Recently, more precise model for *V. parahaemolyticus* behavior based on a real-time reverse transcription-PCR (RT-PCR) assay has been developed (Liao et al. 2017). While previous predictive models established data based on plate counting methods or on DNA-based PCR can underestimate or overestimate the number of surviving cells, Liao et al. (2017) developed and validated RNA-based molecular predictive models to describe the survival of *V. parahaemolyticus* in oysters during low-temperature storage (0, 4, and 10°C). The RNA-based predictive models show the advantage of being able to count all of the culturable, nonculturable, and stressed cells. The new method would enable to evaluate the total surviving *V. parahaemolyticus* population, as well as differentiate the pathogenic ones from the total population.

2. Growth Modelling of *Listeria Monocytogenes* Contamination Minced Tuna

2.1 Background

A previous study also reported that raw seafood may be contaminated with *L. monocytogenes*, and minced tuna in particular was contaminated with *L. monocytogenes* in 14.3% of Japanese retail stores (Handa et al. 2005). In addition, the cell numbers of *L. monocytogenes* in minced tuna increased by 10^2/g at refrigerated temperatures within 3 days, using a most probable number (MPN) method (Miya et al. 2010). As raw minced tuna is one of the most popular sushi ingredients, consuming minced tuna may have the potential risk of listeriosis food poisoning. Although there has been no report on the outbreak of *L. monocytogenes* caused by minced tuna, the dose-response models for *L. monocytogenes* illustrated an infection probability of 10^{-6} to 10^{-1} by ingesting ~ 10^6 CFU, according to the FAO/WHO report (Food and Agriculture Organization of the United NationsWorld Health Organization FAOWHO 2004). If the number of *L. monocytogenes* increased by 10^3~10^4 CFU/g in minced tuna, total ingestion dose per meal might reach 10^5~10^6 CFU. Although the total ingested dose depends on the consumption quantity, the risk is not negligible. Thus, accurate prediction of *L. monocytogenes* growth in minced tuna would contribute to safer raw tuna consumption.

2.2 Competition growth modelling of L. monocytogenes

Non-pathogenic microorganisms are present in most ready-to-eat foods. The interaction and/or competition of a particular pathogen with the natural flora (NF) has been investigated and modelled (Gimenez and Dalgaard 2004). Competing flora affect the maximum population density (N_{max}) of co-cultured pathogenic bacteria. The "Jameson effect" suggests that the growth of both competing members of the flora stops when one member has reached its maximum level in the environment. For realistic growth predictions of the targeted pathogenic bacteria, it is important to take into account the effect of competing natural members of the flora. Mejlholm and Dalgaard (2007a) developed the interaction model of growth of *L. monocytogenes* and lactic acid bacteria in lightly preserved seafood, such as cold-smoked salmon. The developed model is available in the Seafood Spoilage and Safety Predictor (SSSP ver. 3.1) (Dalgaard et al. 2002). Although this model is quite useful for simulating the simultaneous growth of *L. monocytogenes* and lactic acid bacteria in lightly preserved seafood, the model might not be applicable to the *L. monocytogenes* strains in minced tuna, due to the different characteristics of the food. Furthermore, as minced tuna could be exposed to room temperature in real distribution situations before consumption, the SSSP model may not be applicable because of its limited temperature range (2 to 15°C).

Here, we show the growth kinetics of *L. monocytogenes* and *NF* in minced tuna from refrigeration temperature to room temperature. The inhibitory effect of *NF* on the growth of *L. monocytogenes* was also examined to attain a more realistic prediction of *L. monocytogenes* growth in minced tuna (Koseki et al. 2011).

2.3 Model development procedure

The maximum specific growth rate (μ_{max}), lag time (λ), and maximum population density (N_{max}) were modelled as follows. The μ_{max} (1/h) values, determined at different temperatures and different inoculation levels of natural flora (NF), were fitted to the modified square root model as follows (Ratkowsky et al. 1982):

$$\sqrt{\mu_{max}} = a_0 + a_1 \times T + a_2 \times 10^{IC} + a_3 \times (T \times 10^{IC}) \tag{1}$$

where a_i are coefficients to be estimated, T is the temperature (°C), and IC is the inoculation level of *NF* ($\log_{10}$ CFU/g). Furthermore, the N_{max} ($\log_{10}$ CFU/g) was modelled as a multivariate linear model as follows:

$$N_{max} = b_0 + b_1 \times T + b_2 \times IC + b_3 \times (T \times IC) \tag{2}$$

where b_i are coefficients to be estimated, and other abbreviations are same as mentioned above.

To simulate the simultaneous growth of *L. monocytogenes* and *NF* under different temperatures, the modified Baranyi model, incorporating the effect of interspecies competition as reported by Gimenez and Dalgaard (2004), was used in this study.

$$\frac{dq_{Lm}}{dt} = \mu_{max}^{Lm} q_{Lm} \tag{3}$$

$$\frac{dLm}{dt} = \frac{q_{Lm}}{1+q_{Lm}} \times \mu_{max}^{Lm} \times \left(1 - \frac{Lm_t}{Lm_{max}}\right) \times \left(1 - \frac{NF_t}{NF_{max}}\right) \times Lm_t \tag{4}$$

$$\frac{dq_{NF}}{dt} = \mu_{max}^{NF} q_{NF} \tag{5}$$

$$\frac{dNF}{dt} = \frac{q_{NF}}{1+q_{NF}} \times \mu_{max}^{NF} \times \left(1 - \frac{NF_t}{NF_{max}}\right) \times \left(1 - \frac{Lm_t}{Lm_{max}}\right) \times NF_t \tag{6}$$

where Lm_t and NF_t denote the bacterial cell concentration (CFU/g) of *L. monocytogenes* and natural flora at time t, respectively. The q_{Lm} and q_{NF} are dimensionless quantities related to the physiological state of the cells, μ_{max} is the maximum specific growth rate (1/h), Lm_{max} and NF_{max} represent the N_{max} of *L. monocytogenes* and natural flora, respectively. For the initial value of q_0, which is a measure of the initial physiological state of the cells, a geometric mean value for the physiological state parameter α_0 was estimated from the constant temperature experimental data. The relationship between lag time (λ) and α_0 could be shown as follows (Baranyi and Roberts 1994):

$$\mu_{max} \lambda = \ln\left(1 + \frac{1}{q_0}\right) = -\ln(\alpha_0) \tag{7}$$

The models for μ_{max} and N_{max}, along with α_0, were substituted into the above differential equations (Eq. 3 to 6), and the temperature was allowed to be dependent on time. The system was solved numerically by the fourth-order Runge-Kutta method, as a means of obtaining predictions of the bacterial concentration.

2.4 Model performance

Above 25°C, the higher level of competitive natural flora reduced the μ_{max} of *L. monocytogenes* (Fig. 8.1). While we could model the μ_{max} corresponding to the level of competitive natural flora and temperature, we do not know the initial natural flora level in real situations prior to the microbiological tests. Although the μ_{max} model for *L. monocytogenes* would be useful for simulating various scenarios, the input of lower levels or absent natural flora may be chosen to manage safety. Other researchers have also reported the influence of competitive micro-flora on the growth rate of the intended pathogenic bacteria (Pin and Baranyi 1998). Conversely, some reports show competitive flora have no effect on the growth rate of pathogenic bacteria (Mellefont et al. 2008). These conflicting results might be a consequence of the experimental designs used, the levels of competitive flora, the bacterial strains, and the food matrix.

High numbers of competitive natural flora reduced the N_{max} of *L. monocytogenes*. As the relationship between the initial flora level and the N_{max} of *L. monocytogenes* was described as a function, we could incorporate the effect of the initial natural flora level with the N_{max} of *L. monocytogenes* into the predictive model (Eq. 8 and 10). This would yield a flexible prediction for both the number of natural flora and *L. monocytogenes*. As an example, the model's performance was illustrated under fluctuating temperature conditions, which were assumed to be similar to actual handling in a retail store (Fig. 8.2).

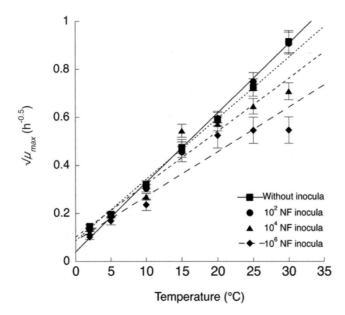

Figure 8.1 Square-root model of specific growth rate (μ_{max}, 1/h) for *L. monocytogenes* in minced tuna. Different symbols represent different level of co-existing natural flora. ●: without inoculation of natural flora, ○: with inoculation of 10^2 CFU/g natural flora, ◆: with inoculation of 10^4 CFU/g natural flora, and Δ with inoculation of 10^6 CFU/g, respectively.

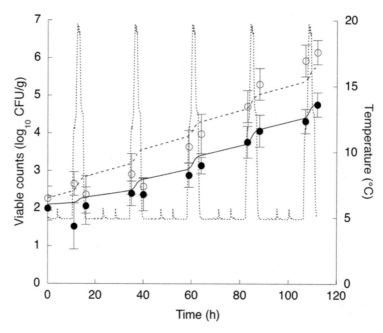

Figure 8.2 Simultaneous prediction of growth of *L. monocytogenes* and natural flora in minced tuna during storage under fluctuating temperatures. Solid and dashed lines represent the prediction using the developed model for *L. monocytogenes* and natural flora, respectively. Observed values of *L. monocytogenes* and natural flora are represented by the symbols ● and ○, respectively.

3. Use of Predictive Tools for Seafood Microbiological Safety

3.1 Overview of web-tools for estimation of microbial behavior

In recent times, there have been many predictive microbiology tools available on the web (Tenenhaus-Aziza and Ellouze 2015). These available tools are classified into four classes according to objectives, such as databases, growth and inactivation predictors, growth and inactivation fitting tools, growth/no growth predictors, and risk assessment modules. Among the tools, databases provide basic information regarding microbial behavior in foods for microbial risk control and assessment. The most representative database is ComBase (www.combase.cc) that consists of microbial growth and inactivation data published in literature. The data and information regarding microbial behavior of vegetables and fruits are allowed to be accessed from some databases, such as ComBase, and tools such as Sym'Previus (www.symprevius.org), MicroHibro (www.microhibro. com), and Baseline (www.baselineapp.com). Although these softwares provide information on bacterial growth kinetics on vegetables and fruits, understanding general trend of bacterial behavior against environmental factors would be useful for food processors. Accordingly, a distinct type database, which is named Microbial Responses Viewer (MRV), has been developed.

3.2 Concept of development of Microbial Responses Viewer (MRV)

ComBase is a large database of microbial responses to food environments, and has attracted the attention of many researchers and food processors. Although ComBase contains a vast amount of data, it is not necessarily easy to obtain desired information from the retrieved data. In order to utilize ComBase to the maximum, we developed a new ComBase-derived database (Microbial Responses Viewer, MRV; http://mrviewer.info (Koseki 2009)) consisting of microbial growth/no growth data. The response was defined as representing "growth" if a significant increase in bacterial concentration (> 0.5 log) was observed. Alternatively, "growth" was defined as a positive value of the specific growth rate. The growth/no growth data of 19 different microorganisms were extracted from all the data in ComBase, comprising 29 kinds of microorganism. Furthermore, the specific growth rate of each microorganism was modeled as a function of temperature, pH, and water activity (a_w) using a Poisson log-linear model, which is a family of generalized linear models (GLM). For 16 of the 19 microorganisms, the specific growth/death rate was successfully modeled as a function of temperature, pH, and a_w using GLM. The specific growth rate was illustrated using a two-dimensional contour plot with growth/no growth data. MRV provides information concerning growth/no growth boundary conditions, and the specific growth rates of queried microorganisms.

In the present initiative, it was important to be able to evaluate data visually and intuitively, and the growth/no growth and μ_{max} data were therefore combined to make it easy to retrieve the required information. This innovative database facilitates the retrieval of growth/no growth data for various types of bacteria, and will contribute to ensuring microbiological food safety. Using MRV, food processors can easily find the appropriate food design and processing conditions. This database will contribute to the efficient and safe production and distribution of processed foods.

Whereas ComBase platform consists of microbial kinetic data, MRV will be a platform focused on growth/no growth data. Databases such as ComBase and MRV should be updated on a continuous basis, incorporating data from both published literature and future studies. Although the current MRV uses only three environmental parameters—temperature, pH, and a_w—to retrieve data, other factors, such as the effect of preservatives, should be incorporated. The development of the MRV is expected to continue in the future to make it more convenient.

References

Augustin, J.-C. and Carlier, V. 2000. Mathematical modelling of the growth rate and lag time for *Listeria monocytogenes*. Int. J. Food Microbiol. 56: 29–51.

Augustin, J.-C., Zuliani, V., Cornu, M. and Guillier, L. 2005. Growth rate and growth probability of *Listeria monocytogenes* in dairy, meat and seafood products in suboptimal conditions. J. Appl. Microbiol. 99: 1019–1042.

Baranyi, J. and Roberts, T.A. 1994. A dynamic approach to predicting bacterial-growth in food. Int. J. Food Microbiol. 23: 277–294.

Baty, F., Flandrois, J.P. and Delignette-Muller, M.L. 2002. Modeling the lag time of *Listeria monocytogenes* from viable count enumeration and optical density data. Appl. Environ. Microbiol. 68: 5816–5825.

Dalgaard, P., Buch, P. and Silberg, S. 2002. Seafood spoilage predictor—development and distribution of a product specific application software. Int. J. Food Microbiol. 73: 343–349.

Delignette-Muller, M.L., Baty, F., Cornu, M. and Bergis, H. 2005. Modelling the effect of a temperature shift on the lag phase duration of *Listeria monocytogenes*. Int. J. Food Microbiol. 100: 77–84.

Emborg, J., Laursen, B.G. and Dalgaard, P. 2005. Significant histamine formation in tuna (Thunnus albacares) at 2 degrees C—effect of vacuum- and modified atmosphere-packaging on psychrotolerant bacteria. Int. J. Food Microbiol. 101: 263–279.

Emborg, J. and Dalgaard, P. 2006. Formation of histamine and biogenic amines in cold-smoked tuna: An investigation of psychrotolerant bacteria from samples implicated in cases of histamine fish poisoning. J. Food Prot. 69: 897–906.

Emborg, J. and Dalgaard, P. 2008. Modelling the effect of temperature, carbon dioxide, water activity and pH on growth and histamine formation by Morganella psychrotolerans. Int. J. Food Microbiol. 128: 226–233.

Fernandez-Piquer, J., Bowman, J.P., Ross, T. and Tamplin, M.L. 2011. Predictive models for the effect of storage temperature on *Vibrio parahaemolyticus* viability and counts of total viable bacteria in Pacific oysters (*Crassostrea gigas*). Appl. Environ. Microbiol. 77: 8687–8695.

Fernandez-Piquer, J., Bowman, J.P., Ross, T., Estrada-Flores, S. and Tamplin, A.L. 2013. Preliminary stochastic model for managing *Vibrio parahaemolyticus* and total viable bacterial counts in a Pacific oyster (*Crassostrea gigas*) Supply Chain. J. Food Prot. 76: 1168–1178.

Food, Agriculture Organization of the United NationsWorld Health Organization FAOWHO. 2004. Risk assessment of *Listeria monocytogenes* in ready-to-eat foods [WWW Document]. fao.org. URL http://www.fao.org/docrep/010/y5394e/y5394e00.HTM (accessed 5.1.14).

Gimenez, B. and Dalgaard, P. 2004. Modelling and predicting the simultaneous growth of *Listeria monocytogenes* and spoilage micro-organisms in cold-smoked salmon. J. Appl. Microbiol. 96: 96–109.

Handa, S., Kimura, B., Takahashi, H., Koda, T., Hisa, K. and Fujii, T. 2005. Incidence of *Listeria monocytogenes* in raw seafood products in Japanese retail stores. J. Food Prot. 68: 411–415.

Koseki, S. 2009. Microbial Responses Viewer (MRV): a new ComBase-derived database of microbial responses to food environments. Int. J. Food Microbiol. 134: 75–82.

Koseki, S., Takizawa, Y., Miya, S., Takahashi, H. and Kimura, B. 2011. Modeling and predicting the simultaneous growth of *Listeria monocytogenes* and natural flora in minced tuna. J. Food Prot. 74: 176–187.

Lianou, A. and Sofos, J.N. 2007. A review of the incidence and transmission of *Listeria monocytogenes* in ready-to-eat products in retail and food service environments. J. Food Prot. 70: 2172–2198.

Liao, C., Zhao, Y. and Wang, L. 2017. Establishment and validation of RNA-based predictive models for understanding survival of *Vibrio parahaemolyticus* in oysters stored at low temperatures. Appl. Environ. Microbiol. 83: e02765–16.

Mejlholm, O. and Dalgaard, P. 2007a. Modeling and predicting the growth of lactic acid bacteria in lightly preserved seafood and their inhibiting effect on *Listeria monocytogenes*. J. Food Prot. 70: 2485–2497.

Mejlholm, O. and Dalgaard, P. 2007b. Modeling and predicting the growth boundary of *Listeria monocytogenes* in lightly preserved seafood. J. Food Prot. 70: 70–84.

Mejlholm, O., Gunvig, A., Borggaard, C., Blom-Hanssen, J., Mellefont, L., Ross, T., Leroi, F., Else, T., Visser, D. and Dalgaard, P. 2010. Predicting growth rates and growth boundary of *Listeria monocytogenes*—An international validation study with focus on processed and ready-to-eat meat and seafood. Int. J. Food Microbiol. 141: 137–150.

Mejlholm, O., Boknaes, N. and Dalgaard, P. 2015. Development and validation of a stochastic model for potential growth of *Listeria monocytogenes* in naturally contaminated lightly preserved seafood. Food Microbiol. 45: 276–289.

Mejlholm, O. and Dalgaard, P. 2015. Modelling and predicting the simultaneous growth of *Listeria monocytogenes* and psychrotolerant lactic acid bacteria in processed seafood and mayonnaise-based seafood salads. Food Microbiol. 46: 1–14.

Mellefont, L.A., McMeekin, T.A. and Ross, T. 2008. Effect of relative inoculum concentration on *Listeria monocytogenes* growth in co-culture. Int. J. Food Microbiol. 121: 157–168.

Miles, D.W., Ross, T., Olley, J. and McMeekin, T.A. 1997. Development and evaluation of a predictive model for the effect of temperature and water activity on the growth rate of *Vibrio parahaemolyticus*. Int. J. Food Microbiol. 38: 133–142.

Miya, S., Takahashi, H., Ishikawa, T., Fujii, T. and Kimura, B. 2010. Risk of *Listeria monocytogenes* contamination of raw ready-to-eat seafood products available at retail outlets in Japan. Appl. Environ. Microbiol. 76: 3383–3386.

Pin, C. and Baranyi, J. 1998. Predictive models as means to quantify the interactions of spoilage organisms. Int. J. Food Microbiol. 41: 59–72.

Ratkowsky, D.A., Olley, J., McMeekin, T.A. and Ball, A. 1982. Relationship between temperature and growth rate of bacterial cultures. J. Bacteriol. 149: 1–5.

Tenenhaus-Aziza, F. and Ellouze, M. 2015. Software for predictive microbiology and risk assessment: a description and comparison of tools presented at the ICPMF8 Software Fair. Food Microbiol. 45: 290–299.

Yang, Z.-Q., Jiao, X.-A., Li, P., Pan, Z.-M., Huang, J.-L., Gu, R.-X., Fang, W.-M. and Chao, G.-X. 2009. Predictive model of *Vibrio parahaemolyticus* growth and survival on salmon meat as a function of temperature. Food Microbiol. 26: 606–614.

Yoon, K.S., Min, K.J., Jung, Y.J., Kwon, K.Y., Lee, J.K. and Oh, S.W. 2008. A model of the effect of temperature on the growth of pathogenic and nonpathogenic *Vibrio parahaemolyticus* isolated from oysters in Korea. Food Microbiol. 25: 635–641.

9

Fish Parasite
Infectious Diseases Associated with Fish Parasite
Md. Ali Reza Faruk

1. Introduction

Parasites have a wide range of distribution in all groups of animals. They are more abundant than free-living animals, and may be found in every phylum of animal from protistan to chordates. A large number and diversity of animal species are capable of patrasitising fish, ranging from microscopic protozoans to easily visible crustaceans and annelids. Most of the fishes, either wild or cultured, are infected with parasites. They not only serve as the host to different parasites, but also serve as carriers of many larval parasitic forms that mature and cause serious diseases in many vertebrates, including man. Parasites exhibit marvelous strategies for adaptation to their hosts. Many parasite species are host-specific to some extent and are capable of infecting one or only a limited number of host species. Individual parasite species may also have widely differing effects on different host species (Roberts 2012). Some of them are parasitic in the external surface of fish; others are parasitic in the internal organs. They can infect fishes in different

Department of Aquaculture, Bangladesh Agricultural University, Mymensingh 2202, Bangladesh.
E-mail: hasin96@yahoo.com

stages of their life, as well as different aquatic environmental conditions, and are also considered to be biological indicators of environmental pollution. Parasites interfere with the nutrition, metabolism, and secretory function of the alimentary canal, damage nervous system, and also upset the normal reproduction of the host.

Most parasite species rarely cause problems in the natural environment but in aquaculture, parasites often cause serious outbreaks of disease (Roberts 2012). They play an important role in determining the productivity, sustainability, and economic viability of aquaculture. Parasite infections cause serious socioeconomic, ecological, and welfare consequences in global finfish aquaculture (Menezes et al. 1990, Barber 2007, Shinn et al. 2015a). They are responsible for direct mortality of farmed stock, retarding growth rate and feed conversion ratios, treatment costs incurred, and rejection of product during processing. Besides direct losses, parasites may have considerable impact on the behavior of fish, reduced fecundity, their resistance to other stressing factors, susceptibility to other infections, the potential legislative burdens, and their presence may also reduce marketability of fish (Williams and Jones 1994, Woo 1995, Schäperclaus 1991, Paladini et al. 2017). Parasites generate costs for managing infections as well as prophylactic treatment in finfish aquaculture. The annual global cost of parasites in finfish aquaculture can be tentatively estimated at $1.05 billion to $9.58 billion (Shinn et al. 2015b). Moreover, many of the parasites are also of zoonotic importance, and have public health consequences. Without stringent and appropriate control measures, the impacts of these pathogens can often be significant. It is therefore an important area for proper attention to be given by the scientists for sustainable and safe fish production. This chapter presents an overview on the important parasitic infections in fish.

2. Protozoan Diseases

Most of the commonly encountered fish parasites are protozoans. They are single-celled organisms, many of which are free-living in the aquatic environment. Their ability to multiply on or within their hosts makes them in many instances very dangerous to fish (Chandra 2004). They have a direct life cycle and mostly reproduce by binary fission; some species have cyst form, off the host. Typically, these parasites are present in large numbers either on the surface of the fish, within the gills, or both. The general effect of these parasites is to irritate the epithelial surface, causing an increasing mucus production. There are three main groups of protozoan parasitising the external tissues of fish: ciliates, flagellates, and amoebae.

2.1 Ciliates

Protozoa belonging to the Ciliophora are equipped with cilia (short, fine cytoplasmic outgrowth), or a structure derived from cilia by secondary modifications, or both. Ciliated protozoa are among the most common external parasites which cause mortalities in a number of wild and farmed fishes. Cilitates can be motile, attached, or found within the epithelium. While they often occur as harmless ectocommensals, under poor environmental conditions or stress, some ciliates can rapidly increase in number, leading to morbidity and mortality.

2.1.1 Ichthyophthiriasis

Ichthyophthirius multifiliis, referred to as 'Ich' is the causative agent of Ichthyophthiriasis or white spot disease. It is the most common pathogen of protozoan parasites of freshwater fishes worldwide (Jessop 1995, Dickerson and Dawe 1995, Buchmann et al. 2001, Matthews 2005, Dickerson 2012). It causes particularly devastating infections in farm-raised fish, where it spreads rapidly within dense populations, leading to extensive morbidity and mortality (Dickerson 2006). Ich infection can occur at any of the growth stages of fish, from fry, fingerling, table size, to brood fish. The parasite can cause catastrophic epizootics in warm and temperate-water fish culture, and may even cause losses in wild fish on occasion. Considerable losses have been reported from cultures of carp, rainbow trout, tilapia, eel, channel catfish, as well as ornamental fish (Ling et al. 1991, Matthews 1994). Ich infections in fingerlings affected about 4% of total channel catfish producers in the U.S. in 2009 (USDA 2010). In Europe, annual losses to trout farmers are estimated to be approximately $140 million per year (Dickerson and Findly 2014). Ich epizootics were reported in China as early as 10th century. The first major outbreak in North America was described in 1898. From 1940, it became a serious disease of carp in Russia. The disease has been reported in fry and fingerlings of Indian major carps from nursery and rearing ponds in Bangladesh (Chandra 2004).

Ichthyophthirius multifiliis is spherical in shape and the cilia are evenly distributed over the whole surface. Characteristic feature of the parasite are the horse-shoe shaped nucleus, and its rotating movement. It is an obligate pathogen and has a unique direct life cycle which allows a rapid intensification of infection (Ewing and Kocan 1992). The life cycle consists of an infective theront, a parasitic trophont, and a reproductive tomont (Nigrelli et al. 1976). Infective theronts swim actively in water in search of hosts. After burrowing into fish epithelium, theronts become trophonts and feed on host tissue until they reach maturity (McCartney et al. 1985). The mature tomont drops off the host, attaches to substrates, and undergoes multiple divisions to produce theronts (MacLennan 1935). The parasite spreads rapidly from fish to fish, as a single Ich tomont can produce hundreds to thousands of infective theronts in less than a day (MacLennan 1935, Matthews 2005).

The parasite invades epithelial tissue of gills, skin, or fins, leaving a small wound and visible white spot or nodule where each parasite encysts (Dickerson 2006, Lom and Dyková 1992). Infected fish are extremely lethargic and covered with visible white dots. Mortality can be rapid and catastrophic. Heavy infection by Ich damages fish skin and gills, causes loss of the respiratory, excretory, and osmoregulatory functions, and might serve as a portal of entry for secondary invaders, leading eventually to death of fish (Hines and Spira 1974, Ewing and Kocan 1987, Dickerson 2006).

2.1.2 Trichodiniasis

Trichodiniasis is one of the major protozoan diseases found in fish worldwide. The disease is caused by a large assemblage of peritrichous ciliates of trichodinids. The trichodinid group includes *Trichodina*, *Trichodinella*, and *Tripartiella*, which are important ectoparasites of freshwater and marine fish worldwide (Lom and Dyková 1992). They are capable, in some cases, of causing heavy damage to their hosts, resulting in mortalities. Infestations caused by trichodinids are particularly significant in aquaculture because they are responsible for causing decreased growth (Ekanem and Oblekezie 1996), chronic mortality during cage production (Valladão et al. 2013) and changes in vision and swimming in larvae, culminating in acute mortality (Valladão et al. 2014). Trichodinids have a monoxenic life cycle, and reproduce mainly by binary fission on the host. They can be horizontally transmitted by direct contact or by contaminated water, in which the parasite searches for new hosts. Outbreaks are often associated with poor water quality and stress. Contaminated fish farming utensils are also another important source of transmission of trichodinids.

Trichodinids are up to 100 µm in diameter, and have a basic saucer-shape with a fringe of cilia around the perimeter which are used for locomotion and feeding. Their body is supported by a rigid ring of interconnected discs called a chitinoid or denticular ring. The parasite brows over the surface of gills and skin with spinning motion, damaging the host tissues and consuming the resulting tissue debris. Infected fish often have a grayish sheen due to excess mucus production, and fins may become frayed. Ultimately, erosion of the epithelium will occur (Roberts 2012). The main pathological changes associated with this parasitic infection are related to gill tissue, such as hyperplasia, hypertrophy, oedema, inflammatory infiltration, and necrosis (Yemmen et al. 2011a,b, Valladão et al. 2013, 2014).

2.1.3 Chilodonellosis

Chilodonella is a highly pathogenic holotrich ciliate, ectoparasite on the skin and gill of a wide range of temperate and tropical freshwater fish. The parasite has a flattened, ovoid shape, is up to 80 μm in length, and is covered by rows of cilia which move it in a steady gliding manner over the epithelial cells of its fish host. Heavy infections of *Chilodonella* are often associated with poor water quality. Carps, salmonids, and catfish are the species most commonly affected. *Chilodonella hexasticha* is most likely to be problematic at lower water temperatures, and is reported as a serious pathogen in overwintered carp (Bauer et al. 1973). *Chilodonella piscicola* (=*C. cyprini*) infects cyprinids particularly, but can be found on other fish, where it can cause problems at higher temperatures (Hoffmann et al. 1979). Fingerlings can be especially vulnerable (Urawa and Yamao 1992, Rintamäki et al. 1994).

The mass development of parasite causes a much higher production of mucus and disturbances in the respiratory function of the skin. The fish is restless, and rises to the upper layers of the water. Its entire body is covered with the bluish-white coating, and is particularly noticeable in the head region. A smear in the skin surface reveals numerous individuals of the parasites. Damage to the skin may then be subject to secondary invasion by bacteria, fungi, and other pathogens. It causes localized hyperplasia of the gill epithelium. The thin respiratory epithelium is covered by the hyperplastic epithelium, and this drastically reduces the respiratory surface of the gills.

2.2 Flagellates

Flagellated protozoans are small parasites that can infect fish externally and internally. They are characterized by one or more flagella that cause the parasite to move in a whip-like or jerky motion. Flagella are longer and more powerful than cilia and are always few in number.

2.2.1 Ichthyobodosis

The disease is caused by heavy infections on skin and gills of fish by parasitic flagellates belonging to the genus *Ichthyobodo* (Isaksen 2013). It is an important disease that has caused severe loss among farmed and ornamental fish worldwide for more than a century (Robertson 1985, Isaksen 2013). *Ichthyobodo* is regarded as one of the most damaging parasites among farmed salmon, and is probably the major cause of mortality among salmonid fry and fingerlings (Robertson 1985). The flagellate that is perhaps best known as a serious fish pathogen is *Ichthyobodo necatrix* (=*Costia necator*), which causes the disease known as *costiasis*.

Ichthyobodo spp. spread rapidly between hosts in fish farms, most likely by both direct contact or through free-swimming parasites (Urawa 1996). Heavy infections may occur when conditions favor the parasites. Poor rearing conditions such as low water flow and high crowding densities are considered particularly important (Schäperclaus 1992, Urawa 1995). Massive infections on skin and gills can cause epithelial hyperplasia or hypertrophy, and may result in severe or fatal osmoregulatory or respiratory problems (Urawa et al. 1998). Clinical signs of *Ichthyobodo* spp. are an excess mucus production that forms a blue-grey or white film over the body surface and gills. There have also been described several non-specific clinical signs of severe and prolonged *Ichthyobodo* spp. infections, including flashing, lethargic behavior, listlessness, loss of appetite, and increased mortality (Robertson 1985, Woo and Poynton 1995).

Ichthyobodo spp. are obligate ectoparasites with a direct life-cycle on the host (Becker 1977), and they cannot subsist or multiply without an appropriate host. There are two forms: a parasitic feeding form which attaches to the fish's epithelial cells and a non-feeding, swimming form that exists off the fish (Lom and Dykova 1992). The free-swimming form is kidney-shaped with two pairs of flagella. The attached form is pear-shaped and attaches to the gill and skin (Southgate 1993). The parasites disappear from a dead host, and die after 30–60 minutes in the free-swimming form outside a host.

2.2.2 Hexamitasis

Hexamitiasis is caused by excessive numbers of flagellated protozoa of the genus *Hexamita* in the alimentary tract of farmed and wild freshwater fishes (Southgate 1993). *Hexamitia* sp. is a small (10 μm) pear-shaped, pyriform organism with three anterior and one posterior pair of flagella. *Hexamita truttae* is common in North American trout hatcheries, which cause mass mortality of fish. Clinically, the young fish have anorexia, debilitated with reduced growth, have trailing faecal casts, excessive nervousness, and the abdomen may be distended. The fish develop an acute enteritis, yellowish watery gut contents with numerous organisms present in the faeces or bile from the gall bladder. Transmission is by ingestion of an infective cyst. In farmed Chinook and Atlantic salmon, the disease can become systemic with fish becoming anaemic with swollen kidneys and exophthalmus. *Hexamita salmonis* may be present in vast numbers in the pyloric intestine of cultured salmonids in freshwater, but the degree of damage it causes has been disputed (Uldal and Buchmann 1996). These parasites present a monoxenic life cycle; the pear-shaped trophozoites change to spherical before cellular division (Woo 2006). They can be horizontally transmitted by the releasing of trophozoites and oocysts into the water from the fish feces that will be ingested by other hosts.

2.3 Amoebae

Some free-living amoebae may change their mode of life and become harmful. Several species of amoeba have been implicated in gill disease in salmonids.

2.3.1 Amoebic Gill Disease

Amoebic gill disease (AGD), caused by the free-living, facultative amoeba *Paramoeba* [=*Neoparamoeba*] *perurans*, is a major issue in marine salmon farming, which leads to gill damage and death of infected fish (Mitchell and Rodger 2011). It has become a significant problem in sea-caged Atlantic salmon and rainbow trout in Tasmania, and has been considered the most serious infectious disease (Roubal et al. 1989, Munday et al. 1990, Bryant et al. 1995, Findlay et al. 1995). Cases of gill amoebic infections of fish other than salmonids have also been reported, e.g., in European catfish (*Silurus glanis*) or turbot (Nash et al. 1988, Dyková et al. 1995, 1998, Paniagua et al. 1998).

Clinical AGD most often occurs at water temperatures of 10–20°C, and is sometimes associated with higher than normal temperatures. Gross pathology in infected fish is characterised by raised, multifocal, white mucoid patches on the gills, which represent regions of epithelial hyperplasia of the primary and secondary lamellae. This phase is followed by desquamation of the epithelium, local disturbances of blood circulation, and progressive changes represented by inflammation (Dyková et al. 1995, Adams and Nowak 2001, 2003). All the above mentioned changes result in decrease or loss of gill respiratory surface area. Significant cardiac changes and acid-base disturbances may occur in AGD affected fish, which may result in acute cardiac dysfunction and death (Powell et al. 2002).

3. Metazoan Diseases

Metazoan parasites include the myxozoans, helminths, crustacean, annelids, and mollusks are common in both wild and cultured fish. All have direct life cycle, but the period of their life cycle in which they act as fish parasites varies considerably with the species concerned.

3.1 Whirling Disease

Whirling disease is a chronic inflammatory disease in salmonid fish caused by the myxozoan parasite *Myxobolus cerebralis*, which is characterized primarily by the tight circular movements due to spinal cord constriction and brain stem compression in infected fish (Rose et al. 2000). It occurs worldwide and rainbow trout are most susceptible to disease, although the parasite is able to infect numerous species of salmonid fishes (O'Grodnick 1979, Hoffman 1990). It has been suggested that *M. cerebralis* was originally a parasite of relatively low pathogenicity of native salmonids, such as the brown trout and Atlantic salmon in central Europe. The introduced rainbow trout, on the other hand, is highly susceptible to whirling disease. Through shipments of live and frozen trout, perhaps even with contaminated trout ova, *M. cerebralis* became established in most countries where rainbow trout are cultured. It is now a serious and expanding problem in the native wild populations of rainbow trout and cut-throat trout in the western United States (Bartholomew and Wilson 2002).

Spores of the parasite are oval and have two distinct polar capsules. This protozoan has a complex life cycle. Spores can be shed from infected live fish, as well as from dead and decomposing fish. The spores can also be spread via bird feces. Spores are ingested by an annelid worm intermediate host, *Tubifex tubifex*, which lives in the bottom mud of ponds, streams, and earthen raceways. The spores develop into actinosporeans that penetrate fish (or the fish ingest the actinosporeans when they eat tubifex worms). Plasmodia develop in the fish's cartilage and eventually produce the characteristic spores.

The disease attacks the cartilage. The development and severity of whirling disease depend on the age and size of the salmonid host when exposed to the parasite. Very young fish are most vulnerable to *M. cerebralis*, and susceptibility decreases with age and growth, as bone replaces cartilage in the developing fish (Hoffman and Byrne 1974, Markiw 1991, El-Matbouli et al. 1992, Ryce et al. 2004, 2005). Heavy infections in young fish often result in death. *Myxobolus cerebralis* infections in the spine can cause the fish's tail to turn black and the spine to curve. A blackened tail is caused by pressure on nerves that control pigmentation. Infections in the head cartilage create head and jaw deformities, while infections in the auditory capsule cause young trout to become disoriented and chase their tails in a whirling motion. Permanent deformities of the head, spine, and operculum are caused by cartilage damage, associated inflammation, and interference with normal bone development (Hedrick et al. 1999, MacConnell and Vincent 2002). During the active phase of infection growth rates may be depressed.

3.2 Proliferative Kidney Disease

Proliferative kidney disease (PKD), caused by the malacosporean parasite *Tetracapsuloides bryosalmonae*, is a widespread disease that causes significant losses among farmed salmonids in Western Europe and North America (Hedrick et al. 1984, Morris and Adams 2006). PKD is also the suspected cause for several declines in wild salmonid populations (Burkhardt-Holm et al. 2005, Sterud et al. 2007). The disease is characterized by a chronic inflammation of the anterior and posterior kidney, which is caused by a massive accumulation of lymphocytes and extensive granuloma formation (Ferguson 1981, Clifton-Hadley et al. 1986).

Fish are infected by parasite spores released from bryozoans, which are the invertebrate hosts of the parasite (Feist et al. 2001). The parasite invades the fish through skin and gills (Longshaw et al. 2002), and afterwards invades inner organs, with the kidney being the main target organ for further development (Ferguson 1981, Clifton-Hadley et al. 1986, Kallert et al. 2011). Infection and clinical signs of PKD are dependent on environmental temperature. Water temperatures above 12°C may induce clinical PKD (Morris et al. 2005), but normally, clinical PKD outbreaks occur at water temperatures above 15°C (Tops et al. 2006). It primarily affects fingerling fish, and usually results in 100% morbidity on an affected farm, with up to 20% mortality (Hedrick et al. 1999).

3.3 Diseases Caused by Crustaceans

A tremendous number of crustaceans have evolved to become dependent on certain animals for existence. Those closely associated with fishes can cause disease problems. Parasitic crustaceans are among the most serious gill and skin parasites of fish worldwide. They may be found attached to the external surfaces of both marine and freshwater fish.

3.3.1 Argulosis

Argulids, commonly referred to as "fish lice", are obligatory ectoparasites causing problems in fisheries and aquaculture worldwide (Fryer 1968, Kabata 1970, Ahmed and Sanaullah 1976, Post 1987, Rushton-Mellor 1992). About 150 Argulus species are known from marine, brackish, and freshwater habitats (Kabata 1985). *Argulus foliaceus* L. 1758, is probably the most common and widespread argulid in Western Europe. This species has been reported as a threat to the culture of tilapia (Roberts and Sommerville 1982), rainbow trout (Menezes et al. 1990, Ruane et al. 1995), and common carp (Jafri and Ahmed 1994). *Argulus japonicus* Thiele, 1900, is believed to have originated in Japan, and probably owes wide-spread distribution to the trade in ornamental fish, such as goldfish and koi carp (Rushton-Mellor 1992). Argulosis represents a major problem within the Bangladeshi carp culture industry, with infections causing mortality, morbidity, and growth loss (Rahman 1996, Ahmed 2004).

At 5–10 mm, *Argulus* is a large parasite that is visible with the naked eye. Argulids have a very distinctive oval-shaped, flattened carapace. Other notable physical features include compound eyes, a pair of large suckers, four pairs of branched thoracic swimming limbs, and a small unsegmented abdomen. While most fish lice are effective swimmers, many species tend to move through the water in a loose cork-screw motion or a somersaulting action. *Argulus* has a four stage life-cycle (egg, nauplius, meta-nauplius, adult), and requires a fish host at least once during their life. The parasites complete their life-cycle within 30–100 days, depending on water temperature. They are transmitted via free-swimming meta-nauplii stages (Shimura 1981). Once emerged from eggs, non-feeding meta-nauplii actively search for a host (Mikheev et al. 2004). Without a host fish, lice survive only for a few days, and the juveniles for an even shorter period (< 48 hours).

The lice attach themselves to the skin of the host with their strong suckers, and then feed on its host's blood or mucus by using a modified disk which possesses piercing and sucking mouthparts. During feeding, argulids inject cytolytic enzymes and then draw off blood (Shimura and Inoue 1984). Localised inflammation is often seen at the site. In the early stages of infection, an increase in the frequency of fish jumping and a reduction in feeding have been observed. As the infection progresses, secondary fungal and bacterial infections develop and fish may exhibit shoaling behavior (Northcott et al. 1997, Gault et al. 2002). If the infection continues, there may be large-scale mortalities. In addition to their destructive nature, *Argulus* species are known to act as vectors for other pathogens including viruses (Ahne 1985), bacteria (Shimura 1983), fungi (Bower-Shore 1940) and nematodes (Moravec 1994).

3.3.2 Lernaeasis

Lernaenid copepodes or 'Anchor worms' are important ectoparasite parasites in freshwater aquaculture of cyprinids, and occasionally of salmonids and other fishes. Epizootics in cultured fish are often associated with high mortality. The parasites also cause problems in commercial aquariums. Infection by this parasite has been associated with reduced host weight, growth, and fecundity (Kabata 1982, 1985, Khan et al. 2003). The very common *L. cyprinacea* is considered one of the most invasive ones, having worldwide distribution. A number of species have been reported from Indian region (*L. chackoensis* and *L. bengalensis*) (Chandra 2004). Indian major carps are particularly affected by lernaeasis in Bangladesh. Infestations with Lernaea are most prevalent in the summer months and occur more commonly in stagnant or slow-moving water bodies.

Only the female stage is parasitic and is found embedded in the dermis of the fish, sometimes penetrating the musculature and reaching the peritoneal cavity in small fish. Attachment is by means of a branched anchor formed by modification of the head region. The embedded parasite induces a necrotic ulcer around its insertion, and eventually a connective tissue capsule forms around the head of the parasite. Lernaea can cause intense inflammation, leading to secondary bacterial and fungal infections. These secondary infections sometimes worsen and kill the fish. Larger numbers of parasites on the gill can interfere with respiration, causing death. Fish can survive *Lernaea* infection, but chronic conditions frequently result in poor growth and body condition. Mortalities occur most often in young fish, but even if they survive they lose weight and are unsightly (Kabata 1985).

3.4 Disease Caused by Monogenetic Trematode

Monogenetic trematodes are parasitic flatworms or flukes that generally live on the surface of a fish host. Damage is caused to the host by the penetration of their attachment organ and by browsing action of the mouth at the free end. They have been recognized as a serious pathogen of fish in aquaculture (Ogawa 2002, Ernst et al. 2002, Grau et al. 2003). Monogeneans are able to multiply rapidly in high density aquaculture environments because they have a direct single host life cycle, as they require no intermediate host (Rohde 1993). They produce freely deposited eggs that often become entangled to high re-infection rates amongst fish (Ernst et al. 2002, Ogawa 2002). Severely affected fish may die, as a result of gill pathology and interference with the exchange of respiratory gasses and ions (Stephens et al. 2003). Monogeneans display a high degree of host specificity. They can be viviparous or oviparous; are hermaphrodite and commonly protandrous. They attach to their host by means of a specialized muscular posterior attachment organ or opisthaptor, which houses a variable array of sclerotized hooks, connecting bars, clamps, or epidermal structures (Paladini et al. 2017). Two most common representatives of this group are *Gyrodactylus* and *Dactylogyrus*.

3.4.1 Gyrodactylosis

The disease is caused by the genus *Gyrodactylus*. They are commonly known as 'skin flukes' and are found all over the surface of the body, including fins and occasionally in the gills of both marine and freshwater fish. *Gyrodactylus salaris* is perhaps the most economically impactful ectoparasitic flatworm of fish which has been responsible for mass mortalities of wild Atlantic salmon when transferred to Norwegian rivers. It is the only notifiable flatworm disease listed by the Office International des Epizooties (OIE) (World Organization for Animal Health) (Bakke et al. 2007). Most species of gyrodactylids are very host-specific and may be even be specific to a particular site on the host (Cone 1995).

Gyrodactylids are small worms, about 0.3–1.0 mm in length, and may just be visible to the naked eyes. Adult parasites carry a fully developed embryo identical to the adult, which in turn carry young of next generation (Eissa 2002). They have a pair of anchors with both dorsal and ventral bars and 16 marginal hooks, and do not have eye spots. Attachment to the fish is made with the marginal hooks; the anchors are used as a spring-like device to assist attachment with the marginal hooks. An embryo with its pair of anchors may frequently be seen inside an adult gyrodactylid. As they are viviparous, gyrodactylids are able to reproduce extremely rapidly if conditions are favorable. Transmission is thought to be by direct fish to fish contact, although parasites may survive for some time in the water column if detached from their host. They move by a characteristic looping action, and small puncture wounds occur each time the hooks penetrate the host. Heavily infected fish may have increased mucus production, frayed fins, skin ulcers, and damaged gills. The lesions are caused by the feeding activity of the parasites. The occurrence of gyrodactylids in epizootic proportions in cultured fish is generally a sign of poor environmental conditions and stressed fish.

3.4.2 Dactylogyrosis

Dactylogyrosis is caused by *Dactylogyrus* species, and in general they are known as "Gill fluke". Some species of *Dactylogyrus* have proved to be very dangerous in cyprinid culture, especially to fry, where relatively few parasites can cause severe gill damage. The intensity of infestation rises rapidly to reach several hundred per fish. Dactylogyrid flukes are common in freshwater fishes in Bangladesh. A numbers of species of this genus (*Dactylogyrus mrigali, D. chauhanus, D. yogendrai, D. labei, D. kalyanensis*) have been reported from the Indian major carps and from other several exotic and indigenous fishes (Chandra and Jannat 2002, Chandra and Yasmin 2003).

Dactylogyrids have two to four eyespots, one pair of large anchor hooks, and are egg layers. They are oviparous and relatively small. Anteriorly there is a pair of pigmented light receptors and two cephalic lobes, which contains secretions of adhesive gland cells. They have a muscular pharynx and a tubular, posteriorly confluent intestine. The haptor has two pairs of hamuli and 14 marginal hooks. The eggs hatch into free-swimming larvae and are carried to a new host by water currents and their own ciliated movement. Eggs are laid by dactylogyrids embryonate and hatch in a period which varies greatly with temperature within a species and between species. In some species, in temperate climates, eggs laid in autumn may overwinter and hatch on the return of favorable conditions in the spring, when newly hatched fry of host species are available. Dactylogyrids also have optimal temperatures for egg production, resulting in seasonal peaks of infection, and also different geographical distributions between parasite species (Paperna 1964, Bauer et al. 1973).

Clinical signs of fish infected by gill fluke include increased breathing frequency, whilst the gill coverings are stretched open widely, the gill being expanded and very pale. Parts of the gills often become protuberant and show as a small pale fleece outside the covering. Parts of the gill sheets on which flukes have settled are covered with a cloudy film, consisting of slime and destroyed epithelial cells. The gills often suffer from heavy damage. Hyperplastic changes of gill epithelium often spread to areas not colonized by the worm. Telangiectasis (gill blood blisters) is frequent and widespread. Local tissue erosion at the attachment site is accompanied by vigorous peripheral proliferation. Mucus production is over stimulated, seriously impeding respiration. Gills are discolored and swollen so much that normal closure of the opercula may be impossible.

3.5 Diseases Caused by Digenetic Trematode

Digeneans are endoparasitic flatworms, also known as 'flukes' or 'digenetic trematodes', principally infecting the alimentary canal or associated organs. The principal developmental stages of digeneans are called miracidium, sporocyst, redia, cercaria, metacercaria, transitory migrating larva, and adult (Paladini et al. 2017). They have a complicated indirect life cycle with at least one intermediate host. The stage most commonly encountered in fish is the larval metacercaria, which may be found in the tissue within the cyst or unencysted, depending on species. The miracidium is generally ovoid and possesses cilia covering the body that allow the parasite to swim and search for the first intermediate host, which is usually represented by a snail. The final host is usually a piscivorous bird. Damage can occur to the fish when cercarial larvae first invade through the skin of the fish. Some metacercaria are extremely damaging to their target organ, while others cause very little harm, although their presence may be aesthetically unappealing.

3.5.1 Diplostomiasis

Diplostomum spathaceum referred to as 'eye fluke' is the causative agent of diplostomiasis. The parasite invades the lens of many species of freshwater fish in Europe and North America. The infection of eye fluke can cause cataract and blindness in a range of fish species (Dwyer and Smith 1989, Karvonen et al. 2003, Whyte et al. 1991), of which Cyprinids and rainbow trout appear to be particularly susceptible to *D. spathaceum*. The first intermediate hosts are *Lymnaea* spp. and the final hosts are primarily gulls (*Larus* spp.). The metacercariae in the lens cause cataract formation and eventually blindness so that the fish is unable to feed and loses condition. It also becomes more susceptible to predation by piscivorous birds, in which adult parasite is found. The numbers of metacercariae required to cause blindness obviously vary according to the size of the fish. Although the fish is not killed by blinding, its growth rate may be greatly reduced due to its inability to feed normally, and such fish may become emaciated.

4. Fish Leech

Leeches are clitellate annelids and are the only important fish pathogens in the phylum. Aquatic leeches, both freshwater and marine, occur in a diversity of habitats worldwide. Leeches are segmented worms that feed on blood. Armed with suckers at each end of the body, they are able to loop over the body of the fish to feed. Leeches can potentially affect the health of fishes in a variety of ways, but most involve blood-feeding activities. In freshwater, *Piscicola geometra* can reach epizootic levels in rainbow trout and cyprinids cultured in earthen ponds. In the marine environment *Hemibdella* sp. has been noted as a problem on cultured Dover sole and turbot. Some species, however, can on occasion become important fish ectoparasites. Leeches act as vectors or intermediate hosts of protozoan parasites among like *Cryptobia* (Chandra 2004).

The fish attacked by a leech shows, especially during the early stages of attachment, signs of irritation and restless. It may attempt to dislodge the parasite by rubbing against objects. Its movements may become unsteady and erratic. Moribund conditions caused by leeches are easily identifiable because they are accompanied by the presence of large numbers of these parasites. Secondary effects of infections might be more serious than the pathogenic changes. Leeches by themselves are only rarely associated with serious pathological effects. Generally the lesions associated with their blood meals are clean and heal readily once the bleeding stops. Mortalities are rare, and usually the effects of the leeches are restricted to growth reduction (Andersson 1988).

5. Fishborne Zoonotic Diseases

A large number of parasites infect fish, but only a few cause illnesses in humans. Many marine and freshwater fishes serve as a source of medically important parasitic zoonoses that include trematodiasis, cestodiasis, and nematodiasis. Some of these infections are highly pathogenic. These diseases are mainly acquired through eating raw or under cooked fish. Generally, fish can either be intermediate host of parasites involving man as the definitive host, or harbor larval parasites of other animals which can invade human tissues. However, the larval stages of a few species of parasite can mature both in animals and man. The reported incidence of these fish-borne zoonoses has increased in recent years due to the development of improved diagnosis, increase in raw fish consumption in those countries in which such dishes have commonly been eaten, increased consumption elsewhere of regional fish dishes based on raw or poorly processed fish, the growth in the international market in fish and fish products, and the remarkable development of aquaculture (Keiser and Utzinger 2005, McCarthy and Moore 2000, Nawa et al. 2005, Robinson and Dalton 2009).

5.1 Trematodiasis

Trematodiasis is the infection of humans by trematode parasites. Among the fish-borne parasitic diseases, infections by digenetic trematodes are the most common. A considerable number of digenean metacercaria from fish may infect humans. The disease is important in Southeast Asia and the Far East where many people are dependent on freshwater fish as the major source of protein. The most significant of these digeneans are perhaps *Clonorchis sinensis, Opisthorchis viverrini*, and *Opisthorchis felineus* (Roberts 2012, Lima dos Santos and Howgate 2011). A large number of freshwater fish species can transmit the infective trematode metacercariae with fish belonging to the Cyprinidae (carps) being the most common (WHO 1995, Touch et al. 2009, Chen et al. 2010). Farmed fish of a variety of species have also been shown to be hosts of trematode parasites (Chi et al. 2008, Thien et al. 2007, 2009, Thuy et al. 2010).

Clonorchiasis caused by the liver flukes *C. sinensis* is endemic in South China, Taiwan, South Korea, and North Vietnam (Rim 2005, Dung et al. 2007, Zhang et al. 2007, Cho et al. 2008). It is estimated that 35 million persons globally could be infected by *C. sinensis*, including 15 million in China (Zhou et al. 2008). The disease is being associated with biliary obstruction leading to hepatic necrosis, cirrhosis, and portal hypertension, in heavy infections. The parasite may also locate in pancreatic ducts, causing acute obstructive pancreatitis, a most painful condition. Opisthorchiasis caused by *O. viverrini* is endemic in Thailand, Lao, Cambodia, and Central Vietnam (Andrews et al. 2008). It is estimated 6 million humans in Asia may be infected with *O. viverrini*. Human infection due to *O. felineus* is found in Russia and countries of Central Europe (Yossepowitch et al. 2004).

5.2 Cestodiasis

There are relatively few cases of fish-borne cestode infections in man. The cestodes that mature in the small intestine of man are not pathogenic, and diseases are never fatal. Diphyllobothriasis is the major cestodiasis transmitted by freshwater, marine, and anadromous fishes. The disease is caused by pseudophyllid cestodes belonging to the genus *Diphyllobothrium*. At least 13 species of the cestode genus *Diphyllobothrium* have been recognised from humans. The genus is found in fish, mammal, and avian hosts, and is usually associated with cold-water habitats. The species most often reported from humans is *D. latum*, which is relatively common in the Baltic region, the European Alps, eastern Russia and Japan (Dick et al. 2001). It is considered a mild disease; persons infected with the tapeworm may often be symptomless, in others it may cause diarrhea, abdominal pain, and anemia (Dick 2007, Scholtz et al. 2009). Recent estimates indicate that approximately 20 million individuals could be affected by the disease (Scholtz et al. 2009).

5.3 Nematodiasis

Fish-borne nematodiasis are generally caused by the incidental infestation of man with nematodes whose natural definite hosts are marine mammals, birds, pigs, or other animals. Freshwater, brackish, or marine fishes are the second intermediate host. In most infections, the worms can only survive for a limited period after the initial invasion of the gastrointestinal tract. The method of infection is by ingesting the infective larvae which are located in the muscles, intestine, or viscera of fish.

Capillariasis is caused by nematode *Capillaria philippinensis*. The disease was originally presumed to be an indigeneous disease of the Philippines, where an epidemic was first recorded in 1967. Subsequently, the disease was also found in Thailand, Japan, Taiwan, Indonesia, Korea, Iran, Egypt, and India. Freshwater fish may be important as a source of infection of humans with this nematode (Cross et al. 1972). The adult worms are found in the gut of humans, where they can cause a severe and even fatal illness (Cross 1990).

Gnathostomiasis is caused by members of the genus *Gnathostoma* who undergo visceral larval migration. The disease occurs in Southeast Asia, China, Japan, Korea, the Indian subcontinent, and Middle East. Its life cycle is complex involving intermediate (crustaceans and fishes), paratenic (piscivorous birds, reptilian, and small mammals), and final hosts (wild and domestic animals). Man is considered an accidental host in whom the parasite can cause a wide clinical picture, internal or external, where the condition 'larva migrans' is one of the known symptoms (Waikagul and Diaz Camacho 2007).

Anisakiasis refers to infection by larval ascaridoid nematodes if ingested with raw or lightly cured fish (Smith and Wootten 1978). The genera involved are *Anisakis*, *Pseudoterranova*. and *Contracaecum*. Their normal definitive hosts are marine mammals. Larvae (located in squids and marine fish) can invade the gastrointestinal tract of man, causing an eosinophilic granuloma syndrome. In Europe, it has also been referred to as the 'herring worm' disease. These nematodes cannot mature in humans, but may cause a severe allergic reaction with granulomatosis of the stomach wall.

A. simplex causes an acute or chronic infection that may lead to abdominal pain, nausea, vomiting, and/or diarrhea. Some patients develop syndromes exhibiting clinical manifestations of allergy following infection or following consumption of dead larvae (Audicana and Kenneddy 2008). The incidence of the disease varies widely among countries, with Japan reported as having the highest incidence.

References

Adams, M.B. and Nowak, B.F. 2001. Distribution and structure of lesions in the gills of Atlantic salmon, *Salmo salar* L., affected with amoebic gill disease. J. Fish Dis. 24: 535–42.

Adam, M.B. and Nowak, B.F. 2003. Amoebic gill disease: sequential pathology in cultured Atlantic salmon, *Salmo salar* L. J. Fish Dis. 26: 601–64.

Ahmed, A.T.A. and Sanaullah, M. 1976. Organal and percentage distribution of some metazoan parasites in *Heteropnensies fossilis* and *Clarius batrachus*. J. Asiat. Soc. Bangladesh 2(1): 7–15.

Ahmed, A.T.A. 2004. Development of environment friendly medicant for the treatment of argulosis in carp brood stock ponds. A SUFER Research Project funded by Department for International Development of the United Kingdom and executed by the University Grant Commission of Bangladesh, 68 pp.

Ahne, W. 1985. *Argulus foliaceus* L. and *Piscicola geometra* L. as mechanical vectors of spring viraemia of carp virus (SVCV). J. Fish Dis. 8: 241–242.

Andersson, E. 1988. The biology of the fish leech *Acanthobdella peledina* Grulse. Zool. Beitr. Neue. Folge. 32: 31–50.

Andrews, R.H., Sithithaworn, P. and Petney, T.N. 2008. *Opisthorchis vierrini:* an underestimated parasite in world health. Trends Parasitol. 24(11): 497–501.

Audicana, M.T. and Kenneddy, M.W. 2008. Anisakis simplex: from obscure infectious worm to inducer of immune hypersensivity. Clin. Microbiol. Rev. 21: 360–379.

Bakke, T.A., Cable, J. and Harris, P.D. 2007. The biology of gyrodactylid monogeneans: "the Russian-doll killers". Adv. Parasitol. 64: 161–218.

Barber, I. 2007. Parasites, behaviour and welfare in fish. Appl. Anim. Behav. Sci. 104: 251–264.

Bartholomew, J.L. and Wilson, J.C. 2002. Whirling Disease: Reviews and Current Topics. American Fisheries Society Symposium 29. American Fisheries Society, Bethesda, MD.

Bauer, O.N., Musselius, V.A. and Strelkov, Y.A. 1973. Diseases of Pond Fishes. Israel Programme for Scientific Translations, Jerusalem.

Becker, C.D. 1977. Flagellate parasites of fish. pp. 357–416. *In*: Kreier, J.P. (ed.). Parasitic Protozoa. Vol. 1. Academic Press, New York.

Bower-Shore, C. 1940. An investigation of the common fish louse, *Argulus foliaceus* (Linn.). Parasitol. 32: 361–371.

Bryant, M.S., Lester, R.J.G. and Whittington, R.J. 1995. Immunogenicity of amoebic antigens in rainbow trout, *Oncorhynchus mykiss* (Walbaum). J. Fish Dis. 18: 9–19.

Buchmann, K., Sigh, J., Nielsen, C.V. and Dalgaard, M. 2001. Host responses against the fish parasitizing ciliate *Ichthyophthirius multifiliis*. Vet. Parasitol. 100: 105–116.

Burkhardt-Holm, Giger, P., Guttinger, W., Ochsenbein, H., Peter, U. and Scheurer, A. 2005. Where have all the fish gone? Environ. Sci. Technol. 39: 441A–447A.

Chandra, K.J. and Jannat, M.S. 2002. Monogenean gill parasites of manor carps from different fish farms of Mymensingh. Bangladesh J. Fish Res. 6: 43–52.

Chandra, K.J. and Yasmin, R. 2003. Some rare and new monogenetic trematodes from air-breathing freshwater fishes of Bangladesh. Indian J. Anim. Sci. 73: 113–118.

Chandra, K.J. 2004. Fish parasitology. K.R. Choudhury, 34/A/2 Ram Babu Road, Mymensingh-2200, 196 pp.

Chen, D., Chen, J., Huang, J., Chen, X., Feng, D., Liang, B., Che, Y., Liu, X., Zhu, C., Li, X. and Shen, H. 2010. Epidemiological investigation of *Clonorchis sinensis* infection in freshwater fishes in the Pearl River Delta. Parasitol. Res. 107: 835–839.

Chi, T.K., Dalsgaard, A., Turnbull, J.F., Pham, J.F. and Murrell, K.D. 2008. Prevalence of zoonotic trematodes in fish from Vietnamese fish-farming community. J. Parasitol. 94: 423–428.

Cho, S.-H., Lee, K.-Y., Lee, B.-C., Cho, P.-Y., Cheun, H.-L., Hong, S.-T., Sohn,W.-M. and Kim, T.-S. 2008. Prevalence of clonorchiasis in Southern endemic areas of Korea in 2006. Korean J. Parasitol. 46: 133–137.

Clifton-Hadley, R.S., Richards, R.H. and Bucke, D. 1986. Proliferative kidney disease (PKD) in rainbow trout, *Salmo gairdneri*: Further observations on the effects of temperature. Aquaculture 55: 165–171.

Cone, D.K. 1995. Monogenea (Phylum Platyhelminthes). pp. 289–328. *In*: Woo, T.T.K. (ed.). Fish Disease and Disorders. Volume 1. London: CABI.

Cross, J.H., Banzon, T., Clarke, M.D., Basaca-Sevilla, V., Watten, R.H. and Dizon, J-J. 1972. Studies on the experimental transmission of *Capillaria phillipinensis* in monkeys. Trans, R. Soc. trap. Med. Hyg. 66: 819–27.

Cross, J.H. 1990. Intestinal capillariasis. Parasitol. Today 6: 26–8.

Dick, T.A., Nelson, P.A. and Choudhury, A. 2001. Diphyllobothriasis: update on human cases, foci, patterns and sources of human infections and future considerations. Southeast Asian J. Trop. Med. Public Health 32: 59–76.

Dick, T.A. 2007. Diphyllobothriasis: The *Diphyllobothrium latum* human infection conundrum and reconciliation with a worldwide zoonoses. pp. 151–184. *In*: Murrell, K.D. and Fried, B. (eds.). Food-borne Parasitic Zoonoses: Fish and Plant-borne Parasites. Springer Science, New York.

Dickerson, H.W. and Dawe, D.L. 1995. *Ichthyophthirius multifiliis* and *Cryptocaryon irritans* (Phylum Ciliophora). pp. 181–227. *In*: Woo, P.T.K. (ed.). Fish Diseases and Disorders. CAB International, Wallingford.

Dickerson, H.W. 2006. *Ichthyophthirius multifiliis* and *Cryptocaryon irritans* (Phylum Ciliophora). pp. 116–153. *In*: Woo, P.T.K. (ed.). Fish Diseases and Disorders. Vol. 1, Protozoan and Metazoan Infections, 2nd Edn. CAB International, Wallingford, UK.

Dickerson, H.W. 2012. *Ichthyophthirius multifiliis*. pp. 55–72. *In*: Patrick, K.B. and Woo, P.T.K. (eds.). Fish Parasites Pathobiology and Protection. CABI, Wallingford.

Dickerson, H.W. and Findly, R.C. 2014. Immunity to *Ichthyophthirius* infections in fish: A synopsis. Develop. Comp. Immunol. 43: 290–299.

Dung, D.T., De, N.V., Waikagul, J., Dalsgaard, A., Chai, J.-Y., Sohn, W.M. and Murrell, K.D. 2007. Fishborne intestinal trematodiasis: an emerging zoonosis in Vietnam. Emerg. Infect. Dis. 13: 1828–1833.

Dwyer, W.P. and Smith, C.E. 1989. Metacercariae of *Diplostomum spathaceum* in the eyes of fishes from Yellowstone lake, Wyoming. J. Wildl. Dis. 25: 126–129.

Dyková, I., Figueras, A. and Novoa, B. 1995. Amoebic gill infection of turbot *Scophthalmus maximus*. Folia Parasitol. 42: 91–96.

Dyková, I., Figueras, A., Novoa, B. and Casal, J.F. 1998. *Paramoeba* spp., an agent of amoebic gill disease of turbot Eissa IAM. Text Book of Parasitic Fish Diseases in Egypt. Dar Elanahdda EL-Arabia Publishing, Cairo.

Eissa, I.A.M. 2002. Parasitic fish diseases in Egypt. Dar El-Nahda El-Arabia Publishing, 32 Abd El-Khalek St. Cairo, Egypt.

Ekanem, D.A. and Oblekezie, A.I. 1996. Growth reduction in African catfish fry infected with *Trichodina maritinkae*. J. Aquacult. Trop. 11: 91–96.

El-Matbouli, M., Fischer-Scherl, T. and Hoffmann, R.W. 1992. Present knowledge of the life cycle, taxonomy, pathology, and therapy of some *Myxosporea* spp. important for freshwater fish. Ann. Rev. Fish Dis. 3: 367–402.

Ernst, I., Whittington, I., Corneille, S. and Talbot, C. 2002. Monogenean parasites in sea-cage aquaculture. Austasia Aquac. 46–8 (February/March).

Ewing, M.S. and Kocan, K.M. 1987. *Ichthyophthirius multifiliis* (Ciliophora) exit from gill epithelium. J. Protozool. 34: 309–312.

Ewing, M.S. and Kocan, K.M. 1992. Invasion and development strategies of *Ichthyophthirius multifiliis*, a parasitic ciliate of fish. Parasitol. Today 8: 204–208.

Feist, S.W., Longshaw, M., Canning, E.U. and Okamura, B. 2001. Induction of proliferative kidney disease (PKD) in rainbow trout *Oncorhynchus mykiss* via the bryozoan *Fredericella sultana* infected with *Tetracapsula bryosalmonae*. Dis. Aquat. Organ. 45: 61–68.

Ferguson, H.W. 1981. Effects of temperature on the development of proliferative kidney disease in rainbow trout, *Salmo gairdneri* Richardson. J. Fish Dis. 4: 175–177.

Findlay, V.L., Helders, M., Munday, B.L. and Gurney, R. 1995. Demonstration of resistance to reinfection with *Paramoeba* spp. by Atlantic salmon *Salmo salar* L. J. Fish Dis. 18: 639–642.

Fryer, G. 1968. The parasitic Crustacea of African freshwater fishes; their biology and distribution. J. Zool. 156: 45–95.

Gault, N.F.S., Kilpatric, D.J. and Stewart, M.T. 2002. Biological control of the fish louse in rainbow trout fishery. J. Fish. Biol. 60: 226–237.

Grau, S.C., Pastor, E., Gonzalez, P. and Carbonell, E. 2003. High infection by *Zeuxapta Seriolae* (Monogenea: Heterarxinidae) associated with mass mortalities of amberjack Seriola dumerili Pisso reared in Sea cages in the Balearic Islands (western Mediterranean). Bull. Eur. Assoc. Fish. Pathol. 23: 139–142.

Hedrick, R.P., McDowell, T.S., Gay, M., Marty, G.D., Georgiadis, M.P. and MacConnell, E. 1999. Comparative susceptibility of rainbow trout (*Oncorhynchus mykiss*) and brown trout (*Salmo trutta*) to *Myxobolus cerebralis* the cause of salmonid whirling disease. Dis. Aquat. Organ. 37: 173–183.

Hedrick, R., Kent, M., Rosemark, R. and Manzer, D. 1984. Occurrence of proliferative kidney disease (PKD) among Pacific salmon and steelhead trout. Bull. Eur. Assoc. Fish. Pathol. 4: 34–37.

Hines, R.S. and Spira, D.T. 1974. Ichthyophthiriasis in the mirror carp *Cyprinus carpio* L. IV. Physiological dysfunction. J. Fish Biol. 6: 365–371.

Hoffman, G.L. and Byrne, C.J. 1974. Fish age as related to susceptibility to *Myxosoma cerebralis*, cause of whirling disease. The Prog. Fish-Cult. 36: 151.

Hoffman, G.L., Kazubski, S.L., Mitchell, A.J. and Smith, C.E. 1979. *Chilodonella hexasticha* (Kiernik, 1909) (Protozoa, Ciliata) from North American freshwater fish. J. Fish Dis. 2: 153–157 .

Hoffman, G.L. 1990. *Myxobolus cerebralis* a worldwide cause of salmonid whirling disease. J. Aquat. Anim. Health 2(1): 30–37.

Isaksen, T.E. 2013. *Ichthyobodo* infections on farmed and wild fish. PhD thesis, University of Bargen, Norway.

Jafri, S.I.H. and Ahmed, S.S. 1994. Some observations on mortality in major carps due to fish lice and their chemical control. Pakistan J. Zool. 26: 274–276.

Jessop, B.M. 1995. *Ichthyophthirius multifiliis* in elvers and small American eels from the East River, Nova Scotia. J. Aquat. Anim. Health 7: 54–57.

Kabata, Z. 1970. Diseases of Fish, Book 1: Crustacea as Enemies of Fishes. Snieszko, S.F. and Axelrod, H.R. (eds.). T.F.H. Publications, U.S.A.

Kabata, Z. 1982. Copepoda (Crustacea) parasitic on fishes: problems and perspectives. Adv. Parasitol. 19: 1–7.

Kabata, Z. 1985. Parasites and Diseases of Fish Cultured in the Tropics. Taylor & Francis (eds.). London UK, 318 pp.

Kallert, D.M., Bauer, W., Haas, W. and El-Matbouli, M. 2011. No shot in the dark: Myxozoans chemically detect fresh fish. Int. J. Parasitol. 41: 271–276.

Karvonen, A., Paukku, S., Valtonen, E.T. and Hudson, P.J. 2003. Transmission, infectivity and survival of *Diplostomum spathaceum* cercariae. Parasitology 127: 217–224.

Keiser, J. and Utzinger, J. 2005. Emerging foodborne trematodiasis. Emerg. Infec. Dis. 11(10): 1507–1514.

Khan, M.N., Aziz, F., Afzal, M.M., Rab, A., Sahar, L., Ali, R. and Naqvi, S.M. 2003. Parasitic infestation in different freshwater fishes of mini dams of Potohar region. Pak. J. Biol. Sci. 6: 1092–1095.

Lima dos Santos, C.A.M. and Howgate, P. 2011. Fishborne zoonotic parasites and aquaculture: A review. Aquaculture 318: 253–261.

Ling, K.H., Sin, Y.M. and Lam, T.J. 1991. A new approach to controlling ichthyophthiriasis in a closed culture system of freshwater ornamental fish. J. Fish Dis. 14: 595–598.

Lom, J. and Dyková, I. 1992. Protozoan parasites of fishes. Dev. Aquacult. Fish. Sci. 26, Elsevier, 315 pp.

Longshaw, M., Le Deuff, R.-M., Harris, A.F. and Feist, S.W. 2002. Development of proliferative kidney disease in rainbow trout, *Oncorhynchus mykiss* (Walbaum), following short-term exposure to *Tetracapsula bryosalmonae* infected bryozoans. J. Fish Dis. 25: 443–449.

MacConnell, E. and Vincent, E.R. 2002. The effects of *Myxobolus cerebralis* on the salmonid host. Symposium number 29. *In*: Bartholomew, J.L. and Wilson, J.C. (eds.). Whirling Disease Reviews and Current Topics. American Fisheries Society and the Whirling Disease Foundation.

MacLennan, R.F. 1935. Observations on the life cycle of *Ichthyophthirius*, a ciliate parasitic on fish. Northwest Sci. 9: 12–14.

Markiw, M.E. 1991. Whirling disease: Earliest susceptible age of rainbow trout to the triactinomyxid of *Myxobolus cerebralis*. Aquaculture 92: 1–6.

Matthews, R.A. 1994. *Ichthyophthirius multifiliis* Fouquet, 1876: Infection and protective response within the fish host. pp. 17–42. *In*: Pike, A.W. and Lewis, J.W. (eds.). Parasitic Disease of Fish. Samara Publishing, Tresaith, UK.

Matthews, R.A. 2005. *Ichthyophthirius multifiliis* Fouquet and ichthyophthiriosis in freshwater teleosts. Adv. Parasitol. 59: 159–241.

McCarthy, J. and Moore, T.A. 2000. Emerging helminth zoonoses. Int. J. Parasitol. 30: 1351–1360.

McCartney, J.B., Fortner, G.W. and Hansen, M.F. 1985. Scanning electron microscopic studies of the life cycle of *Ichthyophthirius multifiliis*. J. Parasitol. 71: 218–226.

Menezes, J., Ramos, M.A., Pereira, T.G. and Moreira da Silva, A. 1990. Rainbow trout culture failure in a small lake as a result of massive parasitosis related to careless introductions. Aquaculture 89: 123–126.

Mikheev, V.N., Pasternak, A.F. and Valtonen, E.T. 2004. Tuning host specificity during the ontogeny of a fish ectoparasite: behavioural responses to host-induced cues. Parasitol. Res. 92: 220–224.

Mitchell, S.O. and Rodger, H.D. 2011. A review of infectious gill disease in marine salmonid fish. J. Fish Dis. 34(6): 41–32.

Moravec, F. 1994. Parasitic nematodes of freshwater fishes of Europe. Academia and Kluwer Acad. Publishers, Praha and Dordrecht, Boston, London, 473 pp.

Morris, D.J., Ferguson, H.W. and Adams, A. 2005. Severe, chronic proliferative kidney disease (PKD) induced in rainbow trout *Oncorhynchus mykiss* held at a constant 18°C. Dis. Aquat. Organ. 66: 221–226.

Morris, D.J. and Adams, A. 2006. Transmission of *Tetracapsuloides bryosalmonae* (Myxozoa: Malacosporea), the causative organism of salmonid proliferative kidney disease, to the freshwater bryozoan *Fredericella sultana*. Parasitology 133: 701–709.

Munday, B.L., Foster, C.K., Roubal, F. and Lester, R.J.G. 1990. Paramoebic gill infection and associated pathology of Atlantic salmon, *Salmo salar*, and rainbow trout, *Salmo gairdneri*, in Tasmania. pp. 215–222. *In*: Perkins, F.O. and Cheng, T.C. (eds.). Pathology in Marine Science. Academic Press, San Diego, California.

Nash, G., Nash, M. and Schlotfeldt, H.J. 1988. Systemic amoebiasis in cultured European catfish, *Silurus glanis* L. J. Fish Dis. 11: 57–71.

Nawa, L., Hatz, Y. and Blum, J. 2005. Sushi delights and parasites: the risk of fishborne and foodborne parasitic zoonoses in Asia. Clin. Infect. Dis. 41: 1297–1303.

Nigrelli, R.F., Pokorny, K.S. and Ruggieri, G.D. 1976. Notes on *Ichthyophthirius multifiliis*, a ciliate parasitic on freshwater fishes, with some remarks on possible physiological races and species. Trans. Am. Micro. Soc. 95: 607–613.

Northcott, S.J., Lyndon, A.R. and Campbell, A.D. 1997. An outbreak of fish lice, *Argulus foliaceus* (L.) (Argulidae, Branchiura, Crustacea). Bangladesh J. Zool. 23: 77–86.

O' Grodnick, J.J. 1979. Susceptibility of various salmonids to whirling disease (Myxosoma: *cerebralis*). Trans. Am. Fish. Soc. 108: 187–190.

Ogawa, K. 2002. Impacts of diclidophorid monogenean infections on fisheries in Japan. Intl. J. Parasitol. 32(3): 373–380.

Paladini, G., Longshaw, M., Gustinelli, A. and Shinn, A.P. 2017. Parasitic diseases in aquaculture: Their biology, diagnosis and control. pp. 37–107. *In*: Austin, B.A. and Newaj-Fyzul, A. (eds.). Diagnosis and Control of Diseases of Fish and Shellfish. First Edition. John Wiley & Sons Ltd.

Paniagua, E., Fernández, J., Ortega, M., Paramá, A., Sanmart´ın, M.L. and Leiro, J. 1998. Effects of temperature, salinity and incubation time on *in vitro* survival of an amoeba infecting the gills of turbot, *Scophthalmus maximus* L. J. Fish Dis. 21: 77–80.

Paperna, I. 1964. Adaptation of *Dactylogyrus extensus* (Mueller and Van Cleave, 1932) to ecological conditions of artificial ponds in Israel. J. Parasit. 50: 90–3.

Post, G. 1987. Textbook of Fish Health. T.F.H. Publications, Canada.

Powell, M.D., Nowak, B.F. and Adams, M.B. 2002. Cardiac morphology in relation to amoebic gill disease history in Atlantic salmon, *Salmo salar* L. J. Fish Dis. 26: 60–64.

Rahman, M. 1996. Effects of a freshwater fish parasite, *Argulus foliaceus* Linn. infection on common carp, *Cyprinus carpio* Linn. Bangladesh J. Zool. 24(1): 57–63.

Rim, H.J. 2005. Clonorchiasis: an update. J. Helminthol. 79: 269–281.

Rintamäki, P., Torpström, H. and Bloigu, A. 1994. *Chilodonella* spp. at four fish farms in northern Finland. J. Eukar. Microbiol. 41: 602–7.

Roberts, R.J. and Sommerville, C. 1982. Diseases of tilapias. *In*: Pullin, R.S.V. and Lowe-McConnell, R.H. (eds.). The Biology and Culture of Tilapias. ICLARM Conf. Proc. 7: 247–263.

Roberts, R.J. 2012. The parasitology of teleosts. *In*: Fish Pathology, Fourth Edition. Blackwell Publishing Ltd.

Robertson, D.A. 1985. A review of *Ichthyobodo necator* (Henneguy, 1883) an important and damaging fish parasite. pp. 1–30. *In*: Muir, J.F. and Roberts, R.J. (eds.). Recent Advances in Aquaculture. Croom Helm, London.

Robinson, M.W. and Dalton, J. 2009. Zoonotic helminth infections with particular emphasis on fasciolosis and other trematodiases. Philos. Trans. R. Soc. Lond. B. Biol. Sci. 364: 2763–2776.

Rohde, K. 1993. Ecology of Marine Parasites. 2nd Edn. Wallingford, UK: CAB International, 1993.

Rose, J.D., G.S. Marrs, C. Lewis and G. Schisler. 2000. Whirling disease behavior and its relation to pathology of brain stem and spinal cord in rainbow trout. J. Aquat. Anim. Health 12: 107–118.

Roubal, F.R., Lester, R.J.G. and Foster, C.K. 1989. Studies on cultured and gill-attached *Paramoeba* spp. (Gymnamoebae: Paramoebidae) and the cytopathology of paramoebic gill disease in Atlantic salmon, *Salmo salar* L. from Tasmania. J. Fish Dis. 12: 481–493.

Ruane, N., McCarthy, T.K. and Reilly, P. 1995. Antibody response to crustacean ectoparasites in rainbow trout, *Oncorhynchus mykiss* (Walbaum), immunized with *Argulus foliaceus* L. antigen extract. J. Fish Dis. 18: 529–537.

Rushton-Mellor, S.K. 1992. Discovery of the fish louse, *Argulus japonicus* Thiele (Crustacea: Branchiura), in Britain. Aquacult. Fish Manage. 23: 269–271.

Ryce, E.K.N., Zale, A.V. and MacConnell, E. 2004. Effects of fish age and development of whirling parasite dose on the disease in rainbow trout. Dis. Aquat. Organ. 59: 225–233.

Ryce, E.K.N., Zale, A.V., MacConnell, E. and Nelson, M. 2005. Effects of fish age versus size on the development of whirling disease in rainbow trout. Dis. Aquat. Organ. 63: 69–76.

Schäperclaus, W. 1991. Fish Diseases. Vol. 2. Oxonian Press Pvt. Ltd. New Delhi.

Schäperclaus, W. 1992. Fish Diseases, 5th Edn. A.A. Balkema, Rotterdam.

Scholtz, T., Garcia, H.H., Kuchta, R. and Wicht, B. 2009. Update on the human broad tapeworm (Genus *Diphyllobothrium*), including clinical relevance. Clin. Microbiol. Rev. 22: 146–160.

Shinn, A.P., Pratoomyot, J., Bron, J.E., Paladini, G., Brooker, E.E. and Brooker, A.J. 2015a. Economic costs of protistan and metazoan parasites to global mariculture. Parasitology 142: 196–270.

Shinn, A.P., Pratoomyot, J., Bron, J. and Brooker, A. 2015b. Economic impacts of aquatic parasites on global Finfish production. Global Aquaculture Advocate, Sept/Oct, 28 pp.

Shimura, S. 1981. The larval development of *Argulus coregoni* Thorell (Crustacea: Branchiura). J. Nat. Hist. 15: 331–348.

Shimura, S. 1983. Seasonal occurrence, sex ratio and site preference of *Argulus coregoni* Thorell (Crustacea: Branchiura) parasitic on cultured freshwater salmonids in Japan. Parasitology 86: 537–552.

Shimura, S. and Inoue, K. 1984. Toxic effect of extract from the mouth parts of *Argulus oregoni*, Thorell (Crustacean Branchiura). Bull. Jap. Soc. Sci. Fish 50(4): 729.

Smith, J.W. and Wootten, R. 1978. Anisakis and anisakiasis. Adv. Parasitol. 16: 93–163.

Southgate, P. 1993. Diseases in aquaculture. pp. 91–130. *In*: Brown, L. (ed.). Aquaculture for Veterinarians. Pergamon Press, Oxford.

Stephens, E.J., Cleary, J.J., Jenkins, G., Jones, J.B., Raidal, S.R. and Thomas, J.B. 2003. Treatment to control *Haliotrema abaddon* in the west Australian dhufish, *Glaucomsoma hebraicum*. Aquaculture 3(215): 1–10.

Sterud, E., Forseth, T., Ugedal, O., Poppe, T.T., Jørgensen, A. and Bruheim, T. 2007. Severe mortality in wild Atlantic salmon *Salmo salar* due to proliferative kidney disease (PKD) caused by *Tetracapsuloides bryosalmonae* (Myxozoa). Dis. Aquat. Org. 77: 191–198.

Thien, P.C., Dalsgaard, A., Thanh, B.N., Olsen, A. and Murrell, K.D. 2007. Prevalence of fishborne zoonotic parasites in important cultured fish species in the Mekong Delta, Vietnam. Parasite Res. 101: 1277–1284.

Thien, C.P., Dalsgaard, A., Nhan, N.N., Olsen, A. and Murrell, K.D. 2009. Prevalence of zoonotic trematode parasites in fish fry and juveniles in fish farms of the Mekong Delta, Vietnam. Aquaculture 295: 1–5.

Thuy, D.T., Kania, P. and Buchmann, K. 2010. Infection status of zoonotic trematode metacercariae in Sutchi catfish (*Pangasianodon hypophthalmus*) in Vietnam: associations with season, management and host age. Aquaculture 302: 19–25.

Tops, S., Lockwood, W. and Okamura, B. 2006. Temperature-driven proliferation of *Tetracapsuloides bryosalmonae* in bryozoans hosts portend salmonid decline. Dis. Aquat. Organ. 70: 227–236.

Touch, S., Komalamisra, C., Radomyos, P. and Waikagul, J. 2009. Discovery of *Opisthorchis viverrini* metacercariae in freshwater fish in southern Cambodia. Acta Trop. 111: 108–113.

Uldal, A. and Buchmann, K. 1996. Parasite host relations: *Hexamita salmonis* in rainbow trout *Oncorhynchus mykiss*. Dis. Aquat. Organ. 25: 229–231.

Urawa, S. and Yamao, S. 1992. Scanning electron microscopy and pathogenicity of *Chilodonella piscicola* (Ciliophora) on juvenile salmonids. J. Aquat. Anim. Health 4: 188–97.

Urawa, S. 1995. Effects of rearing conditions on growth and mortality of juvenile chum salmon (*Oncorhynchus keta*) infected with *Ichthyobodo necator*. Can. J. Fish. Aquat. Sci. 52: 18–23.

Urawa, S. 1996. The Pathobiology of ectoparasitic protozoans on hatchery-reared Pacific salmon. Scientific Reports of the Hokkaido Salmon Hatchery 50: 1–99.

Urawa, S., Ueki, N. and Karlsbakk, E. 1998. A review of *Ichthyobodo* infection in marine fishes. Fish Pathol. 33(4): 311–320.

USDA. 2010. Catfish 2010 Part III: Changes in Catfish Health and Production Practices in the United States, 2002–2009. USDA-APHIS-VS-CEAH-NAHMS.

Valladão, G.M.R., Pádua, S.B., Gallani, S.U., Menezes-Filho, R.N., Dias-Neto, J., Martins, M.L. and Pilarski, F. 2013. *Paratrichodina africana* (Ciliophora): a pathogenic gill parasite in farmed Nile tilapia. Vet. Parasitol. 197: 705–710.

Valladão, G.M.R., Gallani, S.U., Pádua, S.B., Martins, M.L. and Pilarski, F. 2014. Trichodna heterodentata (Ciliophora) infestation on *Prochilodus lineatus* larvae: a host-parasite relationship study. Parasitology 141: 662–669.

Waikagul, J. and Diaz Camacho, S.P. 2007. Gnasthostomiasis. pp. 235–262. *In*: Murrell, K.D. and Fried, B. (eds.). Food-borne Parasitic Zoonoses: Fish and Plant-borne Parasites. Springer Science, New York.

WHO. 1995. Control of foodborne trematode infections. Report of a WHO Study Group: WHO Tech. Rep. Ser. 849.

Whyte, S.K., Secombes, C.J. and Chappell, L.H. 1991. Studies on the infectivity of *Diplostomum spathaceum* in rainbow trout (*Oncorhynchus mykiss*). J. Helminthol. 65: 169–78.

Williams, H.H. and Jones, A. 1994. Parasitic worms of fish. Taylor and Francis, London, 593 pp.

Woo, P.T.K. 1995. Fish diseases and disorders. Vol. I. Protozoan and metazoan infections. CAB International.

Woo, P.T.K. and Poynton, S.L. 1995. Diplomonadida, Kinetoplastida and Amoebida (Phylum Sarcomastigophora). pp. 27–96. *In*: Woo, P.T.K. (ed.). Fish Diseases and Disorders. Vol. 1: Protozoan and Metazoan Infections. CAB International, Wallingford, UK.

Woo, P.T.K. 2006. Diplomonadida (Phylum Parabasalia) and Kinetoplastea (Phylum Euglenozoa). pp. 116–153. *In*: Woo, P.T.K. (ed.). Fish Diseases and Disorders 2nd Ed. Wallingford: CABI Publishing, 2006.

Yemmen, C., Ktari, M.H. and Bahri, S. 2011a. Seasonality and histopathology of *Trichodina puytoraci* Lom, 1962, a parasite of flathead mullet (*Mugil cephalus*) from Tunisia. Acta Adriat. 52: 15–20.

Yemmen, C., Quilichini, Y., Ktari, M.H., Marchand, B. and Bahri, S. 2011b. Morphological, ecological and histopathological studies of *Trichodina gobii* Raabe, 1959 (Ciliophora: Peritrichida) infecting the gills of Solea aegyptiaca. Protistology 6(4): 258–263.

Yossepowitch, O., Gotesman, T., Assous, M., Marva, E., Zimilichman, R. and Dan, M. 2004. Opisthorchiosis from imported raw fish. Emerg. Infect. Dis. 10: 2122–2126.

Zhang, R.-L., Gao, S.-T., Geng, Y.-J., Huang, D.-A., Yu, L.-I., Zhang, S.-X., Cheng, J.-Q. and Fu, Y.-C. 2007. Epidemiological study on Clonorchis sinensis infection in Shenzhen area of Zhujiang delta in China. Parasitol. Res. 101: 179–183

Zhou, P., Chen, N., Zhang, E.-L., Lin, R.-Q. and Zhu, X.-Q. 2008. Food-borne parasitic zoonoses in China: perspective for control. Trends Parasitol. 24: 190–196.

10

Detection and Control of Fish-borne Parasites

Yoshihiro Ohnishi

1. Introduction

Human diseases are caused by 1,415 species of infectious organisms, including 217 virus and prions, 538 (38.0%) bacteria and rickettsia, 307 (2.7%) fungi, 66 (4.7%) protozoa, and 287 (20.3%) helminthes [157]. Out of these species, 868 (61%) from 313 different genera are zoonotic, in that these diseases can be transmitted between humans and animals.

On the other hand, food-borne parasites are divided into meat-borne (Toxoplasmosis, taeniasis, and trichinellosis), reptile/amphibian-borne (Sparganosis), fish-borne, arthropod-borne (Paragonimiasis), mollusk-borne (Angiostrongyliasis), plant-borne (Fasciolosis), and water-borne parasites (Cryptosporidiosis and giardiasis) [42]. Fish-borne parasites include protozoa (*Kudoa septempunctata*), cestodes (*Diphyllobothrium* spp. and *Diplogonoporus* spp.), trematodes (*Clonorchis sinensis*, *Opisthorchis* spp., and minute intestinal flukes), nematodes (Anisakidae, *Capillaria philippinensis*, and *Gnathostoma* spp.), pentastomids, and so on [26, 42]. Fish-borne parasites infect humans when they consume raw or undercooked fish. Therefore, fish-borne parasitic diseases are important diseases in the public health.

WHO estimated the global burden of food-borne diseases [169]. Briefly, about 56.2 million people were infected with food-borne

Kansai University of Health Sciences, Faculty of Health Sciences, Japan.
E-mail: sski83882@iris.eonet.ne.jp

trematodes in 2005: 7.9 million had severe sequelae, and 7,158 died, most from chlangiocarcinoma and cerebral infection [55]. Among food-borne trematodiasis, the total number of patients infected with clonorchiasis, opisthorchiasis, and intestinal fluke infections as fish-borne trematodiasis were estimated to be 15.3 million, 8.3 million, and 6.7 million in 2005 [55], and resulted in 31,620, 16,315, and 18,924 in 2010, respectively [160].

In Japan, the consumption of raw olive flounder infected with *K. septempunctata* caused the new food poisoning in humans, recently [73, 101, 171, 176]. Diphyllobothriasis nihonkaiense had been reported almost exclusively until recently, but Arizono et al. (2009) retrospectively reported 149 cases of diphyllobothriasis nihonkaiense during 1988 to 2008 in 2 main institutes [14]. There had been many endemic areas of liver flukes, but the number of these patients was recently decimated by measures of control and prevention. Nineteen cases with clinostomiasis had been reported until 2014 [10]. Approximately 50 cases infected with *Crassicauda giliakiana* had been reported after eating raw squid or fish [91]. The number of patients with anisakiasis was estimated to be 7,147 per year from 2005 to 2011, and the majority of patients were infected with *Anisakis simplex* s. s. [150].

This chapter describes fish-borne parasites and parasitic diseases. The main fish-borne parasites are shown in Table 10.1.

2. Life Cycle of Fish-borne Parasites

There are four types of life cycles in fish-borne parasites (Fig. 10.1). Type A is a two-host life cycle of *Kudo septempunctata* involving a fish and an annelid [47, 175]. A fish is an intermediate host, and a human is a paratenic host. Type B is a three-host life cycle of cestode and trematode parasites. A fish is a second intermediate host, and a human is a final host. Types C and D are two-host life cycles of nematode parasites. A fish is an intermediate host or a paratenic host. A human is a final host in type C, and a paratenic host in type D.

3. Classification and Characteristics of Fish-borne Parasites

3.1 Protozoa

Kudoa septempunctata is only infective to humans in protozoa.

- *Kudoa septempunctata* **Matsukane et al. 2010 [95]**

  Life cycle: Type A in Fig. 10.1 [175]. *K. septempunctata* is one of the myxosporean parasites in olive flounder, *Paralichthys olivaceus* [95].

  Distribution: Korea and Japan [95, 101, 102].

Table 10.1 Human parasaitoses caused by fish food.

	Diseases	Main causative parasites	Main causative fish		Geographic distributions
			Scientific name	English name	
Protozoa	Kudoasis	*Kudoa septempunctata*	*Paralichthys olivaceus*	Olive flounder	Japan
Cestoda	Diphyllobothriasis	*Diphyllobothrium nihonkaiense*	*Oncorhynchus masou* *Oncorhynchus keta*	Cherry salmon Chum salmon	Japan, Korea, China, Russia and North America
		Diphyllobothrium latum	*Esox lucius* *Perca fluviatilis*	Pike perch	Europe (Fennoscandia, Western Russia), North and South America, Korea, China and Cuba
	Diplogonoporiasis	*Dipologonoporus grandis* (*D. balaenopterae*)	*Sardinops melanostictus* *Engraulis japonica*	Japanese sardine Japanese anchovy	Japan, Korea and Spain
Trematoda	Clonorchiasis	*Clonorchis sinensis*	*Cyprinus carpio* *Carassius* spp. *Pseudorasbora parva*	Carp Crucian Topmouth gudgeon	China, Korea, Vietnam, Taiwan, Philippines, Thailand, India, Russia and Japan
	Opisthorchiasis	*Opisthorchis viverini*	*Dangila lineata* *Cyclocheilichthys apagon* *Hampala dispar* *Cyclocheilichthys armatus* *Henicorhynchus lineatus*	Beardless barb Spotted Hampala barb Freshwater river-dwelling carp	Thailand, Lao PDR, Cambodia, Vietnam, Philippines and India

Table 10.1 contd. …

...Table 10.1 contd.

Diseases	Main causative parasites	Main causative fish		Geographic distributions
		Scientific name	English name	
Trematoda (continued)	*Opisthorchis felineus*	*Tinca tinca* *Rutilus rutilus* *Scardinius erythrophthalmus* *Alburnus alburnus* *Abramis brama*	Tench Common rouch Rudd Common bleak Bream	Eastern Europe (Spain, Italy, Albania, Greece, France, Macedonia, Switzerland, Germany, Poland and Russia), Turkey, Ukaraine and ndia
Metagonimiasis	*Metagoninus yokogawai*	*Plecoglossus altivelis* *Tribolodon* spp. *Lateolabrax japonicus*	Sweetfish Dace Perch	Korea, China, Taiwan, Japan, Russia, Indonesia, Israel, Spain, Lao PDR, Thailand, Hawaii, Balkans, Philippines, Turkey and Siberia
Heterophyiasis	*Heterophyes nocens*	*Mugil cephalus* *Acanthogobius flavimanus*	Flathead grey mullet yellowish goby	Korea, Japan, China, Taiwan, Lao PDR, Thailand, Hawaii, Balkans, Philippines, Turkey and Siberia
Haplorchiasis	*Haplorchis taichui*	*Labiobarbus leptocheila* *Cyclocheilichthys repasson* *Onychostoma elongatum* *Hampala dispar* *Puntioplites falcifer*	Actinopterygii Spotted Hampala barb	Taiwan, Philippines, Bangladesh, India, Palestine, Egypt, Malaysia, Thailand, Lao PDR, Vietnam and China
Echinostomiasis	*Echinostoma hortense*	*Misgurnus anguillicaudatus*	Weather loach	Korea, Japan, China, Philippines, Indonesia, Malaysia and Thailand
Clinostomiasis	*Clinosotomum complanatum*	*Carassius carassius* *Cobitis anguillicaudatus* *Pseudogobio esocinus* *Carassius cuvieri* *Carassius gibelio langsdorfi*	Crucian carp Asin pond loach Pike gudgeon Deepbodies crucian carp Silver crucian carp	Japan, Korea, Europe, North America and East Asia

Nematoda	Anisakiasis	*Anisakis simplex s.s.* *Scomber japonicus* *Gadus macrocephalus* *Trachurus japonicus*	Mackerel Cod Horse mackerel	Iberian Atlantic coast (France and Spain), Japan Sea (Korea), Sakhlin islands, Baltic Sea, West Atlantic and North-east Pacific Sea (Japan), Bering Sea, Mauritanian coast, Azores islands and Brazil
		Anisakis physeteris *Aphanopus carbo* *Xiphias gladius*	Black scabbardfish Swardfish	Japan, Central Mediterranean Sea, Mauritanian coast, West Mediterranean Sea, Iberian Atlantic coast, Azores islands, East Mediterranean Sea and Brazil
		Pseudoterranova decipiens *Notothenia neglecta* *Gadus morhua* *Trachurus trachurus* *Merluccius merluccius*	Cod Atlantic cod Atlantic horse mackerel European hake	Japan, Korea, Chile, USA, Canada, Greenland, UK, France, Spain, Denmark, Baltic Sea, Brazil and Antartic Ocean
	Spiruriasis	*Crassicauda giliakiana* *Watasenia scintillans* *Arctoscopus japonicas* *Gadus microcephalus*	Firefly squid Sailfin sandfish Pacific cod	Japan
	Gnathostomiasis	*Gnathostoma spinigerum* *Channa (Ophicephalus) argus* *Channa striata* *Clarias batrachus* *Anguilla japonica* *Misgurnus anguillicaudatus*	Northern snakehead murrel Snakehead murrel Walking catfish Japanese eel Weather loach	China, Japan, Southeast Asia, India, Central America, South America, and East America (Bostswana and Zambia)
		Gnathostoma nipponicum *Misgurnus anguillicaudatus*	Japanese loach	Japan and Korea

Table 10.1 contd. ....

...Table 10.1 contd.

Diseases	Main causative parasites	Main causative fish		Geographic distributions
		Scientific name	English name	
Nematoda (continued)	*Gnathostoma hispidum*	*Misgurnus anguillicaudatus*	Japanese loach	China, Korea, Japan, Southeast Asia, Australia, and Central America
	Gnathostoma doloresi	*Oncorhynchus masou* *Lepomis macrochirus*	Brook trout Blue-gill	Japan, Southeast Asia and Mexico
	Gnathostoma binucleatum	*Petenia splendida* *Cichlasoma managuense*	Bay snook Jaguar cichlid	Central and South America (Mexico)
Capillariasis	*Capillaria philippinensis*	*Hypseleotris bipartita* *Ambassis commersoni* *Elotris melanosoma*	Bagsit Bagsang Birut	Philippines, Thailand, Japan, Taiwan, Iran, Egypt, Indonesia, Korea and Colombia

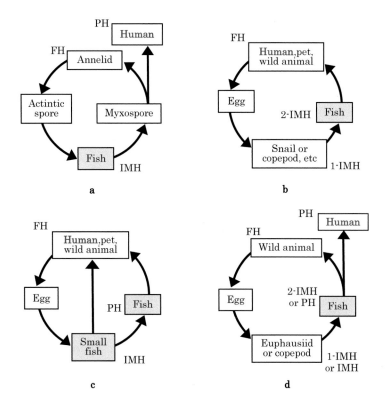

Figure 10.1 Life cycles of fish-borne parasites. a: Type A of life cycle in *Kudoa septempunctata*; b: Type B of life cycles in the cestodes (*Diphyllobothrium* spp.) and trematodes (liver and intestinal flukes); c: Type C of life cycle in *Capillaria philippinensis*; d: Type D of life cycles in *Anisakis* spp. and *Gnathostoma* spp. FH, Final host; IMH, Intermediate host; 1-IMH, First intermediate host; 2-IMH, Second intermediate host; PH, Paratenic host.

Final host: Ragworm [47, 175].

Intermediate host: Olive flounder

Causative fish: Olive flounder (*P. olivaceus*)

Symptoms in humans: In Japan, the consumption of raw olive flounder infected with *K. septempunctata* caused the food poisoning in humans recently [73, 101, 171, 176]. Metalloprotease may be related to the release of seroplasm from spores in *K. septempunctata* [139, 140]. The released seroplasm forms a large hole of 10 μm or more in the intestinal cell, and causes the consequent diarrhea as food poisoning [123]. The main symptoms are diarrhea, emesis, abdominal pain, and fever after consumption with spores of *K. septempunctata* [171].

3.2 Cestoda

3.2.1 Family Diphyllobothriidae

Fourteen species of the genus *Diphyllobothrium* and 4 species of the genus *Diplogonoporus* were reported from humans [26, 136]. Only three species are described here.

• *Diphyllobothrium nihonkaiense* **Yamane et al. 1986 [172]**

Type of life cycle: Type B in Fig. 10.1

Distribution: Japan [14, 26, 100, 102, 152, 172, 173], Korea [56, 76], China [180], Russia [14] and North America [14, 168].

Final host: Humans and brown bear

First intermediate host: Copepods

Second intermediate host: Pacific salmon, including cherry salmon (*Oncorhynchus masou*), chum salmon (*O. keta*), pink salmon (*O. gorbuscha*) and Japanese huchen (*Hucho perryi*) [136].
In Japan, the detection rates of chum salmon, pink salmon, and cherry salmon were 51.1% (24/47), 18.5% (5/27), and 12.2% (10/82), respectively [152].

Causative fish: Cherry salmon, chum salmon, pink salmon and Japanese huchen [136].

Symptoms in humans: The symptoms are generally mild or asymptomatic, and sometimes severe diarrhea, and abdominal pain. The presence of tapeworm can be noticed by the expulsion of strobila.

• *Diphyllobothrium latum* **(Linnaeus 1758)**

Type of life cycle: Type B in Fig. 10.1

Distribution: Europe (Fennoscandia, and West Russia [26]), North America (Alaska and Great Lakes) [26, 136], South America [26], Korea [26, 33, 79, 83, 138], China [180], and Cuba [26]. In Korea, 43 cases of diphyllobothriasis latum had been reviewed until 2007 [83, 138].

Final host: Humans and terrestrial mammals

First intermediate host: Copepods

Second intermediate host: Mainly pike (*Esox lucius*), perch (*Perca fluviatilis*), burbot (*Lota lota*), char (*Salvelinus alpinus*); less frequently ruff (*Gymnocephalus cernuus*), pikeperch (*Sander vitreum*) and yellow perch (*Perca flavescens*) [136].

Causative fish: Pike, perch, burbot, char, ruff, pikeperch and yellow perch [136].

Symptoms in humans: The symptoms are similar to diphyllobothoriasis nihonkaiense. In addition, this parasite also causes Vitamin B_{12}-dependent pernicious anemia [136].

- *Diplogonoporus grandis* **(Blanchard 1894)**

  Three species of the genus *Diplogonoporus* (*D. grandis, D. brauni* and *D. fukuokaensis*) may represent a synonym of *D. balaenopterae* [136].

  Type of life cycle: Type B in Fig. 10.1

  Distribution: Circumpolar (Japan [13, 78, 100], South Korea [35] and Spain).
  Until 2002, there had been more than 180 cases of human diplogonoporiasis recorded in Japan [78].

  Final host: Whales and occasionally humans

  First intermediate host: Copepods

  Second intermediate host: Probably Japanese anchovy (*Engraulis japonica*), Japanese sardine (*Sardinops melanostictus*), skipjack tuna (*Katsuwonus pelamis*), and *Thynnus* spp. [78, 136].

  Causative fish: Japanese anchovy, Japanese sardine, skipjack tuna and *Thynnus* spp. [78, 136].

  Symptoms in humans: The symptoms are generally mild. The presence of tapeworm is noticed by the expulsion of segments in the stool.

3.3 Trematoda

3.3.1 Liver flukes

The oriental liver flukes, including *Clonorchis sinensis, Opisthorchis viverrini, O. felineus*, and *Metrorchis conjunctus* are important causes of human fish-borne parasitic zoonoses [26]. The liver and intestinal fish-borne trematodes (flukes) are estimated to infect more than 35 million people globally, of which, 15 million are distributed in China [84, 88, 89].

Family Opisthorchiidae

- *Clonorchis sinensis* **(Looss 1907)**

  Type of life cycle: Type B in Fig. 10.1

  Distribution: China [26, 88], Korea [26, 77, 84, 138], Vietnam [26], Taiwan [26], Philippines, Thailand, India, Russia [26], and Japan [26, 100].

  Final host: Humans, dogs, cats, rats, pigs, badgers, weasels, camels and buffalos [26].

First intermediate host: Snails

Second intermediate host: 132 species of Cyprinoid freshwater fish [88] and 3 species of freshwater shrimp [26].

Causative fish: In Korea, Japan, and China, *Pseudorasbora parva* is the most commonly infected fish [26]. In China, grass carp (*Ctenopharyngodon idellus*), *Pseudorasbora parva, Saurogobio dabryi, Cyprinus carpio, Carassius auratus,* white Amur bream (*Parabramis pekinensis*), *Abbottina sinensis* (*A. rivularis*), *Gnathopogon timberbis, Parapelecus argenteus, Carassius carassius, Cultriculus* spp., *Opsariichthys* spp., *Rhodeus* spp., *Sarcocheilichthys* spp., *Zacco* spp. and *Hypomesus olidus* [26, 88].

Symptoms in humans: Liver flukes cause chronic inflammation around the biliary tree, chlolangitis, periductal fibrosis, liver cirrhosis, and liver cancer as cholangiocarcinoma in the worst case [77, 84].

- *Opisthorchis viverrini* **(Poirier 1886) Stiles & Hassal 1896**

Type of life cycle: Type B in Fig. 10.1

Distribution: Thailand [12, 26, 71, 72, 146, 169], Lao PDR [26, 92, 132, 146], Cambodia [26, 162], Vietnam [26], Philippines and India.

Final host: Fish-eating mammals (Humans, dogs, cats, rats and pigs) [26].

First intermediate host: Snails

Second intermediate host: more than 40 species of Cyprinid fish *Cyclocheilichthys siaja, Hampala dispar, Puntius* spp., *Labiobarbus lineatus* (*Dangila lineata*), *Esomus metallicus,* and *Osteochilus* sp. [26].

Causative fish: In Lao PDR, infection rates were as follows: *Cyclocheilichthys repasson* (58.5%), *C. armatus* (43.1%), *C. enoplos* (10.0%), *Dangila lineata* (69.6%), *Henicorhynchus lineatus* (42.9%), *Hampala dispar* (44.4%), *Puntioplites proctzysron* (26.8%), and *Osteochilus waandersii* (30.5%) from April 2007 to March 2008. [92].

Symptoms in humans: The symptoms are similar to clonorchiasis [84, 146].

- *Opisthorchis felineus* **(Rivolta 1884) Blanchard 1895**

Type of life cycle: Type B in Fig. 10.1

Distribution: Eastern Europe (Spain [26], Italy [26, 130], Albania [26], Greece [26, 130], France [26], Macedonia [26], Switzerland [26], Germany [26, 89, 130], Poland [26, 89], Russia [26, 89]), and Turkey [26], Ukraine [26], Caucasus [26] and India.

Final host: Fish-eating mammals (Humans, dogs, foxes, cats, rats, rabbits, seals, lions, wolverines, martens and polecats) [26].

First intermediate host: Snails

Second intermediate host: at least 23 species of carp: *Idus melanotus, Tinca tinca, Tinca vulgaris, Abramis brama, Abramis sapa, Barbus barbus, Cyprinus carpio, Blicca bjorkna, Leuciscus idus, Alburnus lucidus, Aspius aspius,* and *Scardinius erythophthalmus* [26].

Causative fish: Tench (*Tinca tinca*), *Rutilus rutilus, Scardinius erythrophthalmus, Alburnus alburnus, Abramis brama, A. ballerus,* and *Blicca bjoerkna*.

Symptoms in humans: In Italy, 72 (34.1%) out of 211 patients were asymptomatic. In symptomatic persons, one or more of the following signs and symptoms were observed: fever (<39°C; 64–75%), abdominal pain (49–56%), headache (36–51%), asthenia (31–46%), arthralgia (27–35%), diarrhea (23–34%), nausea (18–27%), diffuse myalgia (8–15%), skin rash (6–11%), lipothymia (4–10%), dry cough (5–8%), constipation (3–8%), and lack of appetite (2–6%) [130]. In addition, *O. felineus* has been associated with bile duct hyperplasia, pancreatic cirrhosis, and liver cancer in naturally infected cats and dogs [130].

3.3.2 Intestinal flukes

Forty-fifty million people worldwide are generally estimated to be infected with food-borne intestinal trematodes [27, 54, 55]. A total of 59 species of food-borne intestinal flukes have been known to occur in humans [27].

Family Heterophyidae

In the fish-borne parasites, four species of the genus *Metagonimus*, namely *M. yokogawai, M. takahashii, M. minutus,* and *M. miyatai*, three species of the genus *Heterophyes*, namely *H. heterophyes, H. dispar,* and *H. nocens* (syn. *H. katsuradai*), five species of the genus *Haplorchis*, namely *H. taichui, H. pumilio, H. yokogawai, H. pleurolophocerca,* and *H. vanissimus*, are responsible for human infections [26].

- *Metagonimus yokogawai* **(Katsurada 1912)**

  Type of life cycle: Type B in Fig. 10.1

  Distribution: Korea [25, 26, 89, 138, 178], China [26, 89, 179], Taiwan [26, 89, 178], Japan [26, 89, 100, 178], Russia [26], Indonesia [26, 178], Israel [26, 178], Spain [26, 178], Lao PDR [89], Thailand [89], Hawaii [89], Balkans [89], Philippines [89, 178], Turkey [89] and Siberia [89, 178].

  Final host: Humans, dogs, cats and rats

  Intermediate host: Sweetfish (*Plecoglossus altivelis*), dace (*Tribolodon* spp.), and perch (*Lateolabrax japonicus*) [25–27].

  Causative fish: In Japan, sweetfish is the most important fish.

Symptoms in humans: In light infections, epigastric pain, diarrhea and anorexia are present; in heavy infections, abdominal cramps, malabsorption, and weight loss may occur [26].

• *Heterophyes nocens* (Onji & Nishio 1916)

Type of life cycle: Type B in Fig. 10.1

Distribution: Korea [25, 26, 138, 178], Japan [26, 89, 178], China [26, 89], Taiwan [89], Lao PDR [89], Thailand [89], Hawaii [89], Balkans [89], Philippines [89], Turkey [89], and Siberia [89].

Final host: Humans and cats

Intermediate host: Flathead grey mullet (*Mugil cephalus*) and yellowish goby (*Acanthogobius flavimanus*) [25, 26, 178].

Causative fish: *Mugil* sp., and *Acanthogobius* sp. [27].

Symptoms in humans: Gastroenteritis (abdominal pain and diarrhea).

• *Haplorchis taichui* (Nishigori 1924)

Type of life cycle: Type B in Fig. 10.1

Distribution: Taiwan [26, 178], Philippines [26, 178], Bangladesh [26, 178], India [26], Palestine [26], Egypt [26], Malaysia [26], Thailand, [26, 178] Lao LDR [26, 132, 178], Vietnam [26], and China [26].

Final host: Humans, cats, dogs, foxes, and egrets [26].

Intermediate host: In Lao LDR, *Puntius brevis* (2/25, 8.0%), *Puntioplites falcifer* (4/14, 28.6%), *Mystacoleucus marginatus* (1/44, 25.0%), *Cyclocheilichthys repasson* (6/10, 60.0%), *C. armatus* (2/20, 10.0%), *Onychostoma elongatum* (4/9, 44.4%), *Hampala dispar* (4/13, 30.8%), *Labiobarbus leptocheila* (8/11, 72.7%) and *Cirrhinus molitorella* (1/8, 12.5%) [132].

Causative fish: *Cyprinus carpio, C. auratus, Zacco platypus, Pseudorasbora parva, Rodeus ocellatus, Gambusia affinis, Puntius orphoides, Puntius leicanthus, Puntius goniontus, Puntius binotatus,* and *Puntius palata* [26, 27, 166] , and Ctenopharyngodon idellus [26], and *Raiamas guttatus, Mystacoleucus, marginatus* and *Henichoryhnchus siamensis* [27].

Symptoms in humans: Gastroenteritis.

Family Echinostomatidae

About 15 species out of more than 200 species of the genus *Echinostoma* infect humans [174].

• *Echinostoma hortense* (Asada 1926)

Type of life cycle: Type B in Fig. 10.1

Distribution: Korea [25, 27, 89, 138, 178], Japan [27, 100, 178] and China [27, 178], Philippines [89], Indonesia [89], Malaysia [89] and Thailand [89].

Final host: Humans, rats, dogs, cats and mice

Intermediate host: *Misgurnus anguillicaudatus, Misgurnus mizolepis, Odontobutis obscura interrupta, Moroco oxycephalus, Coreoperca kawamebari,* and *Squalidus coreanus* [25–27, 178].

Causative fish: 69.7% of loach, *Misgurnus anguillicaudatus,* were infected in China [178].

Symptoms in humans: The major symptoms are abdominal pain, diarrhea, and fatigue.

Family Clinostomatidae

• *Clinostomum complanatum* **(Rudolphi 1814)**

Type of life cycle: Type C in Fig. 10.1

Distribution: Japan [9, 10, 58, 75], Korea [34, 127], Europe, North America (Mexico [167]) and East Asia.

Final host: Fish-eating birds and humans

Intermediate host: Deepbodies crucian carp (*Carassius cuvieri,* 18/41), silver crucian carp (*Carassius gibelio langsdorfi,* 114/378), carp (*Cyprinus carpio,* 31/110), slender bitterling (*Rhodeus lanceolatus,* 53/529), topmouth gudgeon (*Pseudorasbora parva,* 9/134), rose bitterling (*Rhodeus ocellatus,* 2/233) [10], and perch (*Lateolabrax japonicus* and *Leuciscus hakonensis*) in Japan [9]. *Acheilognathus koreensis* (19/40, 47.5%), *A. rhombea* (8/9, 88.9%), *A. yamatsutae* (20/48, 41.7%), *Carassius auratus* (77/298, 25.5%), *Cobit sinensis* (2/7, 38.6%), *Microphysogobio yaluensis* (8/9, 88.9%), *Pseudorasbora parva* (45/147, 32.6%), *Pungtungia herzi* (6/12, 50%), *Rhodeus uyekii* (7/12, 58.3%), *Squalidus chankaensis tsuchigae* (2/4, 50%), *Squalidus gracilis majimae* (4/28, 14.3%) and *Zacco temminckii* (1/61, 1.6%) in Korea [34]. *Eleotris picta, Cichlasoma trimaculatum, Dormitator latifrons, Mugil curema, H. guatemalensis* and *Galaxis maculatus* in Mexico [167].

Causative fish: Fresh-water fishes, brackish fish [167] and marine fish [75].

Symptoms in humans: In Japan, there were 19 cases with laryngitis until 2014 [10]. These symptoms are violent cough, irritable sensation (15/19), and throat pain (5/19) [58].

3.4 Nematodes

3.4.1 Family Anisakidae

There are nine species of the genus *Anisakis: Anisakis pegreffii, A. simplex* sensu stricto, *A. simplex* complex, *A. typica, A. ziphidarum, Anisakis* sp., *A. physeteris, A. brevispiculata,* and *A. paggiae* [96]. The previous 5 species are

Type I larval morphotype having a relative long ventriculus, and the last 3 species are Type II larval morphotype having a short ventriculus.

- *Anisakis simplex* **sensu stricto (Nascetti et al. 1986)**

  Anisakis Type I larvae in human infections were subdivided to *Anisakis simplex* s. s. in chub mackerel of Pacific ocean, *Anisakis pegreffii* in chub mackerel of Japan Sea, and *Anisakis typica* in hairtail (*Trichiurus* spp.) in coastal areas of Taiwan [96, 164, 165].

  Type of life cycle: Type D in Fig. 10.1
  Humans are paratenic hosts.

  Distribution: This species is widespread between 35° N and the Arctic Polar Circle: Iberian Atlantic coast (France [20] and Spain [15]), Japan Sea (Korea [145]), Sakhalin islands, Baltic Sea, West Atlantic and North-east Pacific Sea (Japan [5, 68, 89, 100, 102, 107, 148, 150, 151, 159, 164]), Bering Sea, Mauritanian coast, Azores islands [96] and Brazil [44]. In Japan, the number of patients with anisakiasis was estimated to be 7,147 per year from 2005 to 2011, and the majority of patients was infected with *Anisakis simplex* s. s. [150]. In Korea, 645 cases of anisakiasis were reported from 1971 to 2010 and later [145].

  Final host: Nine species of cetacean hosts (*Balaenoptera acutorostrata, Delphinapterus leucas, Delphinus delphis, Globicephala melaena, Lagenorhynchus albirostris, Pseudorca crassidens, Stenella coeruleoalba,* and *Phocoena phocoena*) [96].

  Intermediate host: Euphausiid

  Paratenic host: Four squid and 26 fish species: *Sepia officinalis, Todaropsis eblanae, Ommastrephes sagittatus, Illex coindettii, Hippoglossus hippoglossus, Lepidorhombus boscii, Trachurus picturatus, T. trachurus,* chub mackerel (*Scomber japonicus*), *S. scombrus, Thunnus thynnus, Spondyliosoma cantharus, Scorpaena scrofa, Eutrigla gurnardus, Belone belone, Scomberesox saurus, Boreogadus saida, Gadus morhua, Micromesistius poutassou, Theragra chalcogramma, Trisopterus luscus, Molva dypterygia, Brosme brosme, Merluccius merluccius, Lophius piscatorius, Oncorhynchus gorbuscha, O. keta, Salmo salar, Clupea harengus,* and *Conger conger* [96].

  In Japan, chum salmon (*O. keta*, 46/50, 92%), cherry salmon (*O. masou*, 24/34, 71%), pink salmon (*O. gorbuscha*, 12/34, 35%), imported chum salmon (3/3, 100%), imported king salmon (*O. tshawytscha*, 10/28, 35.7%), imported silver salmon (*O. kisutsh*, 1/13, 7.7%), and imported sockeye salmon (*O. nerka*, 10/17, 58.8%) from 1996 to 2001 [151], and Arabesque greenling (*Pleurogrammus azonus*), Alaska pollack (*Theragra chalcogramma*), chub mackerel and surf smelt (*Spirinchus lanceolatus*) from North Pacific Ocean [164], seven hairtail (*Trichiurus* spp., 10/28, 35.7%) [165], and chub mackerel (162/218, 74.4%) from 2007 to 2009 [159].

In Korea, *Todarodes pacificus* (1/5, 20%) [80], and common conger (*Conger myriaster*), croaker (*Pseudosciaena* spp.), yellowtail (*Seriola* spp.), flatfish, Pacific cod (*Gadus macrocephalus*) and so on. [145].

In Brazil, *Cynoscion guatucupa*, *Engraulis anchoita*, *Genypterus blacodes*, *Lophius gastrophysus*, *Paralichthys isosceles*, and *Trachurus lathami* [44].

In Baltic Sea, Baltic cod (*Gadus morhua*, 4.3–15.6%) [97].

In the North African coasts of the Mediterranean Sea, *Scomber scombrus* (12/282, 4.3%) and *Merluccius merluccius* (1/282, 0.4%) [48].

Causative fish: Pacific bluefin tuna and chum salmon in Japan [107], and common conger (120/311, 38.6%), croaker (36/311, 11.6%), squid (32/311, 10.3%), yellowtail (30/311, 9.7%), flatfish (17/311, 5.5%) and so on in Korea [145].

Symptoms in humans: Anisakiasis was divided into three types, depending on the migrate site of the larvae: gastric, intestinal, and extragastrointestinal anisakiasis [15, 68, 148, 150]. The patients with gastric or intestinal anisakiasis have abdominal pain, nausea, and vomiting, occasionally gastrospasm or ileus within several hours after the ingestion of a living *Anisakis* larvae in raw or undercooked fish. These symptoms are associated with the production of Th2 cytokines, mastocytosis, IgE response, and eosinophilia as allergic reactions [7, 15, 16, 149].

- *Anisakis physeteris* (Baylis 1920)

This species is Type II larval morphotype.

Type of life cycle: Type D in Fig. 10.1

Distribution: Japan [148], Central Mediterranean Sea, Mauritanian coast, West Mediterranean Sea [22], Iberian Atlantic coast, Azores island, East Mediterranean Sea [96] and Brazil [44].

Final host: Sperm whale (*Physeter macrocephalus*) and *Ommastrephes sagittatus*

Intermediate host: Euphausiid

Paratenic host: Squid and marine fish (*Aphanopus carbo*, *Xiphias gladius*, *Trachurus trachurus* and *Merluccius merluccius*). *Merluccius merluccius*, *Phycis blennoides*, *Scomber scombrus* and *Phycis phycis* from the North African coasts of Mediterranean Sea, Tunisia, and Algeria [48]. Four percent in 250 larvae from spotted markerel (*Scomber autralasicus*) from Taiwanese waters [29]. Splendid alfonsino (*Beryx spenders*, 41/44, 93.2%) from Japanese waters [108]. Two (22.2%) out of nine larvae from the frigate tuna (*Auxis thazard*) from the Atlantic Ocean off Brazil: *A. typica* (7/9, 77.8%) [67]. All of 15 larvae from opah fish (*Lampris guttatus*) in the Tyrrnenian Sea (NW Mediterranean) [22]. *Auxis thazard* and *Genypterus blacodes* in Brazil [44].

Causative fish: *Aphanopus carbo*, *Xiphias gladius*, *Trachurus trachurus* and *Merluccius merluccius* [96].

Symptoms in humans: These symptoms are similar to anisakiasis infected with *A. simplex* s. s.

- *Pseudoterranova decipiens* **(Krabbe 1978)**

Type of life cycle: Type D in Fig. 10.1

Distribution: Japan [68, 89, 100, 102, 109, 148, 154], Korea [144, 145, 179], Chile [98, 99, 161], USA [94], Canada [21, 94], Greenland [94], U.K. [94], France [20], Spain [15], Denmark [97], Baltic Sea [97], Antarctic Ocean [23], and Brazil [44]. In Korea, 24 out of 203 anisakiasis cases reported were confirmed to be due to *Pseudoterranova* type A larvae until 2015 [145].

Final host: Earless seal, Steller's sea lion (*Eumetopias jubata*) [154] and walruses [94]

Intermediate host: Euphausiid

Paratenic host: Cod (*Notothenia neglecta*) [23], Atlantic or Baltic cod (*Gadus morhua*) [21,97] and *Genypterus blacodes* [44].

In Japan, halibut, cod (Alaska Pollack), sailfin sand fish, nurt smelt and arctic smelt [109]. In Baltic Sea, Baltic cod (*Gadus morhua*) (14.8% and 45.8% in 2013 and 2014) [97]. In Chile, *Schoroeder ichthys chilensis*, *Raja chilensis*, *Genypterus chilensis*, *G. blacodes*, *G. maculatus*, *Cilus gilbertti*, *Trachurus murphyi*, *Mugil cephalus*, *Merluccius gayi*, *Macrouronus magellanicus*, *Hippoglossina montemaris*, *H. macrops*, *Paralichthys microps*, *P. adspersus*, *Aphos porosus*, *Helicolenus lengerichi* [98], and hakes (*Merluccius australis* or *M. gayi*), pomfret (*Brama australis*), Inca scad (*Trachurus murphyi*), and corvina (*Cilus gilberti*) [161]. *Sebastes inermis* in Korea [144]. In Brazil, *Genypterus blacodes* [44].

Causative fish: Cod, haddock, halibut, long rough dab, and sculpin

Symptoms in humans: These symptoms are the same as anisakiasis with *A. simplex* s. s. Elimination of the larva, a pharyngeal tickling sensation, coughing, vomiting, or a foreign body in the mouth or throat [161].

3.4.2 Tetrameridae

- *Crassicauda giliakiana* **(Skjabib and Andreeva 1934)**

Type of life cycle: Type D in Fig. 10.1

The previous study identified the type X larva as *Crassicauda giliakiana* [149].

Distribution: Japan [4, 11, 91, 100, 102, 148, 149]. In Japan, approximately 50 cases had been reported after eating raw squid or fish until 2014 [91].

Final host: Marine mammal (Baired beaked whale, *Berardius bairdii*)

Intermediate host: Small squid (Firefly squid, *Watasenia scintillans*; Japanese flying squid, *Todarodes pacificus*) and fish (Sailfin sandfish,

Arctoscopus japonicas; Pacific cod, *Gadus microcephalus*; Mackerel, *Scomber japonicas*) [4].

Causative fish: *Arctoscopus japonicas* and *Gadus microcephalus*

Symptoms in humans: Clinical symptoms are abdominal pain, nausea/ vomiting, and occasionally diarrhea, ileus, ascites and creeping eruption [11, 91].

3.4.3 Gnathostomaidae

Six species of the 13 known species within the genus *Gnathostoma* have been known to infect humans: *G. spinigerum*, *G. nipponicum*, *G. hispidum*, *G. doloresi*, *G. binucleatum* and *G. malaysiae* [19, 41, 119].

- **Gnathostoma spinigerum (Owen 1836), type species of the genus**

  Type of life cycle: Type D in Fig. 10.1

  Distribution: Southeast Asia [119] (Thailand [40, 70, 81, 89, 117, 137, 142], Lao PDR [40, 41, 70, 142], Myanmar [40, 117, 142], Indonesia [40, 142], Malaysia [40, 142], Vietnam [40, 142, 170] and Philippines [40, 142], China, Korea, Japan [89, 100, 104, 105, 112, 117, 190, 163], Bangladesh [117], and India [117, 129]), Central and South America (Mexico [89, 122], Colombia [40, 117, 142], Ecuador [89, 103, 125, 126], Brazil [44], Argentina [89], and Peru [89]), East Africa (Botswana and Zambia [41, 61]) and Spain.

  Final host: wild and domestic cats (tigers, felids) and dogs [70], rodents, domestic and wild swine, otters, raccoons, marsupials and weasels [19, 104, 105].

  First intermediate host: Copepods

  Second intermediate host: Brackish water fish (*Channa (Ophicephalus) argus* (1002/1246, 80.4%), *Anguilla japonica* (10/33, 30.3%), *Parasilurus asotus* (12/48, 25%), *O. tadianus* (16/111, 14.4%), *Misgurnus anguillicaudatus* (21/291, 7.2%), *Mogurnda obscura* (4/95, 4.2%), *Acanthogobius hasta* (1/94, 1.1%) and *Cyprinus auratus* (1/104, 0.96%) in 1962 in Japan [105]; *Channa striata* (39/55, 70.9%), *Fluta alba* (103/223, 46.2%) and *Rana rugulosa* (11/24, 45.8%) in Thailand [137]), frogs, snakes *Ptyas koros* in Lao PDR [70]), birds and humans.

  Causative fish: Northern snakehead (*Channa argus*) and snakehead murrel (*Channa striata*), walking catfish (*Clarias batrachus*), white ricefield eel (*Monopterus albus*), tank goby (*Glossogobius giurus*), *Therapon argenteus*. and *Cichla* sp. in Brazil [44].

  Symptoms in humans: After infection, the third-stage larvae migrate into the body and cause cutaneous manifestations (intermittent migratory swellings, erythema [119] or subcutaneous nodules) [61], visceral or gastrointestinal manifestations [61], cerebral manifestations (meningoencephalitis [41, 61, 81, 141]), ocular manifestations [41, 61, 117, 129, 170] with eosinophilia (20–72%) [41, 61, 81], and death [61].

- *Gnathstoma nipponicum* (Yamaguti 1941)

  Type of life cycle: Type B in Fig. 10.1

  Distribution: Japan [8, 41, 100, 102, 112, 119, 155] and Korea [41].

  Final host: Weasels [8].

  First intermediate host: Copepods

  Second intermediate host: Japanese loaches (*Misgurnus anguillicaudatus*) (7/3098, 0.23% from July 1986 to April 1987) [8].

  Causative fish: Loaches

  Symptoms in humans: Creeping eruption [8, 119, 155].

- *Gnathstoma hispidum* (Fedchenko 1972)

  Type of life cycle: Type B

  Distribution: China [32, 41], Korea [41, 143], Japan [3, 41, 100, 102, 112, 119, 156], Southeast Asia [41], Australia [41], Central America [41] and Europe [119].

  Final host: Pigs [142]

  First intermediate host: Copepods

  Second intermediate host: Loaches (*Misgurnus anguillicaudatus*), bullfrogs, snakes (*Dinodon rufozonatum rufozonatum* (18/80, 22.5%) and *Elaphe davidi* (1/50, 2.0%) in China [32] and *Agkistrondon* spp. (12/87, 13.6%) in Korea [143]).

  Causative fish: Loaches [3]

  Symptoms in humans: Fever [3] and creeping eruption [3, 119].

- *Gnathostoma doloresi* (Tubangui 1925)

  Type of life cycle: Type B

  Distribution: Japan [41, 66, 100, 102, 112, 113, 115, 116, 118, 119, 121, 153], Southeast Asia [41], Ecuador and Mexico.

  Final host: Wild boars (*Sus scrofa leucomystax*, 7/10, 70%) [113], and pig [66, 142].

  First intermediate host: Copepods

  Second intermediate host: Freshwater fishes and snakes (*Agkistrodon halys*, 6/6, 100% in Japan) [66].

  Causative fish: Brook trout (*Oncorhynchus masou* [100]) and blue-gill (*Lepomis macrochirus*) [115, 118, 121].

  Symptoms in humans: Creeping eruption [119] with eosinophilia (6–67%) [112, 115, 121].

- *Gnathostoma binucleatum* (Almeyda-Artigas 1991)

  Type of life cycle: Type B

  Distribution: Central and South America [41] (Mexico [19, 74, 117, 119] and Ecuador [19, 119]).

  Final host: Wild and domestic cats and dogs

  First intermediate host: Copepods

  Second intermediate host: Freshwater fishes (*Petenia splendida*, *Cichlasoma managuense* and *Gobiomorus dormitor* [74]).

  Causative fish: Freshwater fishes

  Symptoms in human: Erythema [117, 119].

- *Gnathostoma malaysiae* (Miyazaki and Dunn 1965)

  Type of life cycle: Type B

  Distribution: Japan [41, 120] and Southeast Asia (Myanmar [41] , Malaysia [119], and Thailand [119]).

  Final host: Rats [142]

  First intermediate host: Copepods

  Second intermediate host: Freshwater shrimp [120].

  Causative fish: Unknown.

  Symptoms in humans: Two Japanese men had eaten raw freshwater shrimp in Myanmar, and were first reported by Nomura et al. 2000 [120]. They had a creeping eruption and Quincke's edema.

3.4.4 Capillariidae

- *Capillaria philippinensis* (Velasquez, Chirwood and Salazar 1968)

  Type of life cycle: Type C in Fig. 10.1

  Distribution: Philippines [31, 36–39, 89, 128, 158], Thailand [36, 37, 39, 89, 131, 133, 134], Japan [36, 37, 63, 100, 106, 114, 134], Taiwan [18, 28, 37, 65, 87], Iran [36, 37, 62], Egypt [1, 2, 6, 17, 30, 36, 37, 45, 46, 89, 93, 177], Indonesia [30, 64, 128], Korea [64, 82] and Colombia [43].

  Final host: Fish-eating birds (*Nycticorax nycticorax*, *Bubulcus ibis* and *Ixobrychus sinensis* in Egypt [93]), human, monkey, rat and Mongolian gerbil (*Meriones unguiculatus*, experimentally [38]).

  Intermediate host: Bagsit (*Hypseleotris bipartita*) in Philippines [37]; Bagsang (*Ambassis commersoni*), birut (*Elotris melanosoma*), *Apagon* sp., and the ipon [158]; tilapia (*Oreochromis nilotica niloticus*) (experimentally) [45]; the silver carp (*Hypophthalmichthys molitrix*), the common carp (*Cyprinus carpio*), and the grass carp (*Ctenopharyngodon idella*) in Egypt [93]), crabs, snails, clams, shrimp, squid, goat and water buffalo.

  Causative fish: Bagsit [37], bagsang [158].

Symptoms in humans: Gastroenteritis (abdominal pain, gurgling stomach (borborygmus), vomiting, and chronic diarrhea [1, 2, 6, 17, 18, 28, 30, 36, 37, 39, 43, 46, 62, 64, 65, 82, 87, 93, 114, 131, 133, 158, 177]), ascites [2, 134], hypoalbuminemia [2, 17, 18, 46, 65, 93, 114, 128, 133, 134 ,177], body weight loss [1, 2, 17, 18, 28, 36, 37, 62, 64, 65, 82, 87, 93, 114, 158, 177], anemia [18, 87], eosinophilia [28, 30, 62, 87, 134], lower limb edema [6, 28, 46, 93] or death [6, 31, 36, 37, 46]. Eggs, larvae, adults, or Charcot-Leyden crystals were detected in the feces or intestinal mucosa [2, 6, 17, 18, 28, 30, 36, 37, 43, 45, 62, 64, 82, 87, 93, 114, 131, 134, 177].

4. Detection of Fish-borne Parasites

4.1 Detective methods in fish

In Japan, the inspection method for fish-borne parasites [53] and *Kudoa* spp. [90] from fish has been officially decided.

4.1.1 Dissection

Larvae of *Anisakis* spp. and plerocercoids of *Diphyllobothrium* spp. in fish are so large that they can be detected with the naked eye. *Anisakis* spp. are detected in a distinctive "watch-spring coil" shape (4–6 mm width) and in uncoiled shape (about 2 cm long) in fish.

4.1.2 Compressive method

Larvae of *Anisakis* spp., plerocercoids of *Diphyllobothrium* spp., and metacercariae of trematodes in the sliced fish meat can be detected by a stereoscopic microscope after compression using two glass plates (5 × 7 mm) [53].

4.1.3 Artificial digestion method

Larvae of *Anisakis* spp., plerocercoids of *Diphyllobothrium* spp., and metacercariae of trematodes can be recovered from fish using artificial digestion fluid: HCl 7 ml, pepsin (1:10,000)1 g, and water 1 L [53].

4.1.4 Microscopic examination

For detection of *K. septempunctata*, samples are taken from the muscle incision of olive flounder, smeared on a slide glass, stained with Loffelr's methylene blue solution and followed by a microscopic examination [90, 176]. Most of the spores of *K. septempunctata* have 5–7 shell valves and polar capsules per spore [73, 90, 95, 101, 176].

Anisakis larva type 1 (*A. simplex*), *Anisakis* larva type 2 (*A. physeteris*), *Pseudoterranova* larva and *Contraceacum* larva can be morphologically distinguished by several characters of ventriculus, intestinal cecum, and ventricular appendage.

4.1.5 Immunological method

For detection of *K. septempunctata*, ARK Checker IC *Kudoa septempunctata* M kit (Ark Resource Co., Japan) is available for immunochromatographic method.

4.1.6 DNA detection

Polymerase chain reaction (PCR) [57, 59, 90, 176], random amplified polymorphic DNA analysis [124], quantitative real-time PCR [60], loop mediated isothermal amplification (LAMP) [69, 147], nucleic acid sequence based amplification-nucleic acid chromatography [147] were used as screening method for detection of *K. septempunctata* from olive flounder. EasyAmp *Kudoa septempunctata* detection kit (Nippon Gene, co) is available for LAMP method.

LAMP targeting internal transcribed spacer 1 (ITS1-LAMP) and microsatellite (OVMS6-LAMP) was used for diagnosis of *Opisthorchis viverrini* [12].

By PCR-RFLP analysis of the ribosomal DNA internal transcribed spacer region, *Anisakis* type I larva from North Pacific Ocean were identified as *A. simplex* s. s., while those from the southern Sea of Japan were *A. pegreffii*, and those from the coast of Taiwan were *A. typica* [164, 165]. In Korea, 47 (78.3%) out of 60 *Anisakis* Type I larvae, isolated from chub mackerel (*Scomber japonicus*), ribbon fish (*Trichiurus lepturus*), and Pacific squid (*Todarodes pasificus*), were identified as *A. pegreffii* by PCR-RFLP, but one (1.7%) as *A. simplex* s. s. in *T. pacificus* [80]. In Taiwanese waters, *Anisakis* third-stage larva were identified as *A. pegreffii* (57.2%), a recombinant genotype of *A. simplex s.s.* × *A. pegreffi* (25.3%), *A. typica* (10%), *A. physeteris* (4.0%), *A. paggiae* (3.0%), and *A. brevispiculata* (0.5%) by PCR-RELP [29].

In the North African Mediterranean Sea, *Anisakis* larvae were identified as *A. pegreffii* (223/282, 79.1%), *A. typica* (26/282, 9.2%), *A. physeteris* (18/282, 6.4%) and *A. simplex* s. s. (13/282, 4.6%) by PCR-RFLP [48].

PCR was available for distinguishing six species of the genera *Gnathostoma* from fish [70]. Jongthawin et al. (2016) reported the molecular-phylogenic identification of *Gnathostoma* larvae from a snake, *Ptyas koros* in Lao PDR, and adult worms from the stomach of a dog in Thailand [70].

4.2 Detection method from human as diagnosis

4.2.1 Stool examinations

Intestinal parasites (*Diphyllobothrium* spp., *Metagonimus* spp., and *Capillaria philippinensis*) parasitize in the intestine of human, produce eggs, and discharge eggs in feces. Eggs of liver flukes are also detected in the feces. Therefore, microscopic detection after stool examination is used for conventional diagnosis. Eggs of tapeworm can be easily detected by direct smear method using feces. Eggs, larvae, and adults of *C. philippinensis* were detected in the feces: Eggs were peanut-shaped with flattened bipolar plugs and measured 40–48 × 15–23 μm [6, 18, 30, 36, 37, 43, 45, 64, 82, 114, 131, 134, 177].

4.2.2 Artificial digestion method

Taniguchi et al. (1999) identified the causative parasite in the skin as a third stage larva of *G. hispidum* with the artificial digestion method of biopsied skin specimen [156].

4.2.3 Immunological diagnosis

Immunodiagnosis is available, due to the increase of specific antibody in the chronic infection.

The RAST may be useful for immunodiagnosis (specific IgE) of anisakiasis using *Anisakis* larvae antigen in spite of having cross-reaction with previous *Ascaris* or *Toxocara* infections [135].

Tada et al. (1987) reported immunodiagnosis of gnathostomiasis by ELISA and double diffusion (Ouchterlony's method) using *G. doloresi* antigen and detected specific IgG antibodies in 22 (73.3%) out of 30 gnathostomiasis cases [153]. An interdermal test had also been useful for gnathostomiasis [3, 163].

Abdel-Rahman et al. (2005) reported immunodiagnosis of capillariasis by western blot using coproantigen and egg antigen [1].

4.2.4 DNA detection

After treatment of diphyllobothriasis, the proglottids of *D. nihonkaiense* and *D. latum* can be distinguished by PCR using DNA extraction from the proglottids [14, 56, 76, 110, 168, 180].

ITS1- and OVMS6-LAMP for specific diagnosis of *Opisthorchis vierrini* was used for distinguishing from *C. sinensis*, *O. felineus*, *Centrocestus caninus*, *Haplorchis taichui*, *Fasciola gigantica*, and *Haplorchoodes* sp. [12].

In Egypt, the prevalence of *C. philippinensis* in diarrheic patients was 11.6% (14/121) by copro-nested PCR targeting the small subunit ribosomal DNA gene [6].

In Japan, 84 out of 100 *Anisakis* type 1 larvae recovered from 85 patients are identified as *A. simplex* s. s. by PCR-RFLP [164].

4.2.5 Endoscopy

Since 1968 until today, gastric anisakiasis [5, 111, 135] or terranovasis [109] can be diagnosed and treated by endoscopy. Clinostomiasis can also be diagnosed and treated by laryngeal endoscopy [58, 75, 127].

4.2.6 X-ray, CT or MRI

In x-rays, pseudotumor formation was observed in 36% (64/178) of the patients with anisakiasis, and swelling of gastric folds was present in 35.4% (63/178) of the cases [135].

Roentgenologic examinations of the small intestine performed on 14 patients with capillariasis revealed a consistent malabsorption pattern characterized by scattering of barium and mucosal fold alteration [128].

Computer tomography (CT) or magnetic resonance imaging (MRI) have been used for diagnosis of clonorchiasis [88].

4.2.7 Biopsy or autopsy

Intestinal capillariasis can be diagnosed by worms or eggs in the biopsied specimen of intestinal mucosa [18, 28, 36, 37, 64, 82, 87, 106, 114].

For diagnosis of human gnathostomiasis, skin biopsies were effectively performed: *G. hispidum* larva in cross section had an intestinal canal which consisted of 25–35 cells, and a large nucleus was observed at the center of each intestinal cell [3]; 3–7 nuclei per intestinal cell in *G. spinigerum*, mainly 2 nuclei in *G. doloresi*, and 0–4 nuclei in *G. nipponicum* [8, 116, 121, 155].

4.2.8 Treatment

Worms, collected after treatment by albendazole [17, 30, 46, 64, 82, 87], mebendazole [36, 37, 43, 87, 158] or thiabendazole [37, 62, 114, 131, 158], were identified as *C. pilippinensis*.

5. Control or Prevention of Fish-borne Parasites

5.1 Control: Break of the life cycles

The way of control fish-borne parasitic diseases must be to break the life cycle of the parasite. Any point of the life cycle can be theoretically attacked.

In practice, there are three principal points: (1) measures towards sources of pathogen, (2) measures towards infective routes, and (3) measures towards humans as hosts.

5.1.1 Measures towards sources of pathogen

First of all, it is necessary to diagnose fish-borne parasitic diseases earlier and more correctly.

Next, the patients must be treated as soon as possible. Reservoir hosts as pets and wild animals must be also treated or controlled in fish-borne parasitic diseases. Moreover, the patient's feces have to be treated properly with sewage for preventing water contamination.

5.1.2 Measures towards infective route

To avoid food poisoning with *K. septempunctata*, the Fishery Agency must be taking measures towards *Kudoa*-free flatfish aquaculture, such as, elimination of *Kudoa*-carrying flatfish fry, cleaning of aquaculture environment of flatfish, and pre-market monitoring of aquacultured flatfish [101]. Shirakashi et al. (2014) reported that the ultraviolet (UV) treatment of culture seawater prevented infections with *Kudoa yasunagai* and *K. amamiensis* in *Seriola* fish, respectively [141]. The culture seawater was treated with UV (manufacturer's specification; 253.7 nm, 68 mJ/cm^2), and water flow was maintained at 200 L/h for all tanks.

Recently, *Anisakis*-free mackerels were successfully cultivated and grown using deep-seawater in Japan (personal communication).

5.1.3 Measures towards humans as hosts

At the elementary schools or posters, the life cycles of fish-bore parasitic diseases must be taught to the public. The knowledge about the source of infection can prevent these diseases.

Campaigning can prevent or control the fish-borne parasitic diseases: Don't eat raw or undercooked fish!

5.2 Prevention

Consumers should continue to follow four food safety precautions: clean, separate, cook, and chill, with hazard assessment at critical control points (HACCP) [52]. In fish-borne parasites, however, separation and cooking are effective means to prevent infection of fish-borne parasites. So, the best ways to prevent these diseases are to avoid eating raw or undercooked fish [40, 142].

5.2.1 Heating

Food poisoning with *K. septempunctata* can be prevented by heating at 90°C for 5 min [101].

Anisakis larvae can be killed by temperatures higher than 60°C for at least 1 min when cooking or smoking, or more than 74°C for at least 15 seconds when microwave cooking [16].

5.2.2 Freezing

Food poisoning with *K. septempunctata* can be prevented by freezing at –16 to –20°C for 4 hours [101].

It is necessary to be frozen for 24 hours at –20°C (–4° F) in order to kill the metacercariae of *C. complanatum* [58].

Anisakis larvae have been found to survive at –20°C for short periods. Therefore, freezing, at –20°C or below for 7 days [99], or at –35°C or below until solid, and storing at –35°C or below for 15 hours, or at –35°C or below until solid, and storing at –20°C or below for 24 hours, is recommended by the FDA [14, 44, 52, 150].

5.2.3 Irradiation

Food irradiation may offer an alternative control method for the major food-borne parasites [24, 49–51, 85, 86]. A minimum dose of 0.1 kGy is effective without changing physiochemical properties of the fish flesh to control infection by liver flukes, *C. sinensis*, and *O. viverrini* [49, 86]. Irradiation of the sweetfish by 200 Gy is effective to control infectivity of *M. yokogawai* metacercariae in rats [24].

6. Conclusion

Fishes have become intermediate or paratenic hosts in the life-cycle of parasites. Therefore, consumers can prevent food-borne parasitic diseases by not eating raw or undercooked fish.

References

1   Abdel-Rahman, S.M., Moneib, M.E.M., Shahin, M.S. and Aziz, L.A.A. 2005. Immunodiagnosis of *Capillaria philippinenesis* by western blot using coproantigen and egg antigen. El-Minia Med. Bull. 16: 9–17.
2   Ahmed, L., el-Dib, N.A., el-Boraey, Y. and Ibrahim, M. 1999. *Capillaria philippinensis*: an emerging parasite causing severe diarrhea in Egypt. J. Egypt Soc. Parasitol. 29: 483–493.
3   Akahane, H., Sano, M. and Kobayashi, M. 1998. Three cases of human gnathostomiasis caused by *Gnathostoma hispidum*, with particular reference to the identification of parasitic larvae. Trop. Med. 29: 611–614.

4   Akao, N., Ando, K., Nakamura, F. and Kawanaka, M. 2004. Annual prevalence of Spiruroid larva migrans caused by eating raw firely squid, *Watasenia scintillans*, 1995–2003. IASR. 25: 116–117. (In Japanese).

5   Akasaka, Y., Kizu, M., Aoike, A. and Kawai, K. 1979. Endoscopic management of acute gastric anisakiasis. Endoscopy 2: 158–162.

6   Ali, M.I., El-Badry, A.A., Rubio, J.M., Ghieth, M.A. and El-Dib, N.A. 2016. Prevalence of *Capillaria philippinensis* in diarrheic patients using the small subunit ribosomal DNA (ssur DNA) gene. Sci. Parasitol. 17: 93–100.

7   Alonso, A., Moreno-Ancillo, A., Daschner, A. and Lopez-Serrano, M.C. 1999. Dietary assessment in five cases of allergic reactions due to gastroallergic anisakiasis. Allergy 54: 517–520.

8   Ando, K., Tanaka, H., Taniguchi, Y., Shimizu, M. and Kondo, K. 1988. Two human cases of gnathostomiasis and discovery of a second intermediate host of *Gnathostoma nipponicum* in Japan. J. Parasitol. 74: 623–627.

9   Aohagi, Y., Shibahara, T. and Kagota, M. 1995. Metacercariae of *Clinostomum complanatum* found from new fish hosts, *Lateolabrax japonicus* and *Leuciscus hakonensis*. Jpn. J. Parasitol. 44: 340–342.

10 Aohagi, Y., Shibahara, T., Machida, N., Yamaga, Y. and Kagota, K. 1992. *Clinostomum complanatum* (Trematoda: Clinostomatidae) in five new fish hosts in Japan. J. Wildflife Dis. 28: 467–469.

11 Aoyama, S., Hinoue, Y., Takahashi, H., Yoshimitsu, Y., Kusajima, Y., Hirono, T., Takayanagi, N., Akao, N. and Kondo, K. 1996. Clinical study of ten cases with acute abdomen after eating raw firefly squid (*Watasenia scintillans*, Hotaruika), which are probably due to type X larvae of the suborder spirurina. Niponsyoukakizasshi 93: 312–321. (In Japanese).

12 Arimatsu, Y., Kaewkes, S., Laha, T. and Sripa, B. 2015. Specific diagnosis of *Opisthorchis viverrini* using loop mediated isothermal amplification (LAMP) targeting parasite microsatellites. Acta Trop. 141: 368–371.

13 Arizono, N., Fukumoto, S., Tademoto, S., Yamada, M., Uchikawa, R., Tegoshi, T. and Kuramochi, T. 2008. Diplogonoporasis in Japan: Genetic analyses of five clinical isolates. Parasitol. Int. 57: 212–216.

14 Arisono, N., Yamada, M., Nakamura-Uchiyama, F. and Ohnishi, K. 2009. Diphyllobothriasis associated with eating raw Pacific salmon. Emerg. Infect. Dis. 15: 866–870.

15 Audicana, M.T., Del Pozo, M.D., Iglesias, R. and Ubeira, F.M. 2003. *Anisakis simplex* and *Pseudoterranova decipiens*. Food, Science and Technology. New York Dekker, pp. 613–636.

16 Audicana, M.T. and Kennedy, M.W. 2008. *Anisakis simplex*: from obscure infectious worm to inducer of immune hypersensitivity. Clinic. Microbiol. Rev. 21: 360–379.

17 Austin, D.N., Mikhail, M.G., Chiodini, P.L. and Murray-Lyon, I.M. 1999. Intestinal capillariasis acquired in Egypt. Eur. J. Gastroenterol. Hepathol. 11: 935–936.

18 Bair, M.-J., Hwang, K.-P., Wang, T.-E., Liou, T.-C., Lin, S.-C., Kao, C.-R., Wang, T.-Y. and Pang, K.-K. 2004. Clinical features of human intestinal capillariasis in Taiwan. World Gastroenterol. 10: 2391–2393.

19 Bertoni-Ruiz, F., Lamothe-Argumedo, M.R., Garcia-Prieto, L., Osorio-Sarabia, D. and Leon-Regagnon, V. 2011. Systematics of the genus *Gnathostoma* (Nematoda: Gnathostomatidae) in the Americas. Rev. Mex. Biodiv. 82: 453–464.

20 Bouree, P., Paugam, A. and Petithory, J.-C. 1995. Anisakidosis: Report of 25 cases and review of the literature. Comp. Immun. Microbiol. Infect. Dis. 18: 75–84.

21 Brattey, J., Bishop, C.A. and Meyers, R.A. 1990. Geographic distribution and abundance of *Pseudoterranova decipiens* (Nematoda: Ascaridoidea) in the musculature of Atlantic cod, *Gadus morhua*, from Newfoundland and Labrador. *In*: Bowen, W.D. (ed.). Population

biology of sealworm (*Pseudoterranova decipiens*) in relation to its intermediate and seal hosts. Can. Bull. Fish. Aquat. Sci. 222: 67–82.

22 Cavallero, S., Ligas, A., Bruschi, F. and D'Amelio, S. 2012. Molecular identification of *Anisakis* spp. from fishes collected in the Tyrrhenian Sea (NW Mediterrannean). Vet. Parasitol. 187: 563–566.

23 Chai, J.-Y., Guk, S.-M., Sung, J.-J., Kim, H.-C. and Park, Y.-M. 1995. Recovery of *Pseudoterranova decipiens* (Anisakidae) larvae from codfish of the Antarctic Ocean. Korean J. Parasitol. 33: 231–234.

24 Chai, J.-Y., Kim, S.-J., Kook, J. and Lee, S.-H. 1995. Effects of gamma-irradiation on the survival and development of *Metagonimus yokogawai* metacercariae in rats. Korean J. Parasitol. 33: 297–303.

25 Chai, J.-Y. and Lee, S.-H. 2002. Food-borne intestinal trematode infections in the Republic of Korea. Parasitol. Int. 51: 129–154.

26 Chai, J.-Y., Murrell, K.D. and Lymbery, A.J. 2005. Fish-borne parasitic zoonoses: Status and issues. Int. J. Parasitol. 35: 1233–1254.

27 Chai, J.-Y., Shin, E.-H., Lee, S.-H. and Rim, H.-J. 2009. Foodborne intestinal flukes in Southeast Asia. Korean J. Parasitol. 47: S69–S102.

28 Chen, C.-Y., Hsieh, W.-C., Lin, J.-T. and Liu, M.-C. 1989. Intestinal capillariasis: report of a case. J. Formosa Med. Assoc. 88: 617–620.

29 Chen, H.-Y. and Shih, H.-H. 2015. Occurrence and prevalence of fish-borne *Anisakis* larvae in the spotted mackerel *Scomber australasicus* from Taiwanese waters. Acta Tropica 145: 61–67.

30 Chichino, G., Bernuzzi, A.M., Bruno, A., Cevini, C., Atzori, C., Malfitano, A. and Scaglia, M. 1992. Intestinal capillariasis (*Capillaria philippinensis*) acquired in Indonesia: A case report. Am. J. Trop. Med. Hyg. 47: 10–12.

31 Chitwood, M.B., Valesquez, C. and Salazar, N.G. 1968. *Capillaria philippinensis* sp. n. (Nematode: Trichinellida), from the intestine of man in the Philippines. J. Parasitol. 54: 368–371.

32 Cho, S.-H., Kim, T.-S., Kong, Y., Na, B.-K. and Sohn, W.-M. 2007. Larval *Gnathostoma hispidum* detected in the red banded odd-tooth snake, *Dinodon rufozonatum rufozonatum*, from China. Korean J. Parasitol. 45: 191–198.

33 Choi, H.-J., Lee, J. and Yang, H.-J. 2012. Four human cases of *Diphyllobothrium latum* infection. Korean J. Parasitol. 50: 143–146.

34 Chung, D.-I., Kong, H.-H. and Moon, C.-H. 1995. Demonstration of the second intermediate hosts of *Clinostomum complanatum* in Korea. Korean J. Parasitol. 33: 305–312.

35 Chung, D.-I., Kong, H.-H., Moon, C.-H., Choi, D.-W., Kim, T.-H., Lee, D.-W. and Park, J.-J. 1995. The first human case of *Diplogonoporus balaenopterae* (Cestoda: Diphyllobothriidae) infection in Korea. Korean J. Parasitol. 33: 225–230.

36 Cross, J.H. 1990. Intestinal capillariasis. Parasitol. Today 6: 26–28.

37 Cross, J.H. 1992. Intestinal capillariasis. Clinic. Microbiol. Rev. 5: 120–129.

38 Cross, J.H., Banzon, T. and Singson, C. 1978. Further studies on *Capillaria philippinensis*: Development of the parasite in the Mongoloan gerbil. J. Parasitol. 64: 208–213.

39 Cross, J.H. and Bhaibulaya, M. 1983. Intestinal capillariasis in the Philippines and Thailand. pp. 103–136. *In*: Croll, N. and Cross, J.H. (eds.). Human Ecology and Infectious Disease. Academic Press Inc., New York.

40 Daengsvang, S. 1981. Gnathostomiasis in Southeast Asia. Southeast Asian J. Trop. Med. Public Health 12: 319–332.

41 Diaz, J.H. 2015. Gnathostomiasis: An emerging infection of raw fish consumers in *Gnathostoma* nematode-endemic and nonendemic countries. J. Travel Med. 22: 318–324.

42 Dorny, P., Praet, N., Deckers, N. and Gabriel, S. 2009. Emerging food-borne parasites. Vet. Parasitiol. 163: 196–206.

43   Dronda, F., Chaves, F., Sanz, A. and Lopez-Velez, R. 1993. Human intestinal capillariasis in an area of nonendemicity: Case report and review. Clin. Infect. Dis. 17: 909–912.

44   Eiras, J.C., Pavanelli, G.C., Takemoto, R.M., Yamaguchi, M.U., Karkling, L.C. and Nawa, Y. 2016. Potential risk of fish-borne nematode infections in humans in Brazil--Current status based on a literature review. Food and Waterborne Parasitol. 5: 1–6.

45   El-Dib, N.A., Ahmed, J.A., El-Arousy, M., Mahmoud, M.A. and Garo, K. 1999. Parasitological aspects of *Capillaria philippinensis* recovered from Egyptian patients. J. Egypt Soc. Parasitol. 29: 139–147.

46   El-Karaksy, H., El-Shabrawi, M., Mohsen, N., Kotb, M., El-Koofy, N. and El-Deeb, N. 2004. *Capillaria philippinensis*: A cause of fatal diarrhea in one of two infected Egyptian sisters. J. Trop. Pediatr. 50: 57–60.

47   El-Matbouli, M. and Hoffmann, R.W. 1998. Light and electron microscopic studies on the chronological development of *Myxobolus cerebralis* to the actinosporean stage in *Tubifex tubifex*. Int. J. Parasitol. 28: 195–217.

48   Farjallah, S., Slimane, B.B., Busi, M., Paggi, L., Amor, N., Belel, H., Said, K. and D'Amelio, S. 2008. Occurrence and molecular identification of *Anisakis* spp. from the North African coasts of Mediterranean Sea. Parasitol. Res. 102: 371–379.

49   Farkas, J. 1998. Irradiation as a method for decontaminating food: A review. Int. J. Food Microbiol. 44: 189–204.

50   FDA. 2000a. Food irradiation: A safe measure, pp. 6.

51   FDA. 2000b. Food irradiation: Health physics fact sheet, pp. 2.

52   FDA. 2011. Fish and Fishery products hazards and controls guidance, 4th ed-Chapter 5: Parasites. pp. 91–98.

53   Food hygiene inspection guideline for microorganism: Detailed examination method for parasites. 2004. Japanese food hygiene association (ed). Tokyo. pp. 517–564. (In Japanese).

54   Fried, B., Graczyk, T.K. and Tamang, L. 2004. Food-borne intestinal trematodiases in humans. Parasitol. Res. 93: 159–170.

55   Furst, T., Keiser, J. and Utzinger, J. 2012. Global burden of human food-borne trematodiasis: a systemic review and meta-analysis. Lancet Infect. Dis. 12: 210–221.

56   Go, Y.B., Lee, E.H., Cho, J., Choi, S. and Chai, J.-Y. 2015. *Diphyllobothrium nihonkaiense* infections in a family. Korean J. Parasitol. 53: 109–112.

57   Grabner, D.S., Yokoyama, H., Shirakashi, S. and Kinami, R. 2012. Diagnostic PCR assays to detect and differentiate *Kudoa septempunctata*, *K. thyrsites* and *K. lateolabracis* (Myxozoa, Multivalvulida) in muscle tissue of olive flounder (*Paralichthys olivaceus*). Aquaculture 338-341: 36–40.

58   Hara, H., Miyauchi, Y., Tahara, S. and Yamashita, H. 2014. Human laryngitis caused by *Clinostomum complanatum*. Nagoya J. Med. Sci. 76: 181–185.

59   Harada, T., Kawai, T., Jinnai, M., Ohnishi, T., Sugita-Konishi, Y. and Kumeda, Y. 2012. Detection of *Kudoa septempunctata* 18S ribosomal DNA in patient fecal samples from novel food-borne outbreaks caused by consumption of raw olive flounder (*Paralichthys olivaceus*). J. Clinic. Microbiol. 50: 2964–2968.

60   Harada, T., Kawai, T., Sato, H., Yokoyama, H. and Kumeda, Y. 2012. Development of a quantitive polymerase chain reaction assay for detection of *Kudoa septempunctata* in olive flounder (*Paralichthys olivaceus*). Int. J. Food Microbiol. 156: 161–167.

61   Herman, J.S. and Chiodini, P.L. 2009. Gnathostomiasis, another emerging imported disease. Clinic. Microbiol. Rev. 22: 484–492.

62   Hoghooghi-Rad, N., Maraghi, S. and Narenj-Zadeh, A. 1987. *Capillaria philippinensis* infection in Khoozestan Province, Iran: Case report. Am. J. Trop. Med. Hyg. 37: 135–137.

63   Hong, S.T. and Cross, J.H. 2005. *Capillaria philippinensis* infection in Asia. *In*: Arisono, N. et al. (eds.). Asian Parasitol. Vol. 1 Food-borne helminthiasis in Asia. Chiba: Federation of Asian Parasitologists, pp. 225–229.

64 Hong, S.-T., Kim, Y.-T., Choe, G., Min, Y.I., Cho, S.H., Kim, J.K., Kook, J., Chai, J.-Y. and Lee, S.-H. 1994. Two cases of intestinal capillariasis in Kora. Korean J. Parasitol. 32: 43–48.

65 Hwang, K.P. 1998. Human intestinal capillariasis (*Capillaria philippinensis*) in Taiwan. Zhonghua Min Guo Xiao Er Ke Yi Xue Hui Za Zhi. 39: 82–85.

66 Imai, J., Asada, Y., Horii, Y. and Nawa, Y. 1988. *Gnathostoma doloresi* larvae found in snakes, *Agkistrodon halys*, captured in the central part of Miyazaki Prefecture. Jpn. J. Parasitol. 37: 444–450.

67 Iniguez, A.M., Santos, C.P. and Vicente, A.C.P. 2009. Genetic characterization of *Anisakis typica* and *Anisakis physeteris* from marine mammals and fish from the Atlantic Ocean off Brazil. Vet. Parasitol. 165: 350–356.

68 Ishikura, H., Kikuchi, K., Nagasawa, K., Ooiwa, T., Takamiya, H., Sato, N. and Sugane, K. 1993. Anisakidae and Anisakidosis. pp. 43–102. *In*: Sun, T. (ed.). Progress in Clinical Parasitology. Vol. III. New York, Springer-Verlg.

69 Jeon, C.-H., Wi, S., Song, J.-Y., Choi, H.-S. and Kim, J.-H. 2014. Development of loop-mediated isothermal amplification method for detection of *Kudoa septempunctata* (Myxozoa: Multivalvulida) in olive flounder (*Paralichthys olivaceus*). Parasitol. Res. 113: 1759–1767.

70 Jongthawin, J., Intapan, P.M., Sanpool, O., Janwan, P., Sadaow, L., Thanchomnang, T., Laymanivong, S. and Maleewong, W. 2016. Molecular phylogenetic confirmation of *Gnathostoma spinigerum* Owen, 1836 (Nematoda: Gnathostomatidae) in Laos and Thailand. Folia Parasitol. 63: 002.

71 Kaewpitoon, N., Kaewpitoon, S.J. and Pengsaa, P. 2008. Opisthorchiasis in Thailand: Review and current status. World J. Gastroenterol. 14: 2297–2302.

72 Kaewpitoon, N., Kaewpitoon, S.J., Pengsaa, P. and Sripa, B. 2008. *Opisthorchis viverrini*: The carcinogenic human liver fluke. World J. Gastroenterol. 14: 666–674.

73 Kawai, T., Sekizuka, T., Yahata, Y., Kuroda, M., Kumeda, Y., Iijima, Y., Kamata, Y., Sugita-Konishi, Y. and Ohnishi, T. 2012. Identification of *Kudoa septempunctata* as the causative agent of novel food poisoning outbreaks in Japan by consumption of *Paralichthys olivaceus* in raw fish. Clinic. Infect. Dis. 54: 1046–1052.

74 Kifune, T., Lamothe-Argumedo, R., Garcia-Prieto, L., Oceguera-Figueroa, A. and Leon-Regagnon, V. 2004. *Gnathostoma binuceatum* (Spirurida: Gnathostomatidae) en peces dulceacuicolas de Tabasco, Mexico. Rev. Biol. Trop. 52: 371–376. (In Spanish).

75 Kifune, T., Ogata, M. and Miyahara, M. 2000. The first case of human infection with *Clinostomum* (Trematoda: Clinostomidae) in Yamaguchi Prefecture, Japan. Med. Bull. Fukuoka Univ. 27: 101–105.

76 Kim, H.-J., Eom, K.S. and Seo, M. 2014. Three cases of *Diphyllobothrium nihonkaiense* infection in Korea. Korean J. Parasitol. 52: 673–676.

77 Kim, T.-S., Pak, J.H., Kim, J.-B. and Bahk, Y.Y. 2016. *Clonorchis sinensis*, an oriental liver fluke, as a human biological agent of cholangiocarcinoma: a brief review. BMB Rep. 49: 590–597.

78 Kino, H., Hori, W., Kobayashi, H., Nakamura, N. and Nagasawa, K. 2002. A mass occurrence of human infection with *Diplogonoporus grandis* (Cestoda: Diphyllobothriidae) in Shizuka Prefecture, central Japan. Parasitol. Int. 51: 73–79.

79 Lee, S.-H., Chai, J.-Y., Seo, M., Kook, J., Huh, S., Ryang, Y.-S. and Ahn, Y.-K. 1994. Two rare cases of *Diphyllobothrium latum* parvum type infection in Korea. Korean J. Parasitol. 32: 117–120.

80 Lee, M.H., Cheon, D.-S. and Choi, C. 2009. Molecular genotyping of *Anisakis* species from Korean sea fish by polymerase chain reaction-restriction fragment length polymorphism (PCR-RFLP). Food Control. 20: 623–626.

81 Lee, S.-H., Hong, S.-T. and Chai, J.-Y. 1988. Description of a male *Gnathostoma spinigerum* recovered from a Thai woman with meningoencephalitis. Korean J. Parasitol. 26: 33–38.

82   Lee, S.-H., Hong, S.-T., Chai, J.-Y., Kim, W.H., Kim, Y.T., Song, I.S., Kim, S.W., Choi, B.I. and Cross, J.H. 1993. A case of intestinal capillariasis in the Republic of Korea. Am. J. Trop. Med. Hyg. 48: 542–546.

83   Lee, E.B., Song, J.H., Park, N.S., Kang, B.K., Lee, H.S., Han, Y.J., Kim, H.-J., Shin, E.-H. and Chai, J.-Y. 2007. A case of *Diphyllobothrium latum* infection with a brief review of diphyllobothriasis in the Republic of Korea. Korean J. Parasitol. 45: 219–223.

84   Lim, J.H. 2011. Liver flukes: the malady neglected. Korean J. Radiol. 12: 269–279.

85   Loaharanu, P. and Murrell, D. 1994. A role for irradiation in the control of foodborne parasites. Trend Food Sci. Tech. 5: 190–195.

86   Loaharanu, P. and Sommani, S. 1996. Preliminary estimates of economic impact of liver fluke infection in Thailand and the feasibility of irradiation as a control measure. Southeast Asian. J. Trop. Med. Public Health 22: S384–S390.

87   Lu, L.-H., Lin, M.-R., Choi, W.-M., Hwang, K.-P., Hsu, Y.-H., Bair, M.-J., Liu, J.-D., Wang, T.-E., Liu, T.-P. and Chung, W.-C. 2006. Human intestinal capillariasis (*Capillaria philippinensis*) in Taiwan. Am. J. Trop. Med. Hyg. 74: 810–813.

88   Lun, Z.-R., Gasser, R.B., Lai, D.-H., Li, A.-X., Zhu, X.-Q., Yu, X.-B. and Fang, Y.-Y. 2005. Clonorchiasis: a key foodborne zoonosis in China. Lancet Infect. Dis. 5: 31–41.

89   Macpherson, C.N.L. 2005. Human behaviour and the epidemiology of parasitic zoonoses. Int. J. Parasitol. 35: 1319–1331.

90   MAFF (Ministry of Agriculture, Forestry and Fisheries). 2016. Detection method of *Kudoa septempunctata* in olive flounder, pp. 45. (In Japanese).

91   Makino, T., Mori, N., Sugiyama, H., Mizawa, M., Seki, Y., Kagoyama, K. and Shimizu, T. 2014. Creeping eruption due to Spirurina type X larva. Lancet 384: 2082.

92   Manivong, K., Komalamisra, C., Waikagul, J. and Radomyos, P. 2009. *Opisthorchis viverrini* metacercariae in cyprinoid fish from three rivers in Khammouane Province, Lao PDR. J. Trop. Med. Parasitol. 32: 23–29.

93   Mansour, N.S., Anis, M.H. and Mikhail, E.M. 1990. Human intestinal capillariasis in Egypt. Trans. R. Soc. Med. Hyg. 84: 114.

94   Margolis, L. 1977. Public health aspects of "Codeworm" infection: A review. J. Fisheries Res. Board Canada 34: 887–898.

95   Matsukane, Y., Sato, H., Tanaka, S., Kamata, Y. and Sugita-Konishi, Y. 2010. *Kudoa septempunctata* n. sp. (Myxosporea: Multivalvulida) from an aquacultured olive flounder (*Paralichthys olivaceus*) imported from Korea. Parasitol. Res. 107: 865–872.

96   Mattiucci, S. and Nascetti, G. 2006. Molecular systematics, phylogeny and ecology of Anisakid nematodes of the genus *Anisakis* Dujardin, 1845: An update. Parasite 13: 99–113.

97   Mehrdana, F., Bahlool, Q.Z.M., Skov, J., Marana, M.H., Sindberg, D., Mundeling, M., Overgaard, B.C., Korbut, R., Strom, S.B., Kania, P.W. and Buchmann, K. 2014. Occrrence of zoonotic nematodes *Pseudoterranova decipiens*, *Contracaecum osculatum* and *Anisakis simplex* in cod (*Gadus morhua*) from the Baltic Sea. Vet. Parasitol. 205: 581–587.

98   Mercado, R., Torres, P. and Maira, J. 1997. Human case of gastric infection by a fourth larval stage of *Pseudoterranova decipiens* (Nematoda, Anisakidae). Rev. Saude Publica 31: 178–181.

99   Mercado, R., Torres, P., Munoz, V. and Apt, W. 2001. Human infection by *Pseudoterranova decipiens* (Nematoda, Anisakidae) in Chile: Report of seven cases. Mem Inst. Oswaldo Cruz, Rio de Janeiro 96: 653–655.

100 MHLW (Ministry of Health, Labour and Welfare in Japan). 2004. Food-borne helminthiasis as emerging in Japan. IASR 25: 114–115.

101 MHLW. 2012. *Kudoa* and *Sarcocystis* food poisoning in Japan. IASR 33: 147–148.

102 MHLW. 2017. Food-borne helminthiases in Japan. IASR 38: 69–70.

103 Mimori, T., Tada, I., Kawabata, M., Ollague, L.W., Calero, G.H. and De Chong, Y.F. 1987. Immunodiagnosis of human gnathostomiasis in Ecuador by skin test and ELISA using *Gnathostoma doloresi* antigen. Jpn. J. Trop. Med. Hyg. 15: 191–196.

104 Miyazaki, I. 1960. On the genus *Gnathostoma* and human gnathostomiasis, with special reference to Japan. Exp. Parasitol. 9: 338–370.

105 Miyazaki, I. 1966. *Gnathostoma* and gnathostomiasis in Japan. pp. 529–586. *In*: Morishita, K., Komiya, Y. and Matsubayashi, H. (eds.). Progress in Medical Parasitology in Japan. Vol. 3. Meguro Parasitological Museum, Tokyo, Japan.

106 Mukai, T., Shimizu, S., Yamamoto, M., Horiuchi, I., Kishimoto, S., Kajiyama, G., Kawamura, H., Tsuji, M. and Miyoshi, K. 1983. A case of intestinal capillariasis. Jpn. Arch. Int. Med. 30: 163–169. (In Japanese).

107 Murata, R. and Suzuki, J. 2008. The prevalence of *Anisakis* larvae in bluefin tuna (*Turnus orientalis*) as food poisoning. Monthly Epidemiological Records, Tokyo Metropolitan Institute of Public Health 29: 1–2. (In Japanese).

108 Murata, R., Suzuki, J., Sadamasu, K. and Kai, A. 2011. Morphological and molecular characterization of *Anisakis* larvae (Nematoda: Anisakidae) in *Beryx splendens* from Japanese waters. Parasitol. Int. 60: 193–198.

109 Nagano, K. 1989. Gastric terranovasis. pp. 133–140. *In*: Ishikawa, H. and Namiki, M. (eds.). Gastric Anisakiasis in Japan. Epidemiology, Diagnosis, Treatment. Springer-Verlag, Tokyo, Japan.

110 Nakao, M., Abmed, D., Yamasaki, H. and Ito, A. 2007. Mitochondrial genomes of the human broad tapeworms *Diphyllobothrium latum* and *Diphyllobothrium nihonkaiense* (Cestoda: Dophyllobothriidae). Parasitol. Res. 101: 233–236.

111 Namiki, M. 1968. From the aspect of endoscopy. Gastroentero. Jpn. 3: 398–399.

112 Nawa, Y. 1991. Historical review and current status of gnathostomiasis in Asia. Southeast Asian J. Trop. Med. Public Health 22: 217–219.

113 Nawa, Y. and Imai, J. 1989. Current status of *Gnathostoma doloresi* infection in wild boars captured in Miyazaki Prefecture, Japan. Jpn. J. Parasitol. 38: 385–387.

114 Nawa, Y., Imai, J., Abe, T., Kisanuki, H. and Tsuda, K. 1988. A case report of intestinal capillariasis—The second case found in Japan. Jpn. J. Parasitol. 37: 113–118.

115 Nawa, Y., Imai, J., Horii, Y., Ogata, K. and Otsuka, K. 1993. *Gnathostoma doloresi* larvae found in *Lepomis macrochirus* Rafinesque, a freshwater fish (common name: blue-gill), captured in the central part of Miyazaki Prefecture, Japan. Jpn. J. Parasitol. 42: 40–43.

116 Nawa, Y., Imai, J., Ogata, K. and Otsuka, K. 1989. The first record of a confirmed human case of *Gnathostoma doloresi* infection. J. Parasitol. 75: 166–169.

117 Nawa, Y., Katchanov, J., Yoshikawa, M., Rojekittkhun, W., Dekumyoy, P., Kusolusuk, T. and Wattanakulpanich, D. 2010. Ocular gnathostomiasis: A comprehensive review. J. Trop. Med. Parasitol. 33: 77–86.

118 Nawa, Y., Maruyama, H. and Ogata, K. 1997. Current status of gnathostomiasis doloresi in Miyazaki Prefecture, Japan. Trop. Med. 28: 11–13.

119 Nawa, Y. and Nakamura-Uchiyama, F. 2004. An overview of gnathostomiasis in the world. Southeast Asian J. Trop. Med. Public Health 35: 87–91.

120 Nomura, Y., Nagakura, K., Kagei, N., Tsutsumi, Y., Araki, K. and Sugawara, M. 2000. Gnathostomiasis possibly caused by *Gnathostoma malaysiae*. Tokai J. Exp. Clin. Med. 25: 1–6.

121 Ogata, K., Imai, J. and Nawa, Y. 1988. Three confirmed and five suspected human cases of *Gnathostoma doloresi* infection found in Miyazaki Prefecture, Kyushu. Jpn. J. Parasiol. 37: 358–364.

122 Ogata, K., Nawa, Y., Akahane, H., Camacho, S.P.D., Lamothe-Argumedo, R. and Cruz-Reyes, A. 1998. Short report: Gnathostomiasis in Mexico. Am. J. Trop. Med. Hyg. 58: 316–318.

123 Ohnishi, T. 2012. Epidemiological and micobiological features of *Kudoa septempunctata* causing novel outbreaks. Modenmedia 58: 205–209. (In Japanese).

124 Ohnishi, T., Furusawa, H., Oyama, R., Koike, S., Yoshinari, T., Kamata, Y. and Sugita-Konishi, Y. 2015. Molecular epidemiological analysis of *Kudoa septempunctata* by random amplified polymorphic DNA analysis. Jpn. J. Infect. Dis. 68: 235–238.

125 Ollague, W. 1985. Gnathostomiasis (Nodular migratory eosinophilic panniculitis). J. A. Academy Dermatol. 13: 835–836.

126 Ollague, W., Ollague, J., Guevara de Veliz, A. and Penaherrera, S. 1984. Human gnathostomiasis in Ecuador (Nodular migratory eosinophilic panniculitis): First finding of the parasitic in South America. Int. J. Dermatol. 23: 647–651.

127 Park, C.-W., Kim, J.-S., Joo, H.-S. and Kim, J. 2009. A human case of *Clinostomum complanatum* infection in Korea. Korean J. Parasitol. 47: 401–404.

128 Paulino, G.B. and Wittenberg, J. 1973. Intestinal capillariasis: A new cause of a malabsorption pattern. Am. J. Roentgenol. Radiother. Nucl. Med. 117: 340–345.

129 Pillai, G.S., Kumar, A., Radhakrishnan, N., Maniyelil, J., Shafi, T., Dinesh, K.R. and Karim, S. 2012. Case report: Intraocular gnathostomiasis: Report of a case and review of literature. Am. J. Trop. Med. Hyg. 86: 620–623.

130 Pozio, E., Armignacco, O., Ferri, F. and Morales, M.A.G. 2013. *Opisthorchis felineus*, an emerging infection in Italy and its implication for the European Union. Acta Tropica. 126: 54–62.

131 Pradatsundarasar, A., Pecharanond, K., Chintanawongs, C. and Ungthavorn, P. 1973. The first case of intestinal capillariasis in Thailand. Southeast Asian J. Trop. Med. Public Health 4: 131–134.

132 Rim, H.-J., Sohn, W.-M., Yong, T.-S., Eom, K.S., Chai, J.-Y., Min, D.-Y., Lee, S.-H., Hoang, E.-H., Phommasack, B. and Insisengmay, S. 2008. Fishborne trematode metacercariae detected in freshwater fish from Vientiane municipality and Savannakhet Provine, Lao PDR. Korean J. Parasitol. 46: 253–260.

133 Saichua, P., Nithikathkul, C. and Kaewpitoon, N. 2008. Human intestinal capillariasis in Thailand. World J. Gastroenterol. 14: 506–510.

134 Sakabe, S., Fujinaga, K., Tsuji, N., Ito, N., Umino, A., Taniguchi, M., Mizutani, M., Tamaki, S., Tanigawa, M., Tsuji, K. and Sato, H. 2006. Marked hypoalbuminemia caused by *Capillaria philippinensis*. Nipponnaikagakuzasshi 96: 162–163. (In Japanese).

135 Sakanari, J.A. and McKerrow, J.H. 1989. Anisakiasis. Clinic. Microbiol. Rev. 2: 278–284.

136 Scholz, T., Garcia, H.H., Kuchta, R. and Wicht, B. 2009. Update on the human broad tapeworm (Genus *Diphyllobothrium*), including clinical relevance. Clin. Microbiol. Rev. 22: 146–160.

137 Setasuban, P., Nuamtanong, S., Rojanakittikoon, V., Yaemput, S., Dekumyoy, P., Akahane, H. and Kojima, S. 1991. Gnathostomiasis in Thailand: A survey on intermediate hosts of *Gnathostoma* spp. with special reference to a new type of larvae found in *Fluta alba*. Trop. Med. 22: 220–224.

138 Shin, E.-H., Guk, S.-M., Kim, H.-J., Lee, S.-H. and Chai, J.-Y. 2007. Trends in parasitic diseases in the Republic of Korea. Trends Parasitol. 24: 143–149.

139 Shin, S.P., Zenke, K. and Yokoyama, H. 2015. Characterization of proteases isolated from *Kudoa septempunctata*. Korean J. Vet. Res. 55: 175–179.

140 Shin, S.P., Zenke, K., Yokoyama, H. and Yoshinaga, T. 2015. Factors affecting sporoplasm release in *Kuado septempunctata*. Parasitol. Res. 114: 795–799.

141 Shirakashi, S., Nishimura, T., Kameshima, N., Yamashita, H., Ishitani, H., Ishimaru, K. and Yokoyama, H. 2014. Effectiveness of ultraviolet irradiation of seawater for the prevention of *Kudoa yasunagai* and *Kudoa amamiensis* (Myxozoa: Multuvalvulida) infections in *Seriola* fish. Fish Pathol. 49: 141–144.

142 Soesatyo, M. 1985. *Gnathostoma spinigerum* and human gnathostomiasis. Berkala Ilmu Kedokteran 17: 21–35.

143 Sohn, W.-M. and Lee, S.-H. 1998. The first discovery of laraval *Gnathostoma hispidum* (Nematoda: Gnathostomidae) from a snake host, *Agkistrodon brevicaudus*. Korean J. Parasitol. 36: 81–89.

144 Sohn, W.-M. and Seol, S.Y. 1994. A human case of gastric anisakiasis by *Pseudoterranova decipiens* larva. Korean J. Parasitol. 32: 53–56.

145 Sohn, W.-M., Na, B.-K., Kim, T.H. and Park, T.-J. 2015. Anisakiasis: Report of 15 gastric cases caused by *Anisakis* Type I larvae and a brief review of Korean anisakiasis cases. Korean J. Parasitol. 53: 465–470.

146 Sripa, B., Bethony, J.M., Sithithaworn, P., Kaewkes, S., Mairiang, E., Loukas, A., Mulvenna, J., Laha, T., Hotez, P.J. and Brindley, P.J. 2011. Opisthorchiasis and *Opisthorchis*-associated cholangiocarcinoma in Thailand and Laos. Acta Trop. 120: S158–S168.

147 Sugita-Konishi, Y., Fukuda, Y., Mori, K., Mekata, T., Namba, T., Kuroda, M., Yamazaki, A. and Ohnishi, T. 2015. New validated rapid screening methods for identifying *Kudoa septempunctata* in olive flounder (*Paralichthys olivaceus*). Jpn. J. Infect. Dis. 68: 145–147.

148 Sugiyama, H. 2010. Food-borne parasitic infection as food poisoning. Jpn. J. Food Microbiol. 27: 1–7. (In Japanese).

149 Sugiyama, H. 2010. Food and parasitic infections. Syokuhineiseishi 51: 285–291. (In Japanese).

150 Sugiyama, H. and Morishima, Y. 2014. Anisakiasis. NIID Homepage, pp. 4. (In Japanese). http://www.niid.go.jp/niid/ja/kansennohanashi/314-anisakis-intro.html.

151 Suzuki, J. 2003. Detection rates of *Anisakis simplex* third stage larvae from the genus *Oncorhynchus* in Tokyo. Monthly Epidemiological Records, Tokyo Metropolitan Institute of Public Health 24: 1–2. (In Japanese).

152 Suzuki, J., Murata, R., Sadamasu, K. and Araki, J. 2010. Detection and identification of *Diphyllobothrium nihonkaiense* plerocercoids from wild Pacific salmon (*Oncorhynchus* spp.) in Japan. J. Helminthol. 84: 434–440.

153 Tada, I., Araki, T., Matsuda, H., Araki, K., Akahane, H. and Mimori, T. 1987. A study on immunodiagnosis of gnathostomiasis by ELISA and double diffusion with special reference to the antigenicity of *Gnathostoma doloresi*. Southeast Asian J. Trop. Med. Public Health 18: 444–448.

154 Takahashi, S., Ishikura, H. and Kikuchi, K. 1998. Anisakidosis: Global point of view. pp. 109–120. *In*: Ishikura, H. et al. (eds.). Host Response to International Parasitic Zoonoses. Springer, Tokyo.

155 Taniguchi, Y., Hashimoto, K., Ichikawa, S., Shimizu, M., Ando, K. and Kotani, Y. 1991. Human gnathostomiasis. J. Cutan. Pathol. 18: 112–115.

156 Taniguchi, Y., Ando, K., Sugimoto, K. and Yamanaka, K. 1999. Creeping eruption due to *Gnathostoma hispidum*—one way to find the causative parasite with artificial digestion. Int. J. Dermatol. 38: 873–874.

157 Taylor, L.H., Latham, S.M. and Woolhouse, M.E.J. 2001. Risk factors for human disease emergence. Phil. Trans. R. Soc. Lond. B. 356: 983–989.

158 Tidball, J.S., Aguas, J.P. and Aldis, J.W. 1978. A new concentration of human intestinal capillariasis on western Luzon. Southeast Asian J. Trop. Med. Public Health 9: 33–40.

159 TMIPH (Homepage of Tokyo Metropolitan Institute of Public Health). 2016. Anisakiasis and the prevalence of *Anisakis* larvae in chub mackerel, *Scomber japonicus*. http://www.tokyo-eiken.go.jp /assets/issue/health/webversion/web28.html. (In Japanese).

160 Torgerson, P.R., Devleesschauwer, B., Praet, N., Speybroeck, N., Willingham, A.L., Kasuga, F., Rokni, M.B., Zhou, X.-N., Fevre, E.M., Sripa, B., Gargouri, N., Furst, T., Budke, C.M., Carabin, H., Kirk, M.D., Angulo, F.J. Havelaar, A. and de Silva, N. 2015. World health organization estimates of the global and regional diseases burden of 11 foodborne parasitic diseases, 2010: A data synthesis. Plos Medicine 12: 1–22.

161 Torres, P., Jercic, M.I., Weitz, J.C., Dobrew, E.K. and Mercado, R.A. 2007. Human Pseudoterranovosis, an emerging infection in Chile. J. Parasitol. 93: 440443.

162 Touch, S., Komalamisra, C., Radomyos, P. and Waikagul, J. 2009. Discovery of *Opisthorchis viverrini* metacercariae in freshwater fish in southern Cambodia. Acta Tropica. 111: 108–113.

163 Tsushima, H., Numata, T., Yamamoto, O., Iwasaki, H. and Iwanaga, Y. 1980. Gnathostomiasis cutis probably infected in Hiroshima city. Acta Medica Hiroshima 33: 1183–1187. (In Japanese).

164 Umehara, A., Kawakami, Y., Araki, J., Uchida, A. and Sugiyama, H. 2008. Molecular analysis of Japanese *Anisakis simplex* worms. Southeast Asian J. Trop. Med. Public Health 39: 26–31.

165 Umehara, A., Kawakami, Y., Ooi, H.-K., Uchida, A., Ohmae, H. and Sugiyama, H. 2010. Molecular identification of *Anisakis* type I larvae isolated from hairtail fish off the coasts of Taiwan and Japan. Int. J. Food Microbiol. 143: 161–165.

166 Velasquez, C.C. 1982. Heterophyidiasis. pp. 99–107. *In*: Steele, J.H. (ed.). CRC Handbook Series in Zoonoses, Section C: Parasitic Zoonoses (Trematode Zoonoses). Vol. III. CRC Press, Boca Raton, Florida.

167 Violante-Gonzalez, J., Aguirre-Macedo, M.L. and Mendoza-Franco, E.F. 2007. A checklist of metazoan parasites of fish from Tres Palos Lagoon, Guerrero, Mexico. Parasitol. Res. 102: 151–161.

168 Wicht, B., Scholz, T., Peduzzi, R. and Kuchta, R. 2008. First record of human infection with the tapeworm *Diphyllobothrium nihonkaiense* in North America. Am. J. Trop. Med. Hyg. 78: 235–238.

169 WHO. 2015. WHO estimates of the global burden of foodborne diseases, pp. 268.

170 Xuan, L.T., Rojekittikhun, W., Punpoowong, B., Trang, L.N. and Hien, T.V. 2002. Case report: intraocular gnathostomiasis in Vietnam. Southeast Asian J. Trop. Med. Pub. Health 33: 485–489.

171 Yahata, Y., Sugita-Konishi, Y., Ohnishi, T., Toyokawa, T., Nakamura, N., Taniguchi, K. and Okabe, N. 2015. *Kudoa septempuctata*-induced gastroenteritis in humans after flounder consumption in Japan: a case-controlled study. Jpn. J. Infect. Dis. 68: 119–123.

172 Yamane, Y., Kamo, H., Bylund, G. and Wikgren, B.-J.P. 1986. *Diphyllobothrium nihonkaiense* sp. Nov. (Cestoda: Diphyllobothriidae)-revised identification of Japanese broad tapeworm. Shimane J. Med. Sci. 10: 29–48.

173 Yamane, Y. 2003. *Diphyllobothrium nihonkaiense* and other marine-origin cestodes. Prog. Med. Parasitol. Jpn. 8: 245–259.

174 Yamashita, J. 1964. Echinostome. Prog. Med. Parasitol. Japan 1: 287–313.

175 Yokoyama, H. 2004. Life cycle and evolutionary origin of myxozoan parasites of fishes. Jpn. J. Protozool. 37: 1–9. (In Japanese).

176 Yokoyama, H. 2016. Kudoosis of marine fish in Japan. Fish Pathol. 51: 163–168. (In Japanese).

177 Youssef, F.G., Mikhail, E.M. and Mansour, N.S. 1989. Intestinal capillariasis in Egypt: A case report. Am. J. Trop. Med. Hyg. 40: 195–196.

178 Yu, S.-H. and Mott, K.E. 1994. Epidemiology and morbidity of food-borne intestinal trematode infections. WHO. pp. 1–26.

179 Yu, J.-R., Seo, M., Kim, Y.-W., Oh, M.-H. and Sohn, W.-M. 2001. A human case of gastric infection by *Pseudoterranova decipiens* larva. Korean J. Parasitol. 39: 193–196.

180 Zhang, W., Che, F.C., Tian, S., Shu, J. and Zhang, X. 2015. Molecular identification of *Diphyllobothrium nihonkaiense* from 3 human cases in Heilongjiang Province with a brief literature review in China. Korean J. Parasitol. 53: 683–688.

11

Seafood Allergy

Hiroki Saeki

1. Definition of Food Allergy

This chapter introduces the feature of hypersensitivity induced by seafood, and explains immunological cross-reactivity among causative seafood.

Food allergy is now a serious social problem worldwide, particularly in developed countries. Food allergy refers to an immunological hypersensitivity (allergy symptom) that develops by ingesting a specific food, and is defined as a "phenomenon in which symptoms that are detrimental to the living body are caused by the antigen-specific immunological mechanism after oral intake, skin contact, inhalation of the causative food" [1], and also defined as "an adverse health effect arising from a specific immune response that occurs reproducibly on exposure to a given food" [2].

2. Prevalence of Food Allergy and Causative Foods

According to the European Respiratory Health Survey involving 15 countries in 1991–1994, 12% of respondents reported food allergy (n = 17,280), except for Spain (Fig. 11.1) [3]. However, other epidemiological research showed the high prevalence of food allergy in Spain (8–9%) [4, 5], suggesting that there is no difference in the tendency of allergy development among the countries.

Faculty of Fisheries Sciences, Hokkaido University, Japan.
E-mail : saeki@fish.hokudai.ac.jp

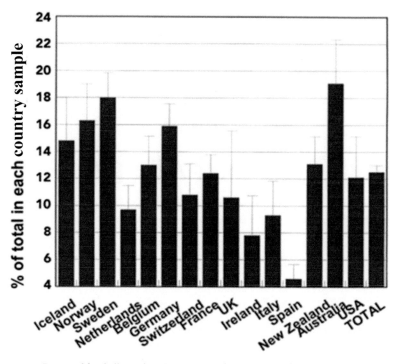

Figure 11.1 Reported food allergy/intolerance prevalence rates and 95% Confidence Intervals by country [3].

Compared to adults, the incidence of food allergy tends to be higher for children and infant groups. The prevalence of food allergy in Japan is presumed to be about 5–10% in infants and 1.5–3% after school age [1]. In the United States, the prevalence is estimated to be about 6–7% in children and 1–2% in adults [6].

Milk, eggs, and wheat are the three major causes of food allergy in any country. In addition, the top eight food allergens, including five kinds of foods (fish, crustacean, shellfish, nuts, peanuts, and soybeans) are called "Big 8". In almost all countries, fish, crustacean, and shellfish are listed as causative foods.

The causative foods tends to vary depending on age. The incidence of eggs and dairy products is high, but the frequency decreases gradually with growth, and the incidence of fish, shrimp, crab, and fish roe increased in Japanese infant and young generations [7, 8] (Table 11.1). In the survey of the United States (2009–2010), the prevalence of each food allergen among food-allergic children was in the order of peanuts (25.2%), followed by milk (21.1%), and shellfish (17.2%) [9].

Table 11.1 Causative foods of new-onset food allergies in Japanese infant and young generations.

< 1 year old (884)	1 year old (314)	2–3 years old (173)	4–6 years old (109)	7–19 years old (123)	≥ 20 years old (100)
Hen's egg 57.6%	Hen's egg 39.1%	Fish roe 20.2%	Fruit 16.5%	Crustacean 17.1%	Wheat 38.0%
Cow's milk 24.3%	Fish roe 12.9%	Hen's egg 13.9%	Hen's egg 15.6%	Fruit 13.0%	Fish 13.0%
Wheat 12.7%	Cow's milk 10.1%	Peanut 11.6%	Peanut 11.0%	Hen's egg Wheat 9.8%	Crustacean 10.0%
	Peanut 7.9%	Nuts 11.0%	Buckwheat Fish roe 9.2%		Fruit 7.0%
	Fruit 6.0%	Fruit 8.7%		Buckwheat 8.9%	

This table lists the causative foods that account for > 5% in each group, ranked from 1st to 5th, from the top N = 1706.

3. Outline of Pathogenic Mechanism of Food Allergy

Food ingredients are not recognized as foreign substances when they are normally ingested and absorbed as nutrients in the intestinal tract. However, food allergy develops when the biological system (oral immune tolerance) causes dysregulation and our living body recognises food ingredients as foreign substances. Food allergy is divided into immunoglobulin E (IgE) mediated and cell-mediated hypersensitivities, and almost all food allergic symptoms are accompanied by IgE involvement [10]. Allergens (Allergenic proteins) have specific sites called epitope, where specific antibodies (IgE) can combine to elicit immune response. The reaction between allergens and IgEs through the epitopes acts as a trigger for the onset of hypersensitivity. Various kinds of seafood, such as fish, crustacean, shellfish, and cephalopod each have specific allergens, and information about IgE-binding epitopes in each allergen is useful to understand the overview of IgE cross-reactivity among seafood (See 7.1).

The outline from food intake to onset of immediate hypersensitivity is as follows [11, 12, 13]: (1) When digested foods containing allergen are recognized as foreign substances, specific antibodies (IgE) to food allergen are produced in our body. (2) The specific IgE to the food allergen binds to the surface of mast cells (establishment of sensitization). (3) When the sensitized allergen in foods is ingested again, it is bound to the specific IgE at the surface of the mast cells. (4) The allergen-IgE reaction triggers release of chemical substances, such as histamine and leukotriene from the mast cells. (5) The irritating substances immediately act on various parts of the body and induce vasodilation, enhancement of vascular permeability, smooth muscle contraction of the bronchi, and peripheral vascular irritation. As a result, (6) we develop undesirable symptoms (urticaria, erythema, itching, conjunctivitis, vomiting, and diarrhea).

The systemic symptoms of food allergy appear in a very short time, and serious shock symptoms (anaphylaxis), such as a sudden decrease in blood pressure and loss of consciousness accompanying organ dysfunction may cause death. We need to realize that food allergy is a serious food risk.

As described above, IgE is involved in the development of food hypersensitivity. On the other hand, some food allergies develop as IgE independent (non IgE-mediated). As a representative example, newborn infants and gastrointestinal allergies (food protein-induced enterocolitis syndrom) are known. In this case, no IgE antibody is produced in the body and the mechanism is discussed as an allergy involved in immune cells [14, 15].

4. Feature of Seafood Allergy

It is known that the type of causative food introducing hypersensitivity is highly dependent on eating habits. Japan is one of the largest seafood consuming countries where more than 500 species of marine bioresources are utilized as food materials, and approximately 40% of the intake of animal protein in the average Japanese is ingested from seafood. Therefore, as shown in Fig. 11.2, seafood (fish roe, crustacean, and fish) accounted for

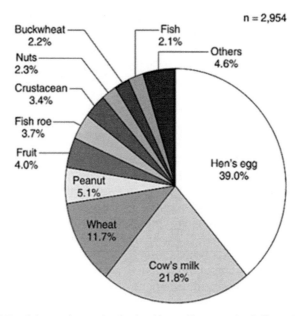

Figure 11.2 Breakdown of causative foods of immediate-type food allergy in Japan [7].

Table 11.2 Examples of seafood causing hypersensitivity.

Fish	*Mackerel, salmon*, cod, tuna, sardine, horse mackerel, flounder, sea bream, eel, yellowtail, Atka mackerel, saury, rockfish, fox jacopever, Pacific herring, red rock fish, greenling
Crustacean	*Shrimp* and *prawn* (black tiger prawn, kuruma prawn, Northern shrimp), lobster, *crab* (King crab, snow crab, hair crab)
Shellfish	Scallop, clam, *abalone*, turban shell, mussel
Cephalopod	*Squid (cattle fish)*, octopus
Fish roe	*Salmon*, walleye pollock, herring, flying fish, sturgeon (caviar)

Underline indicates seafood that are designated food allergen labeling system in Japan.

9.2% of all cases that developed food allergy, and it occupied 23% (crustacean and fish) of causative foods in food allergic patients over 20-years-old [7]. This is one example that the causative foods of food allergy reflect the eating habit of the community.

In Japan, various kinds of seafood induce allergic cases, as listed in Table 11.2. For example, more than 30 kinds of fish have been reported as causative foods of hypersensitivity [16, 17], and many allergens have been isolated in marine bioresources. In addition, IgE cross-reactivity is established among various types of seafood allergens. Therefore, identifying allergen in each seafood and clearly grasping the scope of IgE cross-reactivity among allergens are essential for reducing food risk of seafood.

5. Seafood Registered in Food Allergen Labeling System

Exclusion of harmful substances and reduction of pathogenic microorganisms are basic principles to ensure food safety. However, since allergens are food ingredients, it is impossible to reduce the risk of food allergy only with conventional food hygiene standards. In response to the situation, food allergen labeling regulations have been enforced in over 20 countries and regions (US, Canada, EU, South American countries, Japan, China, Hong Kong, Taiwan, Korea, Thailand, Australia, New Zealand, and South Africa) in order to avoid health damage due to food allergy [18], and seafood (fish, shellfish, crustacean) are designated as mandatory labeling foods in almost all the countries [19].

Japan's food allergen labeling system for specific allergenic ingredients has been mandated in 2002. When allergens are included in foods at concentration levels of several mg/g or several mg/mL as a total protein weight, they are subject to labeling [7, 8]. In this system, the listed causative foods are classified into "mandatory labeling" and "recommended labeling", according to the number of onset and the severity of the case. Seven kinds of

Table 11.3 Allergenic ingredients designated in Japan food labeling system.

Regulation	Processed food ingredient
Food with mandatory labeling (7 ingredients)	Egg, milk, wheat, buckwheat, peanut, shrimp/prawn, crab
Food with recommended labeling (18 ingredients)	Salmon, mackerel, abalone, squid, salmon roe, orange, kiwifruit, beef, walnut, soybean, chicken, banana, pork, peach, yam, apple, Matsutake mushroom, gelatin

"mandatory labeling ingredients" and 20 kinds of "recommended labeling ingredients" have been listed as of the year of 2017, as listed in Table 11.3. The fact that the seven seafood (shrimp, crab, squid, abalone, mackerel, salmon, salmon roe) have been designated in these 27 items reflects the feature of food allergy in Japan.

6. Seafood Allergens

6.1 Outline

Various kinds of allergens contained in seafood are listed in Table 11.4. Parvalbumin, the first-reported food allergen in the world [20], is a major allergen in fish muscle. Collagen, the major structural protein in connective tissues also acts as an allergen [21]. Tropomyosin is the major allergen of

Table 11.4 Allergens contained in seafood.

Allergen	Seafood	Feature	Range of IgE cross-reactivity
Parvalbumin	Fish	Ca-binding protein	Fish, fish and bullfrog
Collagen	Teleost fish, Cartilaginous fish	Connective tissue	Teleost fish, Cartilaginous fish
Aldehyde phosphate dehydrogenase	Fish	Glycolytic enzymes	No information
Tropomyosin	Shrimp, crab, squid, octopus, shellfish	Muscle fiber component	Invertebrate including snail, mite, and cockroach
Arginine kinase	Shrimp, shellfish, octopus	ATP synthesis	No information
Troponin Myosin light chain Sarcoplasmic calcium-binding protein	Shrimp	Muscle contraction	No information
Parbalbumin	Abalone	Muscle component	No information
β'-components	Fish roe	Yolk protein	Fish roes

invertebrates, such as crustacea (shrimp, crab) and mollusks (squid, octopus, shellfish) [22]. Fish roe also contains a allergen in a yolk, whose protein is called β′(beta prime)-component [23] in the field of developmental biology.

Other IgE-reactive proteins with a few onset cases have also been identified in seafood. Enolase and aldolase, glycolytic enzymes have been identified from cod, salmon, and tuna as new fish allergens [24]. Aldehyde phosphate dehydrogenase [25] was also found as a new allergen in cod. In crustacea, arginine kinase [26, 27, 28] troponin [29], myosin light chain [30], and sarcoplasmic calcium binding protein [31] have been identified as novel allergens. Paramyosin was also found as a new abalone allergen [32].

6.2 Parvalbumin

Parvalbumin is a major allergen in fish muscle, which is water-soluble and calcium-binding protein with a molecular mass of about 10 kDa to 12 kDa. Parvalbumin has two isoforms (α and β) with isoelectric points 5.0 and 4.5, respectively. Mammals and birds have the α type, and fish and amphibians have the both types. Parvalbumin is a thermally stable protein, and its characteristic is often utilized for purification process. Parvalbumin is recovered from hot water-extract of fish meat after removal of other proteins by heating at more than 90°C. Parvalbumin is an important Ca-regulatory factor in muscle [33], and is considered to be a water-soluble muscle relaxant factor in fast muscle of vertebrate in muscle [34]. Two molecules of calcium ions are bound in one molecule of parvalbumin [35], and a structure of calcium ion binding site (EF hand motif structure [33]) is conserved in all organisms' parvalbumin.

Parvalbumin of Baltic cod is famous as the first identified fish allergen (allergen M) [20], and is registered as *Gad c 1* in WHO/IUIS allergen database (www.allergen.org) [36]. In addition, 12 types of fish parvalbumin as allergens, such as *Sal s 1* (from Atlantic Salmon) and *The c 1* (from walleye pollock), are registered in the database. All fish parvalbumin are β-type, but α-parvalbumin has been reported as an allergen of an edible frog [37, 38].

Parvalbumin content in muscle depends on fish species. In measuring parvalbumin content in many fishes using the same method, the content difference reached up to 48 times (Fig. 11.3) [17], and IgE reactivity of fish was roughly proportional to parvalbumin content [17]. In addition, the strength of fish muscle allergenicity was explained from the viewpoint of intramuscular expression level of parvalbumin [38].

The homology of amino acid sequence in fish β-parbalbumin is in the broad range of 60–90% among teleost fish, but the primary structure of the calcium binding site in the EF hand motif shows high homology. The amino acid sequence of IgE-binding epitope has been identified in some teleost fish (cod [39, 40], carp [41], Pacific mackerel [42], and Atlantic salmon [43]).

However, the identified sequential IgE-epitopes tend to be different among fish species; for example, the region 21–40 was judged to contain a major IgE-binding epitope of Pacific mackerel parvalbumin (Sco j 1), but no major IgE-binding epitope was found in this region of other fish parvalbumins (sardine, Japanese eel, cod, horse mackerel, crimson sea bream, skipjack, and flounder) [42]. On the other hand, when cod parvalbumin was treated with a chelating agent to remove calcium ions, the IgE-binding ability tended to decrease [44], indicating the importance of tertiary structure recognition in the binding of IgE to parvalbumin in fish allergic patients. Indeed, recent research revealed that IgE-binding epitope of various fish parvalbumins exist in a stereoscopic conformation maintained by calcium binding [45].

6.3 Collagen; another fish meat allergen

Collagen is a major component of the extracellular matrix of animals, and it has been reported as an allergen in various fish such as mackerel, skipjack, and bigeye tuna [46, 47]. Collagen is composed of three polypeptide subunits of 120 kDa, and they assemble into a right-handed triple helical structure. The major part of each collagen polypeptide chain is the helical structure part (atelocollagen), and both ends (telopeptide) have random structures. IgE-binding epitopes in collagen was found in the atelocollagen part [48], which is as the same as that of bovine collagen [49].

Gelatin, heat-denatured collagen with water-solubility is a popular food ingredient, and it is listed as one of the allergic food ingredients in Japan's food allergen labeling system. Although the triple helical structure of collagen is degraded by heating, the IgE-binding ability is highly retained in gelatin [50].

Parvalbumin has been considered a major allergen in fish meat for a long time. On the other hand, recently, specific IgE to fish collagen has been detected in half of Japanese fish allergic patients (18/36 people), and this positive rate was almost equal to that of parvalubumin [51] (Fig. 11.4). This data suggests the necessity of recognizing collagen as a fish pan-allergen in the same way as parvalbumin. Interestingly, such a high positive rate of collagen has not been reported except in the research in Japan, where there is a habit of eating raw fish, "*Sashimi*", and the specific IgE to collagen showed no reactivity to heat-muscle extract containing gelatinized collagen. The high rate of collagen allergy may be attributable to the traditional Japanese food culture [51].

Recently, there are increasing cases of utilizing fish collagen as a medicinal substitute for mammalian collagen. Therefore, it should be noted that fish collagen has a probability to induce hypersensitivity, and it is also necessary to verify contamination of parvalbumin in fish collagen products.

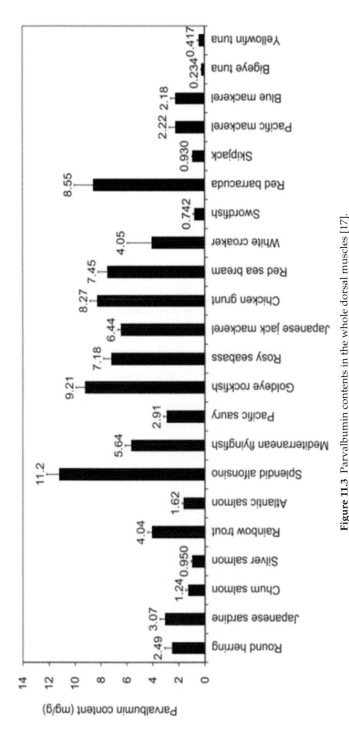

Figure 11.3 Parvalbumin contents in the whole dorsal muscles [17].

Round herring *Etrumeus teres*, Japanese sardine *Sardinops melanostictus*, chum salmon *Oncorhynchus keta*, silver salmon *Oncorhynchus kisutch*, rainbow trout *Oncorhynchus mykiss*, Atlantic salmon *Salmo salar*, splendid alfonsino *Beryx splendens*, Mediterranean flyingfish *Cheilopogon heterurus*, Pacific saury *Cololabis saira*, goldeye rockfish *Sebastes thompsoni*, rosy seabass *Doederleinia berycoides*, Japanese jack mackerel *Trachurus japonicus*, chicken grunt *Parapristipoma trilineatum*, red sea bream *Pagrus major*, white croaker *Pennahia argentata*, swordfish *Xiphias gladius*, red barracuda *Sphyraena pinguis*, skipjack *Katsuwonus pelamis*, Pacific mackerel *Scomber japonicus*, blue mackerel *Scomber australasicus*, bigeye tuna *Thunnus obesus* and yellowfin tuna *Thunnus albacares*.

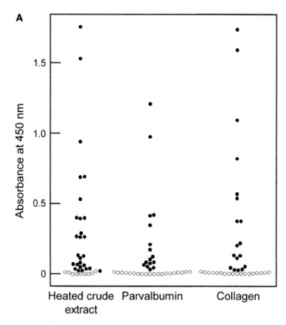

Figure 11.4 Analysis of IgE reactivity to the heated crude extract from muscle, purified parvalbumin, and purified collagen from Pacific mackerel by enzyme-linked immunosorbent assay [51]. Open and closed circles show sera from fish allegeic patients and healthy volunteers, respectively. Fish allegeic patients' sera showed IgE reaction to parvalbumin and collagen.

6.4 Tropomyosin; major allergen of invertebrates

Tropomyosin is a constituent protein of muscle fiber binding to actin filament in muscle, which is composed of two α-helical chains (each molecular mass is 35 kDa) taking a coiled structure. Tropomyosin is contained in the muscle of all organisms, whereas it only acts as a food allergen in invertebrates. Seafood, including invertebrates, such as shrimp, crab, squid, octopus, and shellfish, has a probability of inducing hypersensitivity. On the other hand, vertebrate tropomyosin do not act as a food allergen; only tilapia tropomyosin has been reported as an allergen, and the association with inflammatory bowel disease has been discussed [52].

Tropomyosin is now recognized as a common allergen of marine invertebrates. Tropomyosin as an allergen was initially reported as a squid allergen [53], and was subsequently understood as a common allergen between crustacea and mollusca [54]. IgE contained in sera of shrimp allergic patients, who have no experience of eating octopus, reacted with squid and octopus tropomyosins [53], and IgE reacting with tropomyosin of crustaceans mollusks also responded to octopus and scallop tropomyosins [55].

Information of IgE-binding epitope in marine invertebrate tropomyosin has been fulfilled, and eight sites of IgE-binding epitope (specific amino acid sequences) have currently been identified, as shown in Fig. 11.5 [56]. As described in 7.3, the IgE-binding epitopes have high-homology sequences. There is an ambitious attempt to prepare a monoclonal antibody (MAb) with reference to the IgE-binding epitope sequence located at the C-terminal of shellfish tropomyosin, and to develop a detection system for invertebrate tropomyosin contained in processed foods using the MAb [57]. This MAb had the ability to selectively react with tropomyosin of 15 kinds of marine invertebrates (shrimp, crab, shellfish, squid, and octopus), and also house dust allergens (mite, cockroach). This is one example showing wide-ranged IgE cross-reactivity of invertebrate tropomyosin.

6.5 β'-component; fish roe yolk protein

The major allergen contained in fish roe is a yolk protein called β'-component [23]. In case of salmon roe allergy, the specific IgE to β'-component is detected in almost all salmon-roe-allergic patients' sera [58].

Salted fish roes are popular in areas having fish eating culture, and various kinds of fish roe foods have been consumed worldwide: *ikura* (salmon), *tarako* (walleye pollock), *kazunoko* (herring), and *karasumi* (mullet) in Japan, *wuyuzi* (mullet) in Taiwan, *bottarga* (tuna or mullet) in the Mediterranean sea, and *cavier* (sturgeon). As shown in Table 11.1 and Fig. 11.2, frequent cases of salmon-roe-allergy have been reported in Japan, and Japan's food allergen labeling system has specified that salmon roe was labeled as a potential cause of allergic reactions [7].

Hen eggs consist of egg white and yolk parts, and the major allergen is egg white proteins, such as ovalbumin and ovomucoid. On the other hand, fish roe has no part equivalent to hen egg white. The major proteins in fish yolk are lipovitellin, phosvitin, and β' (beta prime)—component, and they are generated from vitellogenin, which is a precursor of the yolk proteins. The DNA sequences encoding the three major yolk proteins are located in the Vg gene in the following order: NH2–(lipovitellin heavy chain)–(phosvitin)–(lipovitellin light chain)–(β'-component)–(C-terminal peptide)–COOH. Vitellogenin synthesized in liver reaches oocytes through the bloodstream, and is enzymatically cleaved to be the three yolk proteins [59], which are accumulated into the yolk with a progress of oocyte growth.

β'-component of chum salmon roe showed a high content of β-pleated sheets and no α-helices, and consists of 16 kDa- and 18 kDa-subunits (Fig. 11.6) having the same level of IgE-binding ability [60]. Structure of β'-component has strong digestion resistance against pepsin and trypsin [61], suggesting that the ingested β'-component seems to reach the small intestine in the form of high molecular mass components with high

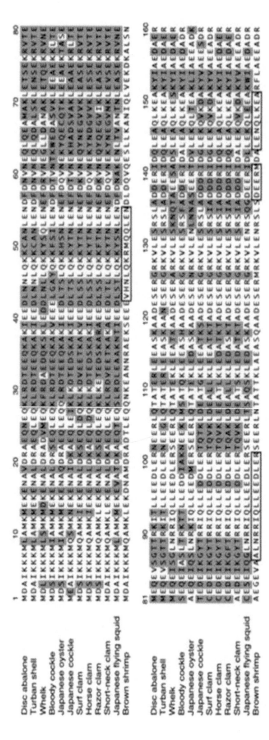

Figure 11.5 contd. ....

...*Figure 11.5 contd.*

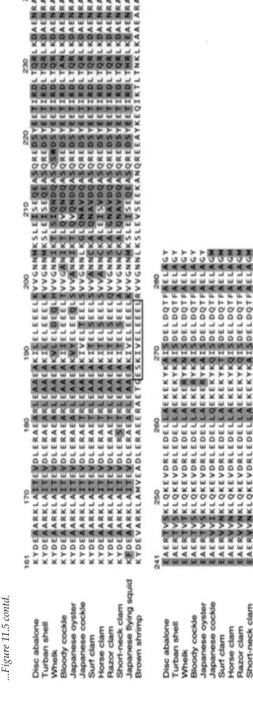

Figure 11.5 Amino acid sequences of the gastropod and bivalve tropomyosins. The residues differing from brown shrimp tropomyosin are shaded. The IgE epitope regions proposed for brown shrimp tropomyosin are boxed [56].

Accession numbers (DDBJ/EMBL/GenBank nucleotide sequence databases): disc abalone *Haliotis discus discus*, AB444939; turban shell *Turbo cornuttus*, AB444940; whelk *Neptunea polycostata*, AB444941; bloody cockle *Scapharca broughtonii*, AB444942; Japanese oyster *Crassostrea gigas*, AB444943; Japanese cockle *Fulvia mutica*, AB444944; surf clam *Pseudocardium sachalinensis*, AB444945; horse clam *Tresus keenae*, AB444946; razor clam *Solen strictus*, AB444947; short-neck clam *Ruditapes philippinarum*, AB444948; Japanese flying squid *Todarodes pacificus*, AB218915; brown shrimp *Penaeus aztecus*, DQ151457.

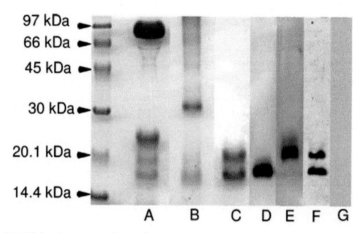

Figure 11.6 Molecular characteristics of chum salmon β'-component [60]. SDS–polyacrylamide gel electrophoresis (PAGE) patterns of yolk protein and β'-component and their IgE reactivity in immunoblotting. Whole yolk proteins extracted with 0.5 M NaCl (pH 8.0) (A), native β'-component. (B, C), 16 kDa and 18 kDa subunits of β'-component (D, E) were electrophoresed. Samples were loaded under reducing (A, C, D, E, F) or non-reducing (B) conditions. β'-component was immunoblotted using salmon-roe-allergic patient's (F) and control (G) sera.

allergenicity. Indeed, chum salmon β'-component orally administered to mice was detected in blood plasma.

There is a structural difference among β'-component of various fishes. For example, rainbow trout β'-component contains two subunits with the same molecular mass of 16 kDa [23], and grey mullet β'-component consists of two subunits with molecular mass of 16 kDa and 30 kDa in SDS-PAGE under reducing conditions [62].

7. Cross Reactivity of Seafood Allergens

7.1 Importance of understanding IgE cross-reactivity among seafood allergens

Allergic patients with sensitization to 'Food α' may complain of hypersensitivity to 'Food β' that they have never ingested so far. This situation is due to the fact that 'Food β' contains some allergic substances which have structural similarity to the allergen in 'Food α', and the patient's specific IgE to allergen in 'Food α' reacted with the allergic substance in 'Food β'. The relationship among different allergens that react with the same IgE is called "IgE cross-reactivity". Knowledge of IgE cross-reactivity among allergens is essential information to plan dietary guidance for allergic patients, and is also important to maintain comfortable eating habits without excess avoidance of foods.

The feature of seafood allergy is existence of IgE cross-reactivity among allegens in various kinds of seafood, reflecting marine bioresources, which include allergen with high structural similarity. In the case of parvalbumin, IgE cross-reactivity is established among various fish species [16, 17, 63]. Marine invertebrates, such as shrimp, crab, squid, octopus, and shellfish, show IgE-cross reactivity via tropomyosin [20, 22]. In countries where various kinds of seafood are readily available through a stable distribution food network, IgE cross-reactivity complicates medical and social correspondence to seafood allergies. Therefore, understanding IgE cross-reactivity among seafood is essential for securing food safety of marine bioresources.

7.2 IgE cross-reactivity among fish meat allergens

There is a high possibility that parvalbumin acts as allergen of processed seafood regardless of fish species. According to the interesting clinical study about fish allergic patients [16], when blood of 38 fish-allergic patients were reacted with 43 kinds of water-extract of fish meat, 33 patients' sera contained parvalbumin-specific IgE that reacted with 4 to 43 kinds of fish extracts. This research indicates that IgE cross-reactivity via parvalbumin is widely established among fish. Indeed, IgE cross-reactivity of parvalbumin has been confirmed in administration of purified or recombinant parvalbumins to fish-allergic patients [64]. In addition, parvalbumin-specific IgE contained in blood of fish allergic patients markedly reacted with parvalbumin of edible frog (bullfrog, *Rana catesbeiana*) [38], indicating the establishment of IgE cross-reactivity between fish and amphibian meat via parvalbumin.

So far, attempts based on fish taxonomy have been carried out to classify fish species that fish-allergic patients can eat, and the strength of IgE cross-reactivity was discussed among some fish species. For example, fish-allergic patients who react only to parvalbumin of limited fish species (e.g., salmonid fish) have been reported [65], and allergic patient's sera reacting with parvalbumin (β type) of teleost fish did not respond to parvalbumin (α type) of cartilage fish [66]. However, at the present, it has not reached the stage that can be used for dietary guidance by defining the range of IgE cross-reactivity in fish.

Collagen, another major fish meat allergen, also showed IgE cross-reactivity among fish species [21, 46, 67]. Pacific mackerel collagen markedly inhibited the reaction between fish meat-allergic patients' sera and 22 types of fish water-extracts containing collagen [67], indicating collagen has highly IgE cross-reactive among fish meat. On the other hand, no IgE cross-reactivity was observed in collagen between teleost fish and cartilaginous fish [68]. Furthermore, fish collagen did not show IgE cross-reactivity with mammals' collagen, such as livestock [21]. Conclusively, the extent of collagen cross-reactivity is limited, depending on the type of organism.

7.3 Tropomyosin: pan-allergen of invertebrates

Invertebrate tropomyosin has widespread IgE cross-reactivity beyond species, and is called "pan-allergen". Figure 11.7 shows a characteristic of tropomyosin as a pan-allergen; Tropomyosin-specific IgE in blood of a scallop-allergic patient reacted with tropomyosin (and its partial digested fragments) contained in six kinds of marine invertebrates. At the end of 2016, 19 kinds of tropomyosin were registered as food allergen in the WHO/IUIS allergen database [36]. 17 of which are shrimp, crab, squid, and octopus, and also edible snail.

The amino acid sequence homology of tropomyosin in shrimps and crabs (10 species of crustacean) and squids and octopuses (four species) is as high as 89.1–98.6% and 91.2–99.6%, respectively [56]. The eight kinds of IgE-binding epitopes (sequential epitope) identified in Brawn Shrimp (Pen a 1) are highly conserved among these invertebrates [69, 70], suggesting that the high homology contributes to the establishment of the IgE cross-reactivity between crustaceans and cephalopods.

In the case of crustacean and mollusk, the amino acid sequence homology is not so high; it was 61.3–64.4% for crustaceans (12 species) and cephalopods (4 species), and 54.6–63.7% for crustaceans and shellfish (19 species) [56], whereas IgE cross-reactivity has been reported among these marine bioresources. The reason is that IgE binding sites are present in highly homologous regions [56].

7.4 Relation between food allergy and living environment via IgE cross-reactivity of allergen

Mites and cockroaches often induce house dust allergy, and seven of their tropomyosin have been registered as allergens in the WHO/IUIS allergen database [36], and IgE cross-reactivity exists between crustacea (shrimp, crab) and arthropod (mite, cockroach) [71, 72], indicating the possibility that the living environment is involved in the development of seafood allergy [73, 74]. Therefore, investigation of living environment as well as the survey of ingested food is important for the cause investigation of food allergy.

7.5 Contamination of tropomyosin in processed seafood

Small shrimp, crab, and krill are major fish feeds and their gastrointestinal digesta containing tropomyosin may contaminate materials of processed

Muscle homogenate

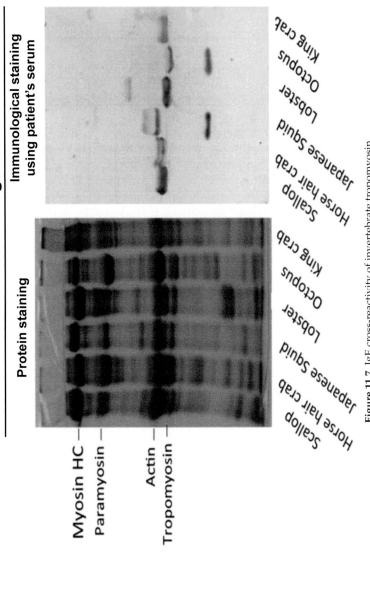

Figure 11.7 IgE cross-reactivity of invertebrate tropomyosin.

Left: SDS-PAGE patterns of invertebrate muscles (Protein staining). Right: The specific IgE contained in scallop allergic patient's serum was reacted with various invertebrate tropomyosin: scallop *Patinopecten yessoensis*, horsehair crab *Erimacrus isenbeckii*, Japanese flying squid *Todarodes pacificus*, lobster *Nephropidae* sp., octopus *Enteroctopus dofleini*, and king crab *Paralithodes camtschaticus*.

seafood during flesh collection process. Indeed, crustacean tropomyosin was detected in frozen walleye pollock surimi, which is a protein material for surimi-based products, such as crab leg analog and fish sausage [75]. Considering the wide range of allergen cross-reactivity of tropomyosin, contamination by small crustaceans cannot be ignored from the viewpoint of food safety. In fish sausages sold in Japan, the package information includes the caution that "This raw material is feeding shrimp and crab."

7.6 IgE cross-reactivity in fish roe allergy

β'-component, the major allergen in fish roe, shows IgE cross-reactivity among various kinds of fish roes [23, 58, 76]. Although the structure of β'-component is different among fish species, the specific IgE contained in salmon-roe-allergic patient's serum reacts with various kinds of fish β'-component. Strong IgE cross-reactivity via β'-component was established between salmonids (chum salmon, rainbow trout, and Sakhalin taimen) [23], and the specific IgE to chum salmon β'-component was reacted with other fish β'-component, such as walleye pollack, Pacfic herring, dusky sole, and grey mullet [62, 76]. On the other hand, there is no report about the establishment of IgE cross-reactivity between chicken egg and fish roe. Kondo et al. [77] explained the absence of IgE cross-reactivity between hen egg and salmon roe by investigating specific IgE in allergic patients' sera.

8. Conclusion

This chapter described that seafood contain a wide variety of allergens, and that the range of the IgE cross-reactivity extends not only between similar organisms, but also beyond species. These features of seafood allergy indicate the importance to read food allergen labeling system in processed seafoods based on the information of IgE cross-reactivity. It is necessary to continually accumulate the information concerning IgE cross-reactivity among seafood allergens, because it contributes to improvement of food allergen labeling systems, as well as diet guidance and ensuring food safety in marine bioresources.

References

1   Japanese society of pediatric allergy and clinical immunology. 2012. Japanese pediatric guideline for food allergy 2012. http://www.jspaci.jp/jpgfa2012/. Accessed 17 Jul 2017.

2   Boyce, J.A., Assa'ad, A., Burks, A.W., Jones, S.M., Sampson, H.A., Wood, R.A., Plaut, M., Cooper, S.F., Fenton, M.J., Arshad, S.H., Bahna, S.L., Beck, L.A., Byrd-Bredbenner, C., Camargo, C.A., Eichenfield, L., Furuta, G.T., Hanifin, J.M., Jones, C., Kraft, M., Levy, B.D., Lieberman, P., Luccioli, S., Mccall, K.M., Schneider, L.C., Simon, R.A., Simons, F.E.R., Teach, S.J., Yawn, B.P. and Schwaninger, J.M. 2010. Guidelines for the diagnosis and management of food allergy in the United States: Summary of the NIAID-

sponsored expert panel report. J. Allergy Clin. Immunol. 126: 1105–1118. Doi: 10.1016/j. jaci.2010.10.008.

3   Woods, R.K., Abramson, M., Bailey, M. and Walters, E.H. 2001. International prevalences of reported food allergies and intolerances. Comparisons arising from the European Community Respiratory Health Survey (ECRHS) 1991–1994. Eur. J. Clin. Nutr. 55: 298–304. Doi: 10.1038/sj.ejcn.1601159.

4   Alvarado, M.I. and Pérez, M. 2006. Study of food allergy in the Spanish population. Allergol. Immunopathol. (Madr.) 34: 185–193.

5   Perez-Sanchez, N.I., Blanca-Lopez, N., Victorio, L., Lopez, J.D., Feo-Brito, F., Villalba, M., Blanca, M., Monteseirin, F., Canto, G. and Cornejo, J.A. 2017. A Population study of sensitisation and allergy to relevant inhaled and food allergens in a Mediterranean region. J. Allergy Clin. Immunol. 139: AB159. Doi: 10.1016/j.jaci.2016.12.523.

6   Boyce, J.A., Assa'ad, A., Burks, A.W., Jones, S.M., Sampson, H.A., Wood, R.A., Plaut, M., Cooper, S.F. and Fenton, M.J. 2010. Guidelines for the diagnosis and management of food allergy in the United States: Report of the NIAID-Sponsored expert panel. J. Allergy Clin. Immunol. 126: S1–58. Doi: 10.1016/j.jaci.2010.10.007.Guidelines.

7   Ebisawa, M., Ito, K. and Fujisawa, T. 2017. Japanese guidelines for food allergy 2017 on behalf of Committee for Japanese Pediatric Guideline for Food Allergy. Allergol. Int. 66: 248–264. Doi: 10.1016/j.alit.2017.02.001.

8   Akiyama, H., Imai, T. and Ebisawa, M. 2011. Japan food allergen labeling regulation-history and evaluation. pp. 139–171. *In*: Taylor, S. (ed.). Adv. Food Nutr. Res. Academic Press.

9   Gupta, R.S., Springston, E.E., Warrier, M.R., Smith, B., Kumar, R., Pongracic, J. and Holl, J.L. 2011. The prevalence, severity, and distribution of childhood food allergy in the United States. Pediatrics 128: e9–e17. Doi: 10.1542/peds.2011-0204.

10 Kumar, S., Verma, A.K., Das, M. and Dwivedi, P.D. 2012. Molecular mechanisms of IgE mediated food allergy. Int. Immunopharmacol. 13: 432–439. Doi: 10.1016/j. intimp.2012.05.018.

11 Singh, M.B. and Bhalla, P.L. 2008. Genetic engineering for removing food allergens from plants. Trends Plant Sci. 13: 257–260. Doi: 10.1016/j.tplants.2008.04.004.

12 Bischoff, S.C. 2007. Role of mast cells in allergic and non-allergic immune responses: Comparison of human and murine data. Nat. Rev. Immunol. 7: 93–104. Doi: 10.1038/ nri2018.

13 Kumar, S., Verma, A.K., Das, M. and Dwivedi, P.D. 2012. Molecular mechanisms of IgE mediated food allergy. Int. Immunopharmacol. 13: 432–439. Doi: 10.1016/j. intimp.2012.05.018.

14 Burks, A.W., Tang, M., Sicherer, S., Muraro, A., Eigenmann, P.A., Ebisawa, M., Fiocchi, A., Chiang, W., Beyer, K., Wood, R., Hourihane, J., Jones, S.M., Lack, G. and Sampson, H.A. 2012. ICON: Food allergy. J. Allergy Clin. Immunol. 129: 906–920. Doi: 10.1016/j. jaci.2012.02.001.

15 Morita, H., Nomura, I., Matsuda, A., Saito, H. and Matsumoto, K. 2013. Gastrointestinal food allergy in infants. Allergol. Int. 62: 297–307. Doi: 10.2332/allergolint.13-RA-0542.

16 Koyama, H., Michiko, K., Kawamura, M., Tokuda, R., Kondo, Y., Tsuge, I., Yamada, K., Yasuda, T. and Urisu, A. 2006. Grades of 43 fish species in Japan based on IgE-binding activity. Allergol. Int. 55: 311–316. Doi: 10.2332/allergolint.55.311.

17 Kobayashi, Y., Yang, T., Yu, C.-T., Ume, C., Kubota, H., Shimakura, K., Shiomi, K. and Hamada-Sato, N. 2016. Quantification of major allergen parvalbumin in 22 species of fish by SDS–PAGE. Food Chem. 194: 345–353. Doi: 10.1016/j.foodchem.2015.08.037.

18 Gendel, S.M. 2012. Comparison of international food allergen labeling regulations. Regul. Toxicol. Pharmacol. 63: 279–285. Doi: 10.1016/j.yrtph.2012.04.007.

19   University of Nebraska–Lincoln Food Allergens—International Regulatory Chart. Food Allergy Research and Resource Program. http://farrp.unl.edu/IRChart. Accessed 29 Jun 2017.

20   Elsayed, S. and Bennich, H. 1975. The primary structure of allergen M from cod. Scand. J. Immunol. 4: 203–208.

21   Hamada, Y., Nagashima, Y. and Shiomi, K. 2001. Identification of collagen as a new fish allergen. Biosci. Biotechnol. Biochem. 65: 285–291. Doi: 10.1271/bbb.65.285.

22   Reese, G., Ayuso, R. and Lehrer, S.B. 1999. Tropomyosin: an invertebrate pan-allergen. Int. Arch. Allergy Immunol. 119: 247–258. Doi: 24201.

23   Shimizu, Y., Nakamura, A., Kishimura, H., Hara, A., Watanabe, K. and Saeki, H. 2009. Major allergen and its IgE cross-reactivity among salmonid fish roe allergy. J. Agric. Food Chem. 57: 2314–2319. Doi: 10.1021/jf8031759.

24   Tong, P., Gao, J., Chen, H., Li, X., Zhang, Y., Jian, S., Wichers, H., Wu, Z., Yang, A. and Liu, F. 2012. Effect of heat treatment on the potential allergenicity and conformational structure of egg allergen ovotransferrin. Food Chem. 131: 603–610. Doi: 10.1016/j.foodchem.2011.08.084.

25   Das Dores, S., Chopin, C., Romano, A., Galland-Irmouli, A.-V., Quaratino, D., Pascual, C., Fleurence, J. and Guéant, J.-L. 2002. IgE-binding and cross-reactivity of a new 41 kDa allergen of codfish. Allergy 57 Suppl. 7: 84–87.

26   Yu, C.-J., Lin, Y.-F., Chiang, B.-L. and Chow, L.-P. 2003. Novel shrimp allergen, Pen m 2 proteomics and immunological analysis of a proteomics and immunological analysis of a novel shrimp allergen, Pen m 2 1. J. Immunol. 170: 445–453. Doi: 10.4049/jimmunol.170.1.445.

27   García-Orozco, K.D., Aispuro-Hernández, E., Yepiz-Plascencia, G., Calderón-de-la-Barca, A.M. and Sotelo-Mundo, R.R. 2007. Molecular characterization of arginine kinase, an allergen from the shrimp Litopenaeus vannamei. Int. Arch. Allergy Immunol. 144: 23–28. Doi: 10.1159/000102610.

28   Shen, H.W., Cao, M.J., Cai, Q.F., Ruan, M.M., Mao, H.Y., Su, W.J. and Liu, G.M. 2012. Purification, cloning, and immunological characterization of arginine kinase, a novel allergen of octopus fangsiao. J. Agric. Food Chem. 60: 2190–2199. Doi: 10.1021/jf203779w.

29   Hindley, J., Wünschmann, S., Satinover, S.M., Woodfolk, J.A., Tim Chew, F., Chapman, M.D. and Pomés, A. 2006. Bla g 6: A troponin C allergen from Blattella germanica with IgE binding calcium dependence. J. Allergy Clin. Immunol. 117: 1389–1395. Doi: 10.1016/j.jaci.2006.02.017.

30   Ayuso, R., Grishina, G., Bardina, L., Carrillo, T., Blanco, C., Ibáñez, M.D., Sampson, H.A. and Beyer, K. 2008. Myosin light chain is a novel shrimp allergen, Lit v 3. J. Allergy Clin. Immunol. 122: 795–802. Doi: 10.1016/j.jaci.2008.07.023.

31   Shiomi, K., Sato, Y., Hamamoto, S., Mita, H. and Shimakura, K. 2008. Sarcoplasmic calcium-binding protein: Identification as a new allergen of the black tiger shrimp Penaeus monodon. Int. Arch. Allergy Immunol. 146: 91–98. Doi: 10.1159/000113512.

32   Suzuki, M., Kobayashi, Y., Hiraki, Y., Nakata, H. and Shiomi, K. 2011. Paramyosin of the disc abalone Haliotis discus discus: Identification as a new allergen and cross-reactivity with tropomyosin. Food Chem. 124: 921–926. Doi: 10.1016/j.foodchem.2010.07.020.

33   Arif, S.H. 2009. A Ca(2+)-binding protein with numerous roles and uses: parvalbumin in molecular biology and physiology. Bioessays 31: 410–421. Doi: 10.1002/bies.200800170.

34   Gillis, J.M. 1985. Relaxation of vertebrate skeletal muscle. A synthesis of the biochemical and physiological approaches. Biochim. Biophys. Acta 811: 97–145. Doi: 10.1016/0304-4173(85)90016-3.

35   Donato, H. and Martin, R.B. 1974. Conformations of carp muscle calcium binding parvalbumin. Biochemistry 13: 4575–4579. Doi: 10.1021/bi00719a016.

36   WHO, IUIS Allergen Nomenclature Home Page. http://www.allergen.org/. Accessed 17 Feb 2016.

37 Hilger, C., Grigioni, F., Thill, L., Mertens, L. and Hentges, F. 2002. Severe IgE-mediated anaphylaxis following consumption of fried frog legs: definition of a-parvalbumin as the allergen in cause. Allergy 57: 1053–1058. Doi: 10.1034/j.1398-9995.2002.23677.x.

38 Hamada, Y., Nagashima, Y. and Shiomi, K. 2004. Reactivity of serum immunoglobulin E to bullfrog Rana catesbeiana parvalbumins in fish-allergic patients. Fish Sci. 70: 1137–1143.

39 Elsayed, S. and Apold, J. 1983. Immunochemical analysis of cod fish allergen M: locations of the immunoglobulin binding sites as demonstrated by the native and synthetic peptides. Allergy 38: 449–459.

40 Perez-Gordo, M., Pastor-Vargas, C., Lin, J., Bardina, L., Cases, B., Ibanez, M.D., Vivanco, F., Cuesta-Herranz, J. and Sampson, H.A. 2013. Epitope mapping of the major allergen from Atlantic cod in Spanish population reveals different IgE-binding patterns. Mol. Nutr. Food Res. 57: 1283–1290. Doi: 10.1002/mnfr.201200332.

41 Swoboda, I., Bugajska-Schretter, A., Verdino, P., Keller, W., Sperr, W.R., Valent, P., Valenta, R. and Spitzauer, S. 2002. Recombinant carp parvalbumin, the major cross-reactive fish allergen: A tool for diagnosis and therapy of fish allergy. J. Immunol. 168: 4576–4584. Doi: 10.4049/jimmunol.168.9.4576.

42 Yoshida, S., Ichimura, A., Shiomi, K. and Vier, E. 2008. Elucidation of a major IgE epitope of Pacific mackerel parvalbumin. Food Chem. 111: 857–861. Doi: 10.1016/j. foodchem.2008.04.062.

43 Perez-Gordo, M., Lin, J., Bardina, L., Pastor-Vargas, C., Cases, B., Vivanco, F., Cuesta-Herranz, J. and Sampson, H.A. 2011. Epitope mapping of Atlantic salmon major allergen by peptide microarray immunoassay. Int. Arch. Allergy Immunol. 157: 31–40. Doi: 10.1159/000324677.

44 Bugajska-Schretter, A., Elfman, L., Fuchs, T., Kapiotis, S., Rumpold, H., Valenta, R. and Spitzauer, S. 2014. Parvalbumin, a cross-reactive fish allergen, contains IgE-binding epitopes sensitive to periodate treatment and Ca2+ depletion. J. Allergy Clin. Immunol. Volume 101: 67–74. Doi: 10.1016/S0091-6749(98)70195-2.

45 Kobayashi, A., Ichimura, A., Kobayashi, Y. and Shiomi, K. 2016. IgE-binding epitopes of various fish parvalbumins exist in a stereoscopic conformation maintained by Ca2+ binding. Allergol. Int. 65: 345–348. Doi: 10.1016/j.alit.2016.02.004.

46 Hamada, Y., Nagashima, Y., Shiomi, K., Shimojo, N., Kohno, Y., Shibata, R., Nishima, S., Ohsuna, H. and Ikezawa, Z. 2003. Reactivity of IgE in fish-allergic patients to fish muscle collagen. Allergol. Int. 52: 139–147. Doi: 10.1046/j.1440-1592.2003.00293.x.

47 Kobayashi, Y., Huge, J., Imamura, S. and Hamada-Sato, N. 2016. Study of the cross-reactivity of fish allergens based on a questionnaire and blood testing. Allergol. Int. 65: 1–8. Doi: 10.1016/j.alit.2016.01.002.

48 Shiomi, K., Yoshida, S., Sawaguchi, T. and Ishizaki, S. 2010. A major IgE epitope of rainbow trout collagen alpha 2 chain. Food Hyg. Saf. Sci. 51: 153–159.

49 Hori, H., Hattori, S., Inouye, S., Kimura, A., Irie, S., Miyazawa, H. and Sakaguchi, M. 2002. Analysis of the major epitope of the aplha 2 chain of bovine type I collagen in children with bovine gelatin allergy. J. Allergy Clin. Immunol. 110: 652–657. Doi: 10.1067/mai.2002.127862.

50 Sakaguchi, M., Toda, M., Ebihara, T., Irie, S., Hori, H., Imai, A., Yanagida, M., Miyazawa, H., Ohsuna, H., Ikezawa, Z. and Inouye, S. 2000. IgE antibody to fish gelatin (type I collagen) in patients with fish allergy. J. Allergy Clin. Immunol. 106: 579–584. Doi: 10.1067/mai.2000.108499

51 Kobayashi, Y., Akiyama, H., Huge, J., Kubota, H., Chikazawa, S., Satoh, T., Miyake, T., Uhara, H., Okuyama, R., Nakagawara, R., Aihara, M. and Hamada-Sato, N. 2016. Fish collagen is an important panallergen in the Japanese population. Allergy 71: 720–723. Doi: 10.1111/all.12836.

52   Liu, R., Holck, A.L., Yang, E., Liu, C. and Xue, W. 2013. Tropomyosin from tilapia (Oreochromis mossambicus) as an allergen. Clin. Exp. Allergy 43: 365–377. Doi: 10.1111/cea.12056.

53   Miyazawa, H., Fukamachi, H., Inagaki, Y., Reese, G., Daul, C.B., Lehrer, S.B., Inouye, S. and Sakaguchi, M. 1996. Identification of the first major allergen of a squid (Todarodes pacificus). J. Allergy Clin. Immunol. 98: 948–953.

54   Leung, P.S., Chow, W.K., Duffey, S., Kwan, H.S., Gershwin, M.E. and Chu, K.H. 1996. IgE reactivity against a cross-reactive allergen in crustacea and mollusca: evidence for tropomyosin as the common allergen. J. Allergy Clin. Immunol. 98: 954–961.

55   Ishikawa, M., Suzuki, F., Ishida, M., Nagashima, Y., Shiomi, K., Masaru Ishikawa, Fumi Suzuki, Masami Ishida, Yuji Nagashima and Kazuo Shiomi. 2001. Identification of tropomyosin as a major allergen in the octopus Octopus vulgaris and elucidation of its IgE-binding epitopes. Fish Sci. 67: 934–942. Doi: 10.1046/j.1444-2906.2001.00344.x.

56   Emoto, A., Ishizaki, S. and Shiomi, K. 2009. Tropomyosins in gastropods and bivalves: Identification as major allergens and amino acid sequence features. Food Chem. 114: 634–641. Doi: 10.1016/j.foodchem.2008.09.100.

57   Zhang, H., Lu, Y., Ushio, H. and Shiomi, K. 2014. Development of sandwich ELISA for detection and quantification of invertebrate major allergen tropomyosin by a monoclonal antibody. Food Chem. 150: 151–157. Doi: 10.1016/j.foodchem.2013.10.154.

58   Shimizu, Y. and Saeki, H. 2010. Characterization of fish roe allergens. pp. 47–59. In: Shiomi, K. and Saeki, H. (eds.). Science of Seafood Allergens. Koseisha-Koseikaku, Tokyo.

59   Hiramatsu, N., Cheek, A.O., Sullivan, C.V., Matsubara, T. and Hara, A. 2005. Vitellogenesis and endcrine distribution. pp. 431–471. In: Mommsen, T.P. and Moon, T.W. (eds.). Biochem. Mol. Biol. fishes. Vol. 6. Elsevier Science Ltd., Amsterdam, The Netherlands.

60   Shimizu, Y., Kishimura, H., Kanno, G., Nakamura, A., Adachi, R., Akiyama, H., Watanabe, K., Hara, A., Ebisawa, M. and Saeki, H. 2014. Molecular and immunological characterization of β′-component (Onc k 5), a major IgE-binding protein in chum salmon roe. Int. Immunol. 26: 139–147. Doi: 10.1093/intimm/dxt051.

61   Fujita, S., Shimizu, Y., Kishimura, H., Watanabe, K., Hara, A. and Saeki, H. 2012. In vitro digestion of major allergen in salmon roe and its peptide portion with proteolytic resistance. Food Chem. 130: 644–650. Doi: 10.1016/j.foodchem.2011.07.099.

62   Li, Z., Shimizu, Y. and Saeki, H. 2016. Grey mullet roe contains yolk protein having IgE cross-reactivity to chum salmon roe major allergen (Onc k 5). Fish Sci. Doi: 10.1007/s12562-016-1057-x.

63   Kuehn, A., Swoboda, I., Arumugam, K., Hilger, C., Hentges, F., Smith, K.A. and Goodman, R.E. 2014. Fish allergens at a glance: variable allergenicity of parvalbumins, the major fish allergens. Doi: 10.3389/fimmu.2014.00179.

64   Van Do, T., Elsayed, S., Florvaag, E., Hordvik, I. and Endresen, C. 2005. Allergy to fish parvalbumins: Studies on the cross-reactivity of allergens from 9 commonly consumed fish. J. Allergy Clin. Immunol. 116: 1314–1320. Doi: 10.1016/j.jaci.2005.07.033.

65   Kuehn, A., Hutt-Kempf, E., Hilger, C. and Hentges, F. 2011. Clinical monosensitivity to salmonid fish linked to specific IgE-epitopes on salmon and trout beta-parvalbumins. Allergy 66: 299–301. Doi: 10.1111/j.1398-9995.2010.02461.x.

66   Yuki, H. 2010. Characterization of fish allergens. pp. 9–20. In: Shiomi, K. and Saeki, H. (eds.). Science of Seafood Allergens. Koseisha-Koseikaku, Tokyo.

67   Kobayashi, Y., Huge, J., Imamura, S. and Hamada-Sato, N. 2016. Study of the cross-reactivity of fish allergens based on a questionnaire and blood testing. Allergol. Int. 65: 272–279. Doi: 10.1016/j.alit.2016.01.002

68   Kobayashi, Y., Kuriyama, T., Nakagawara, R., Aihara, M. and Hamada-Sato, N. 2016. Allergy to fish collagen: Thermostability of collagen and IgE reactivity of patients' sera

with extracts of 11 species of bony and cartilaginous fish. Allergol. Int. 65: 450–458. Doi: 10.1016/j.alit.2016.04.012.

69 Ayuso, R., Lehrer, S.B. and Reese, G. 2002. Identification of continuous, allergenic regions of the major shrimp allergen pen a 1 (Tropomyosin). Int. Arch. Allergy Immunol. 127: 27–37. Doi: 10.1159/000048166.

70 Ayuso, R., Reese, G., Leong-Kee, S., Plante, M. and Lehrer, S.B. 2002. Molecular basis of arthropod cross-reactivity: IgE-binding cross-reactive epitopes of shrimp, house dust mite and cockroach tropomyosins. Int. Arch. Allergy Immunol. 129: 38–48.

71 Witteman, A.M., Akkerdaas, J.H., van Leeuwen, J., van der Zee, J.S. and Aalberse, R.C. 1994. Identification of a cross-reactive allergen (Presumably tropomyosin) in shrimp, mite and insects. Int. Arch. Allergy Immunol. 105: 56–61. Doi: 10.1159/000236803.

72 Lehrer, S.B. and Reese, G. 1998. Cross-reactivity between cockroach allergens and arthropod, nematode and mammalian allergens. Rev Française d'Allergologie d'Immunologie Clin. 38: 846–848. Doi: 10.1016/S0335-7457(98)80154-8.

73 Ayuso, R., Gerald, R., Leong-Kee, S., Plante, M. and Lehrer, S.B. 2002. Molecular basis of arthropod cross-reactivity: IgE-binding cross-reactive epitopes of shrimp, house dust mite and cockroach tropomyosins. Int. Arch. Allergy Immunol. 129: 38–48.

74 Popescu, F.-D. 2015. Cross-reactivity between aeroallergens and food allergens. World J. Methodol. 5: 31–50. Doi: 10.5662/wjm.v5.i2.31.

75 Sakai, S., Adachi, R., Shibahara, Y., Oka, M., Abe, A., Seiki, K., Oda, H., Yoshioka, H., Urisu, K., Akiyama, H. and Teshima, R. 2008. A sunvey of crustacean soluble proteins such as "shrimp" and "crab" content in food ingredients. Japanese J. Food Chem. 15: 12–17.

76 Shimizu, Y., Oda, H., Seiki, K. and Saeki, H. 2015. Development of an enzyme-linked immunosorbent assay system for detecting β'-component (Onk k 5), a major IgE-binding protein in salmon roe. Food Chem. 181: 310–317. Doi: 10.1016/j.foodchem.2015.02.071.

77 Kondo, Y., Kakami, M., Koyama, H., Yasuda, T., Nakajima, Y., Kawamura, M., Tokuda, R., Tsuge, I. and Urisu, A. 2005. IgE Cross-reactivity between fish roe (salmon , herring and pollock) and chicken egg in patients anaphylactic to salmon roe. Allergol. Int. 54: 317–323.

12

Scombroid (Histamine) Poisoning Associated with Seafood

Masataka Satomi

1. Introduction, Scombrotoxin Fish Poisoning (SFP)

Histamine is a biogenic amine, and one of the important chemical compounds that causes food-borne intoxication, which is called an allergy-like food poisoning (EFSA 2011, FAO/WHO 2012, Hungerford 2010, Ladero et al. 2010, Halasz et al. 1994, Silla Santos 1996, Spano et al. 2010, Taylor 1986, Toda et al. 2009). The representative symptoms of histamine poisoning are rash, facial flushing, and low blood pressure, when humans ingest large amounts of histamine at one meal. It is easy to accumulate in marine products, especially scombroid fish treated with unsuitable handlings. In the case of allergy-like food poisoning caused by scombroid fish (EFSA 2011, Emborg et al. 2006, 2008, FAO/WHO 2012, Kanki et al. 2002, 2004, 2007, Lehane et al. 2000, Nei et al. 2012, Okuzumi et al. 1981, 1994, Sato et al. 1994, Takahashi et al. 2003, 2015, Tao et al. 2002, 2009, Torido et al. 2012a,b, Visciano et al. 2012), it is called scombroid fish poisoning (SFP). However, the main causative agent of SFP is histamine, which is derived from bacterial decarboxylation of L-histidine included in scombroid

Japan Fisheries Research and Education Agency, Japan.
E-mail: msatomi@affrc.go.jp

fish body. Histamine accumulation occasionally occurs in variety of foods, except for scombroid fish products, fermented seafood (Harada et al. 2008, Hernandez-Herrero et al. 1999, Mongkolthanaruk et al. 2012, Nakazato et al. 2002, Sato et al. 1995, Satomi et al. 1997, 2008, 2011, 2012, Stute et al. 2002, Yatsunami and Echigo 1991, 1993), cheese (Burdychova and Komprda 2007, Chang and Snell 1968, Chang et al. 1985, Joosten and Northolt 1989), wine (Coton et al. 1998, 2010, Coton and Coton 2005, Landete et al. 2005, Lonvaud-Funel and Joyeux 1994, Lonvaud-Funel 2001, Lucas et al. 2005, 2008), meat products (Landeta et al. 2007, Silla Santos 1998, Suzzi and Gardini 2003), and others (Calles-Enríquez et al. 2010, Hamaya et al. 2014, Ibe et al. 2003, Le Jeune et al. 1995, Tsai et al. 2006, 2007), though fish products are the most important causative food in an allergy-like food poisoning. Therefore, the regulation values are set for an amount of histamine in many countries and organizations such as FDA, CODEX, and others. Here I would like to mention characteristics of histamine and histamine producing bacteria.

2. Toxicological Aspects of Histamine and other Biogenic Amines

2.1 Histamine

Chemically, the molecular mass of histamine ($C_5H_9N_3$) is 111.14, the aqueous solution is odorless, colorless, and transparent, and histamine is a basic compound since it is a biogenic amine. Histamine is stable at elevated temperatures, indicating that histamine is not destroyed by usual cooking method (EFSA 2011, FAO/WHO 2012, Hungerford 2010, Taylor 1986). Histamine present in food is yielded form decarboxylation of L-histidine by certain bacteria (Fig. 12.1). In the mammal body including humans, it is usually stored *in vivo* in mast cells, enterochromaffin-like (ECL) cells, and basophils, and its transient release into the extracellular space can be triggered by an external stimulus, such as the binding of an antigen to the cell surface antibody (FAO/WHO 2012, Hungerford 2010). Histamine exerts its pharmacological effects, such as vasodilation, by acting as a mediator of allergic reaction and inflammation.

2.2 Other biogenic amines

Although the primary causative agent of SFP is histamine, other biogenic amines derived from amino acid by microbial metabolism can also cause allergy-like symptoms cooperatively with histamine. Tyramine, cadaverine, putrescine, and others sometimes act as antagonists for histamine-metabolizing enzymes, such as diamine oxidase and histamine-N-methyltransferase, though small amounts of these amines cannot affect the body (FAO/WHO 2012, Hungerford 2010, Ladero et al. 2010, Taylor

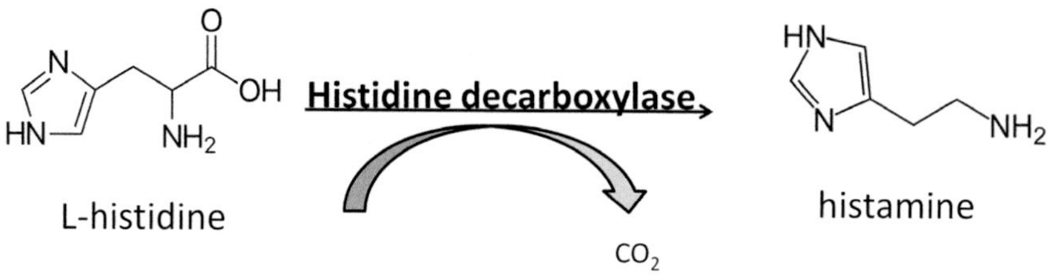

Figure 12.1 Conversion of L-histidine to histamine. Decarboxylation of L-histidine is caused by bacterial enzymes in the food.

1986). Therefore, these amines in foods have also been studied. Since tyramine is sometimes accumulated in fermented food, including fish sauce with significant level (Stute et al. 2002), it should be focused on tyramine producing bacteria (Coton and Coton 2005, 2009, Lucas et al. 2003, Satomi et al. 2014). Some reviews describing biogenic amines are available (FAO/WHO 2012, Hungerford 2010, Ladero et al. 2010, Taylor 1986).

2.3 Dose-response in human

2.3.1 Symptom

Large amounts of histamine can cause hypertension, hypotension, headache, urticaria, nausea, and vomiting, among other symptoms (EFSA 2011, FAO/WHO 2012, Hungerford 2010, Ladero et al. 2010, Taylor 1986). Ingestion of foods containing large amounts of histamine can sometimes trigger histamine poisoning, including SFP or an allergy-like food poisoning, which is clearly different from a food allergy that patients develop as an immune reaction after eating specific causal foods. People do not develop histamine poisoning unless they ingest more than the allowed amount of histamine. In other words, food poisoning does not occur in the absence of a causal agent, such as large amount histamine in fish and shellfish (FAO/WHO 2012).

No observed adverse effect level

The no observed adverse effect level (NOAEL), the appropriate hazard level for healthy individuals as determined by the Joint FAO/WHO Expert Meeting on the Public Health Risks of Histamine and Other Biogenic Amines from Fish and Fishery Products (2012), was a dose of 50 mg of histamine. Moreover, using the available consumption data combined with expert opinions, the meeting agreed that a serving size of 250 g captured the maximum amount eaten in most countries at a single eating event. Based on the hazard level of 50 mg of histamine and a serving size of 250 g, the maximum concentration of histamine in that serving was consequently calculated to be 200 mg/kg. Most of the histamine that is ingested is rapidly eliminated in the urine. After the remaining histamine in the body is degraded, it is released as carbon dioxide. Therefore, histamine is eliminated within a few hours of ingestion and is not accumulated in the body.

2.4 Analytical methods for biogenic amines

Many simple kits for measuring the histamine content in foods are commercially available (FAO/WHO 2012). Measurement principle is also diverse, there is the ELISA method using antihistamine antibody, and colorimetric method using histamine degrading enzyme (Table 12.1). Since these kits do not require special measuring equipment, they are often used as part of quality control at the food production site. The AOAC method which has been used as an official method from ancient times is complicated in operation, and a certain degree of skill is required for measurement. In recent years, high performance liquid chromatography (HPLC) has become widespread, and measurement by this method is mainstream in each inspection institution and research institute. In addition, the LC-MS method combining HPLC with a mass spectrometer (MS) has also been developed. Whereas the simple kit and AOAC method are limited to the measurement of histamine only, HPLC and LC-MS methods have the advantage of being able to simultaneously measure other biogenic amines, such as tyramine and putrescine. Also, the accuracy of the analysis is higher than the former. Common problems of each method include extraction of amines from food

Table 12.1 List of the most commonly used test methods for determination of histamine levels.

	AOAC	HPLC	ELISA	Colorimetric method
Time needed for one test	1–2 h	1–2 h	1 h	1 h
Equipment	Fluorometer	HPLC	Spectrofluorometer	Spectrophotometer
Limit of quantification	1–5 mg/kg	1.5–5 mg/kg	2–5 mg/kg	20 mg/kg
Range	1–150 mg/kg	5–2500 mg/kg	0–500 mg/kg	0.8–300 mg/kg
General advantages	Robust, repeatable, accurate, precise	Quantification of all biogenic amines, accuracy, precise	Easy (kit), fast, low equipment costs and possibility of multiple tests simultaneously	Easy (kit), fast, low equipment costs and possibility of multiple tests simultaneously. Sample calibration and possibility of semiquantitative evaluation by visual colorimetry

specimens. Extraction work is time-consuming and a laborious task, and shortening processing time and reducing workload at the site depends on simplification of this process.

3. Histamine Producing Bacteria

As mentioned above, histamine production is the result of bacterial decarboxylation of L-histidine, and histamine accumulation occurs in many foods other than fish products. Histamine-producing bacteria have histidine decarboxylase (HDC: EC.4.1.1.22), which catalyzes the conversion of L-histidine to histamine. Histamine production in bacteria is one of the stress responses to low pH in the environment surrounding bacteria (Molenaar et al. 1993). Two HDC isozymes are known as bacterial enzymes: pyridoxal phosphate (PLP)-dependent HDC in gram-negatives and pyruvoyl-type HDC in gram-positives (Landete et al. 2008). Interestingly, the bacterial species that produces histamine is somewhat specific in each food (Fig. 12.2).

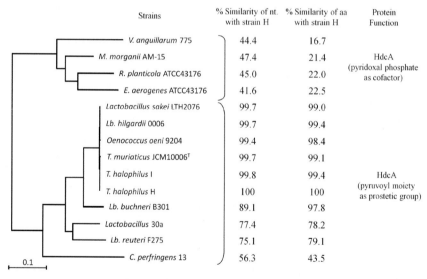

Strains	% Similarity of nt. with strain H	% Similarity of aa with strain H	Protein Function
V. anguillarum 775	44.4	16.7	
M. morganii AM-15	47.4	21.4	HdcA (pyridoxal phosphate as cofactor)
R. planticola ATCC43176	45.0	22.0	
E. aerogenes ATCC43176	41.6	22.5	
Lactobacillus sakei LTH2076	99.7	99.0	
Lb. hilgardii 0006	99.7	99.4	
Oenococcus oeni 9204	99.4	98.4	
T. muriaticus JCM10006ᵀ	99.7	99.1	
T. halophilus I	99.8	99.4	HdcA (pyruvoyl moiety as prostetic group)
T. halophilus H	100	100	
Lb. buchneri B301	89.1	97.8	
Lactobacillus 30a	77.4	78.2	
Lb. reuteri F275	75.1	79.1	
C. perfringens 13	56.3	43.5	

Figure 12.2 Genetic relationships among the hdcA of different bacteria. Phylogenetic tree were constructed based on the nucleotide sequences of hdcA using the Neighbor-Joining method by Saitou and Nei (1987). The bar indicates genetic distance (Knuc). The nucleotide accession numbers are as follows; *Vibrio anguillarum* (AY312585), *Morganella morganii* (J02577), *Klebsiella planticola* (M62746), *Enterobacter aerogenes* (M62745), *Lactobacillus sakei* (AY800122), *L. hilgardii* (AY651779), *Oenococcus oeni* (U58865), *Tetragenococcus muriaticus* (AB040487), *T. halophilus* (AB076394), The strain H (AB362339), *L. buchneri* (AJ749838), *L. reuteri* (CP000705), *Lactobacillus* 30a (J02613), *Clostridium perfringens* (NC_003366).

3.1 Gram-negative histamine producing bacteria

In the case of raw fish, such as scombroid fish, mainly gram-negative bacteria, Enterobacteriaceae, and Photobacterium spp., are known as histamine producers (EFSA 2011, Emborg et al. 2006, 2008, FAO/WHO 2012, Kanki et al. 2002, 2004, 2007, Kimura et al. 2000, Lehane, 2000, Okuzumi et al. 1981, 1994, Sato et al. 1994, Takahashi et al. 2003, 2015, Tao et al. 2002, 2009, Torido et al. 2012a,b, Visciano et al. 2012). Major histamine producing bacteria are listed in Table 12.2. Although the original habitation of both bacterial groups is different, they are problematic organisms as histamine producers in fishery products. In Japan, histamine producing bacteria belonging to the both bacterial groups were isolated from commercial raw fish in retail shops through the season (Yoguchi et al. 1990a,b). Both bacterial groups have pyridoxal phosphate (PLP)-dependent HDC. Bacteriological characteristics of both groups are mentioned as follows.

Table 12.2 List of gram-negative histamine producing bacteria.

Species		Taxonomic characteristics	Original habitat	Growth feature	References
Morganella	*morganii*	Enterobacteriaceae	Mammalial intestine	mesophile	Miyaki 1954
	psychrotolerans			psychrophile	Basby et al. 1998
Enterobacter	*aerogenes*			mesophile	Taylor 1986
Klebsiella	*oxytoca*			mesophile	Kanki et al. 2002
Raoultella	*planticola*			mesophile	Kanki et al. 2002
Photobacterium	*damselae*	Vibrionaceae	Marine environments	mesophile	Okuzumi et al. 1994
	phosphoreum			mesophile	Kanki et al. 2004
	kishitanii			psychrophile	Bjornsdottir-Butler et al. 2017
	angustum			psychrophile	Bjornsdottir-Butler et al. 2017
Vibrio	*anguillarum*			psychrophile	Tolmasky et al. 1995

a) Enterobacteriaceae

Bacteriological characteristics

Histamine-producing Enterobactericeae include common enteric mesophiles, such as *Morganella, Enterobacter, Hafnia, Raoultella,* and others (EFSA 2011, Emborg et al. 2006, 2008, FAO/WHO 2012, Kanki et al. 2002, 2004, 2007, Kimura et al. 2000, Lehane, 2000, Takahashi et al. 2003, Tao et al. 2002, 2009, Visciano et al. 2012), which have been studied for a long time as histamine-producing bacteria in fish and marine products. Phenotypic characteristics of this group bacteria are gram negative, rod, facultative anaerobic, D-glucose catabolizing, oxidase negative, non-halophilic, and others (Brenner and Farmer III 2005). *Morganella morganii* is especially famous as a significant histamine producer owing to its rapid growth and the potent activity of its histamine-producing enzyme. The optimum growth temperature of this group bacteria is around 37°C. Many species of Enterobactericeae are mesophilic, but some species are psychrotolerant. Therefore, it is not a problem when foods contaminated with these bacteria are managed at low temperature (refrigerated). In recent years, *Morganella psychrotolerans* that can grow even at low temperatures have been reported (Emborg et al. 2006, 2008). It should be able to grow and produce histamine in refrigerated food including raw fish or fish materials.

Habitat

Generally, the original habitation of Enterobacteriaceae bacteria is mammalian intestine and their neighboring environments, indicating that their optimal growth condition is nutrient rich, near mammalian body temperature, and so on (Brenner and Farmer III 2005). They are ubiquitously distributed in nature, mainly in terrestrial areas, but even through aquatic or estuarine environments. Since histamine producing strains have common biological features as Enterobacteriaceae bacteria with the exception of histamine production, their habitation is also same as that of Enterobacteriaceae bacteria. These bacteria also exist at fishing ports and foods processing plants, indicating that they have opportunity to adhere to the fish surface considerably frequently (FAO/WHO 2012, Yoguchi et al. 1990a,b). Multiple species of Enterobacteriaceae are isolated as histamine-producing bacteria from food poisoning specimens, and it is sometimes impossible to identify which kind is the causative bacteria. This is due to the extensive existence of this group of bacteria. It cannot be distinguished whether it was accidentally contaminated in the causative food, or whether it was the true causative bacteria. Also, it is sometimes difficult to identify where the histamine-producing bacteria are contaminated and histamine accumulated in the fish, because histamine-producing Enterobacteriaceae are attached frequently even in fish caught in coastal waters (Yoguchi et al. 1990a,b).

b) Marine bacteria

As the important histamine producers, some species belonging to the genus Photobacterium are known. Although certain strains of *Vibrio anguillarum*, which is a fish pathogen, also have been known to produce histamine (Tolmasky et al. 1995), basically oceanic histamine-producing bacteria are mainly bacteria belonging to the genus Photobacterium. Here we describe the genus Photobacterium which has been isolated as a histamine-producing bacterium from food.

Bacteriological characteristics

The genus Photobacterium is a marine bacterium belonging to the family Vibrionaceae; gram negative, rod, facultative anaerobic, D-glucose metabolizing, requires sodium ions for growth and motile by mean flagella (Framer III and Janda 2005). *P. damselae* and *P. phosphoreum* are known as main halophilic histamine producing bacteria for a long time (EFSA 2011, FAO/WHO 2012, Hungerford 2010, Kanki et al. 2002, 2004, 2007, Morii et al. 2006, Okuzumi et al. 1981, 1994, Tao et al. 2002, 2009, Torido et al. 2012a,b). In recent studies, other *Photobacterium* species, *P. iliopiscarium*, *P. kishitanii*, and *P. angstam*, are also recognized as histamine producers (Bjornsdottir-Butler et al. 2017, Takahashi et al. 2015). *P. damselae* is a mesophilic bacteria with 30–37°C for optimal temperature, hence other *Photobacterium* species are psychrophile, which can grow at 4°C, but not at 30°C. These marine bacteria require salt for growth, and are often not counted by the general viable count using standard plate agar. It is preferable to employ a medium consisting of seawater or artificial seawater for cultivation and growth of Photobacterium.

Habitat

The original habitat of *Photobacterium* spp. is marine environment, thus they are strictly oceanic organisms and easily isolated from marine organisms, seawater, and related materials. Some of the species belonging to the genus are known as luminescent symbiotic bacteria in some marine organisms. Therefore, *Photobacterium* spp. including histamine producing strain are universally present in seawater and adhere to seafood before capture (Framer III and Janda 2005, Yoguchi et al. 1990a,b). They can survive in humid environment for a while, but not in aired environment. Therefore,

Table 12.3 Comparative properties of pyridoxal phosphate dependent histidine decarboxylase in gram positive bacteria

Bacterium	Molecular mass		Subunit structure	Optimum pH	V_{max} (µmol· min⁻¹·mg⁻¹)	K_m (mmol/L)	References
	Native	Subunit					
Morganella morganii AM-15	17,000	42,742	α4	6.5	150	1.1	Guirard and Snell (1987)
Raoultella planticola	~ 20,000	~ 42,500	α4 and α2	6.5	142	2.4	Guirard and Snell (1987)
Enterobacter aerogenes	~ 20,000	~ 50,000	α2	6.5	110	2.2	Guirard and Snell (1987)
Photobacterium phosphoreum		42,600		5.5			Morii et al. (2006)

Photobacterium are generally not detected from highly processed food, and terrestrial environment.

c) Histamine producing enzyme

In the many PLP-dependent HDCs, the enzymes are constructed as a homotetramer of 170 kDa that is widely distributed in Enterobacteriaceae, *Photobacterium* spp., mammal liver, and others (Fujii et al. 1994, Guirard and Snell 1987, Kamath et al. 1991, Kanki et al. 2007, Morii et al. 2004, 2006, Tanase et al. 1985). The summary of molecular characteristics in HDC is shown in Table 12.3. Since this enzyme is resistant to freezing, histamine can be produced after thawing without inactivating it even if it is kept at –20°C with the crude enzyme solution. Therefore, when histamine-producing bacteria adhering to fresh fish are frozen together with foods and thawed, the enzyme leaking from the bacterial cells may produces histamine. Generally, since gram negative bacteria are weak against freezing, the number of bacteria after thawing is greatly reduced, and in some cases, histamine producing bacteria cannot be detected from the thawed food. In that case, histamine is produced in the absence of bacteria.

3.2 Gram-positive histamine producing bacteria

Gram-positive histamine-producing bacteria were isolated from fermented or aged foods (Table 12.4), such as wine, cheese, meat products, fish sauce, and others (references are quoted above). Non-halophilic lactic acid bacteria, such as *Oenococcus oeni*, *Lactobacillus* 30a, *L. hilgardii*, *L. buchneri*, and others, or staphylococci were the major sources of histamine accumulation in dairy

Table 12.4 List of major gram positive histamine producing bacteria related to food processing.

Species		Physiological characteristics	Original habitat	Growth feature	References
Tetragenococcus	*halophilus*	Halophile	salted fermented food	mesophile	Satomi et al. 2008
	muriaticus		salted fermented seafood	mesophile	Satomi et al. 1997
Staphylococcus	*epidermidis*	Halotolerant	salted fish paste	mesophile	Yokoi et al. 2011
	capitis		sausage	mesophile	De las Rivas et al. 2008
Lactobacillus	*otakiensis*	Non-halophile	salted marinade fish	psychrotolerant	Hamaya et al. 2014
	buchneri		cheese	mesophile	Martín et al. 2005
	hilgardii		wine	mesophile	Lucas et al. 2005
	30a		cheese	mesophile	Chang et al. 1968
Oenococcus	*oeni*	Alcohol tolerant	wine	mesophile	Coton et al. 1998
Streptococcus	*thermophilus*		milk product	mesophile	Calles-Enríquez et al. 2010

products. Additionally, histamine accumulation occasionally occurred in fishery products, such as fish sauce (Sato et al. 1995, Satomi et al. 1997, 2008, 2011, 2012, Stute et al. 2002), salted fish (Hernandez-Herrero et al. 1999, Mongkolthanaruk et al. 2012, Yatsunami and Echigo 1991, 1993), and others (Harada et al. 2008, Hamaya et al. 2014), and certain bacterial strains, belonging to *Tetragenococcus*, *Staphylococcus*, and others, were identified as the histamine producers. Generally, tetragenococci are causative agents of histamine accumulation in the fermented or aged foods containing large amounts of salt (approximately 15% or more) (Kimura et al. 2001, Satomi et al. 2012). Materials of fermented foods sometimes have a high salinity, a low pH, or contain alcohol, making them an unsuitable environment for bacterial growth, generally. Therefore, histamine producers take more time to produce histamine during fermentation. However, since there is a long fermentation period in the processing of fermented foods, the risk of histamine accumulation may be high in such foods.

a) Halophilic lactic acid producing bacteria

The primary causative agent of histamine accumulation in salted fermented fishery products is *Tetragenococcus* spp. They have been isolated from fish sauce, soy sauce, fermented rice brine products, and others as a histamine producer (Satomi 2016, Satomi et al. 1997, 2008, 2011, 2012).

Bacteriological characteristics

The genus *Tetragenococcus* is gram-positive tetrad cocci, halophile, and lactic acid producer (Collins et al. 1990, Dicks et al. 2009, Garvie 1986, Satomi et al. 1997, Weiss 1992). The strains are typically non-histamine producers. However, certain *Tetragenococcus* strains cause histamine accumulation in the fermented fishery products (Kimura et al. 2001, Konagaya et al. 2002, Mongkolthanaruk et al. 2012, Sato et al. 1995, Satomi et al. 1997, 2008, 2011, 2012, Udomsil et al. 2010). Their physiological and biochemical properties are almost coincident with that of the typical non-halophilic lactic acid bacteria, except for being a halophile, meaning they are facultative anaerobic bacteria without a respiratory pathway. *T. halophilus* strains are the predominant bacteria during the processing of fermented fishery products and soy sauce (Ito et al. 1985a,b, Fukui et al. 2012, Kobayashi et al. 1995, 2000, Röling et al. 1996, Udomsil et al. 2010). They play the important role of producing lactic acid, which reduces pH in the products during fermentation. Therefore, *T. halophilus* is also used as a fermentation starter in these foods (Kimura et al. 2015, Shozen et al. 2012a, Udomsil et al. 2010, Zaman et al. 2011). The optimum growth temperature is around 30°C; it does not grow at temperatures higher than 40°C, and hardly grows below 10°C. Despite being a lactic acid bacterium, they prefer an alkaline environment for growth; their growth is significantly inhibited at pH 5.0 or less. *T. muriaticus* is also isolated from fermented fishery products (Satomi et al. 1997, 2012), and the original isolation source is a squid liver sauce (Satomi et al. 1997), though there is less information about its distribution.

Habitat

The genus involves five species and two of those species, *T. halophilus* and *T. muriaticus*, are commonly isolated from fermented fishery foods, fish sauce, pickled seafood with rice bran, and others. Both species are able to proliferate in salt concentrations greater than 20% in foods. *T. halophilus* is commonly isolated from salted type fermented products including soy sauce, miso paste, and fishery products; moreover, the species is widely distributed in high osmotic pressure environments, such as salted foods, sugar stick juice, and others (Fujii et al. 2008, Fukui et al. 2012, Ito et al. 1985a,b, Justé et al. 2008, 2012, Kobayashi et al. 1995, 2000, Kosaka et al.

2012, Kuda et al. 2001, Röling and van Verseveld 1996, Sato et al. 1995, Taira et al. 2007).

HDC gene

The genetic information of tetragenococcal pyruvoyl-HDC has been reported recently (Mongkolthanaruk et al. 2012, Satomi et al. 2008, 2011, 2012). At present, it is likely that histamine-producing tetragenococci, including *T. halophilus* and *T. muriaticus*, harbor a plasmid encoding *hdcA*. Generally, bacterial amino acid decarboxylase gene systems were located on the chromosome, excepting an aspartate decarboxylase operon detected on a 25-kb plasmid of the *T. halophila* (Abe et al. 2002) and *hdc* cluster on 80-kb plasmid of the *L. hilgardii* (Lucas et al. 2005). From the results of complete sequences of the representative plasmids encoding *hdcA*, several pieces of information have been clarified. The strains harbor an approximately 21–37 kbp plasmid encoding a single copy of the *hdc* cluster (Table 12.5), which consists of four genes, *hdcP, hdcA, hdcB*, and *hdcRS* (Fig. 12.3). The nucleotide sequence of the *hdc* cluster shares a > 99% sequence similarity with that of lactic acid bacteria (Fig. 12.3), including non-halophilic lactic acid bacteria that are phylogenetically distinct from *Tetragenococcus* spp. (Fig. 12.4). Finally, the structure of most tetragenococcal plasmids was the same, and two putative mobile genetic elements, ISLP1 (Nicoloff and Bringel 2003)-like and IS200 (Beuzón et al. 2004)-like, were identified in the up- and downstream regions of the *hdc* cluster (Fig. 12.7). Additionally, the order of genes within the *hdc* cluster was identical to the sequences of lactic acid bacteria reported as histamine producers. The sequence types of the putative plasmid replication processes are divided into two groups: theta type replication and rolling circle replication (RCR), though almost all parts related to the *hdc* cluster including the putative transposons are conserved in both types of plasmids (Benachour et al. 1997, Kantor et al. 1997, Satomi et al. 2008, 2011, 2012).

b) Non-halophilic lactic acid producing bacteria

In agricultural products, such as wine, cheese, sausage, and other dairy products, *Oenococcus oeni, Lactobacillus* spp., and *Streptococcus thermophilus* are isolated as problematic histamine-producing bacteria (references have been quoted above). These bacterial species have been known as typical lactic acid bacteria from ancient times, and both the physiological and biochemical properties are consistent with the nature of the typical lactic acid bacteria. Certain *Lactobacillus* strains have long been known as representative histamine-producing bacteria. In the fishery products, *L. otakiensis* have been reported as a histamine producer (Hamaya et al. 2014, Satomi 2016).

Table 12.5 Comparative properties of pyruvoyl type histidine decarboxylase in gram positive bacteria.

Bacterium	Molecular mass (kDa) Native	α	β	Subunit structure	Optimum pH	V_{max} (μmol·min^{-1}·mg^{-1})	K_m (mmol/L)	ProHDC (π chain) detected	References
Staphylococcus epidermidis TYH1	121	26	9	$(\alpha\beta)_{3(67)}$	5.0–6.0	45.5	1.1	+	Furutani et al. (2013)
Tetragenococcus muriaticus	257	29	13	$(\alpha\beta)_6$	4.5–7.0	16.8	0.74	+	Konagaya et al. (2002)
Oenococcus oeni	190	28	11	$(\alpha\beta)_6$	4.8	17.8	0.33	+	Coton et al. (1998)
Lactobacillus buchneri	203	25	9	$(\alpha\beta)_6$	5.5	69	0.6	+	Recsei et al. (1983) van Poelje and Snell (1990)
Lactobacillus 30a	208	25	9	$(\alpha\beta)_6$	4.8	80	0.4	+	Vanderslice et al. (1986) van Poelje and Snell (1990)
Clostridium perfringens	213	25	10.5	$(\alpha\beta)_6$	4.5	25	0.2	+	Recsei et al. (1983)
Micrococcus sp.	100	29	8	$(\alpha\beta)_{3(67)}$	4.4–5.8	25	0.8	+	Alekseeva et al. (1976, 1986)

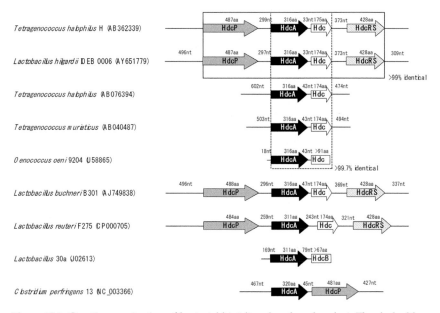

Figure 12.3 Genetic organization of bacterial histidine decarboxylase loci. The dashed box indicates a set of genes that are more than 99% identical. nt, nucleotides; aa, amino acids.

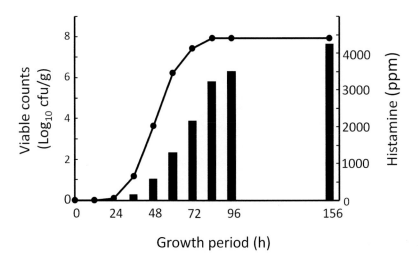

Figure 12.4 Changes in the viable counts of histamine producing bacteria and histamine content during cultivation in histidine broth. Closed circles: bacterial counts, solid bars: histamine content.

Bacteriological characteristics

According to the previous literature (Hammes et al. 1992, Kandler and Weiss 1986), *Lactobacillus* strains are bacilli or cocci, gram-positive, and are facultative anaerobes without a respiratory pathway. These strains do not require salt to grow; growth is inhibited by salt concentrations of 6% or more. The lactic acid-producing ability of these species is great, and they can grow in acidic environments as low as pH 4.0. Their optimal temperature is around 30°C. Histamine-producing lactic acid bacteria are rarely isolated from marine products except for fermented products, but there are some lactobacilli or relatives that were isolated from kamaboko as slime producing bacteria or rancid bacteria.

Habitat

Lactobacilli are widely distributed in nature, mainly in dairy and agricultural products. Histamine producing species have been isolated from cheese, wine, meat products, and others, which are accompany with lactic acid fermentation. In an interesting case, *L. otakiensis* like strains have been isolated as histamine producers from pickled mackerel in Japan (Hamaya et al. 2014, Satomi 2016). Originally, this species was isolated from Japanese vegetable pickles as dominant bacteria (Watanabe et al. 2009), which did not produce histamine at the original description. Physiological characteristics of the bacteria are shown in Table 12.6. Since it has been regarded that a major histamine producer is gram-negative bacteria in the pickled mackerel production owing to contamination from materials (Furutani et al. 2013), it should be studied further. The fact that histamine-producing *Lactobacillus* strains are present in the field where fishery products are manufactured is meaningful when considering the possibility of histamine accumulation in pickled type fermented fishery products.

Table 12.6 Free amino acids content in various fish species (mg/100g).

	Saury	Mackerel	Sardine	Sea bream
Taurine	128	17	45	138
Histidine	677	754	964	4
Alanine	13	10	20	13
Others	53	127	61	235
Total	871	908	1090	390

HDC gene

The histamine formation potential, even in these lactic acid bacteria, is strain-dependent; therefore, it is believed that the histamine-forming enzyme gene is mobile. The homology of histamine-producing enzymes and the amino acids of the aforementioned halophilic lactic acid bacteria are highly conserved, indicating that the origin of this enzyme has been inferred. There is little known about the genetic information of the histamine-producing *L. otakiensis*. At present, it has been reported that histamine-producing strains have *hdcA*, which shares > 99% sequence similarity to that of *L. hilgardii,* and *Tetragenococcus* spp. (Hamaya et al. 2014, Satomi 2016).

c) Other gram-positive bacteria

Stapylococcus strains have been isolated from various food as a histamine producer-salted fish, sausage, fish paste, and others. Since the bacteria have tolerance for more than 10% of salt concentration and can produce large amount of histamine, it is problematic in the food processing field. In addition to the aforementioned bacteria, *Bacillus* spp., *Enterococcus* spp., and *Micrococcus* sp. are known as histamine producers (FAO/WHO 2012), though their behavior in fishery products is unsure. Some Bacillus strains have been isolated from fishery products including fishmeal (Tsai et al. 2006), but it is unclear if the bacteria can produce significant levels of histamine in fermented fishery products with high salt concentrations, or under low water activity conditions. It is thought to make spores in a high salt concentration environment. In the case of soy sauce fermentation, *Enterococcus* strains have been isolated as a histamine producer, which is contrary to expectations (Ibe et al. 2003). However, further information, such as genetic analysis of *hdcA* and taxonomy, has not been reported. In addition, the information on the histamine-producing *Micrococcus* sp. has not been provided; only the N-terminal sequence of HDC is available (Alekseeva et al. 1976, 1986, Prozorovski and Jörnvall 1974, van Poelje and Snell 1990). Since these bacteria have similar growth features to the previously discussed histamine producers in fermented products, future research should include their genetic and biochemical analysis. Here the focus is on histamine producing *Staphylococcus*.

Bacteriological characteristics

The genus *Staphylococcus* is gram-positive cocci, a facultative anaerobe with a respiratory pathway, and encompasses more than 40 species, which includes

S. aureus, famous as food-poisoning bacteria or sometimes clinical bacteria (Kloos and Shileifer 1986, Götz et al. 2006). In addition, staphylococci were frequently reported as histamine-formers in salted fishes (Hernandez-Herrero et al. 1999, Yatsunami and Echigo 1991, 1993), fish paste (Harada et al. 2008), and fermented meat products (Silla Santos 1998, Suzzi and Gardini 2003, Landeta et al. 2007), and soy bean products (Tsai et al. 2007). There are only two histamine-producing staphylococcal species reliably identified, *S. capitis* (De las Rivas et al. 2008) and *S. epidermidis* (Yokoi et al. 2011). The rest of the histamine-producing staphylococci remain unidentified. Therefore, there is little information about the distribution and significance of staphylococci as a histamine producer in the fishery products. Although histamine-producing staphylococci are problematic in dairy or meat products, they are a presence that cannot be ignored in salted fish products, due to their tolerance to a high salt concentration. Staphylococci are infrequently isolated from fermented fish products with a long-term fermentation period. Since most fermented fishery products, like fish sauce, contain more than a 15% salt concentration, it is rare to find staphylococci as the dominant bacteria for the whole fermentation period in these foods. According to previous reports (Kloos and Shileifer 1986, Götz et al. 2006), the growth ranges of staphylococci are usually less than 15%, indicating that the bacteria are halotolerant. Actually, staphylococci are commonly isolated from foods containing less than 10% salt, such as salted dry fish, dry-cured sausage, and others (Hernandez-Herrero et al. 1999, Landeta et al. 2007, Silla Santos 1998, Suzzi and Gardini 2003, Yatsunami and Echigo 1991, 1993). In the study of the transition of bacterial flora in fish sauce fermentation, it was reported that staphylococci, such as *S. xylosus* and its relatives, are the dominant bacteria only during the initial stage of fermentation and then decline (Fukui et al. 2012), suggesting that high salt concentration is not suitable for them to maintain their domination. The niches in fish sauce mashes, in which staphylococci lost, could be occupied by tertagenococci.

HDC gene

The staphylococcal HDCs are also pyruvoyl-type HDCs, as in other gram-positive histamine producers. As is shown in the phylogenetic tree based on HDC amino acid sequences, the phylogenetic position of its HDC is far from that of lactic acid bacteria, but rather close to clostridial HDC (Furutani et al. 2014, Yokoi et al. 2011, De las Rivas et al. 2008). Moreover, since *hdcB* and *hdcRS* are lacking in the *hdc* cluster (De las Rivas et al. 2008, Yokoi et al. 2011), the genetic characteristics are similar to those of *Clostridium*. Previous studies strongly indicate that the HDC gene (*hdcA*) is encoded in genomic DNA and located in mobile genetic element (De las Rivas et al. 2008, Yokoi et al. 2011). In the case of *S. epidermidis* TYH1 isolated from fish-miso, the *hdc* cluster resides in the staphylococcal cassette chromosome elements (Yokoi et al. 2011), which is one of the mobile genetic elements and was first discovered as a mobile genetic element, and is composed of the *mec* gene complex, encoding methicillin resistance (SCCmec) (Ito et al. 1999). Since the SCC gene complex encodes recombinases of the invertase/resolvase family (Ito et al. 2003; International working group on the classification of staphylococcal cassette chromosome elements, IWG-SCC 2009), which mediate the site-specific integration of SCC into the chromosome of a staphylococcal strain, the *hdc* cluster may transfer to other strains by using this mobile genetic element.

d) Histamine producing enzyme

Hence, pyruvoyl-type HDCs are generally constructed as a hexamer $(\alpha\beta)_6$ containing alpha and beta subunits, of about 200 kDa in complete form (Huynh and Snell 1985, Recsei and Snell 1970, 1984, van Poelje and Snell 1990), and the HDC gene is coded on *hdcA*. A summary of the chemical characteristic of each enzyme in histamine producing gram-positive bacteria is shown in Table 12.5. The polypeptide chain translated from *hdcA* is cleaved to form the alpha and beta subunit, and then the N-terminal residue in the alpha subunit is converted to a pyruvoyl group via pyruvate in the process of maturation. The enzyme is thought to specifically distribute in gram positive, mainly lactic acid bacteria (Calles-Enriquez et al. 2010, Coton et al. 1998, Gallagher et al. 1993, Konagaya et al. 2002, Lucas et al. 2005, Martín et al. 2005, Recsei et al. 1983, Recsei and Snell 1984, Vanderslice et al. 1986), staphylococci (De las Rivas et al. 2008, Furutani et al. 2014), clostridia (Recsei et al. 1983, Shimizu et al. 2002), and Micrococci (Alekseeva et al. 1976, 1986, Prozorovski and Jörnvall 1974, van Poelje and Snell 1990). A summary of the chemical characteristic of each enzyme in histamine producing gram-positive bacteria is shown in Table 12.2. As was mentioned above, the *hdc* cluster consists of 4 genes, *hdcP, hdcA, hdcB,* and *hdcRS* (Fig. 12.3). The region related to histamine production is conserved in several lactic acid bacteria (Calles-Enríquez et al. 2010, Lucas et al. 2005, Martín et al. 2005, Satomi 2016). The putative functions of the rest of genes are as follows: *hdcP* encodes a histidine/histamine antiporter (Lucas et al. 2005, Martín et al. 2005), *hdcB* encodes a cleavage factor of immature HDC translated from *hdcA* (Trip et al. 2011), and *hdcRS* encodes histidyl-RNA synthetase (Lucas et al. 2005, Martín et al. 2005). Although some genes are deleted in certain species, *hdcA* and *hdcP* are likely essential genes for producing HDC (Calles-Enríquez et al. 2010, De las Rivas et al. 2008, Yokoi et al. 2011).

4. Histamine Accumulation in Seafood

4.1 Raw fish

4.1.1 Mechanisms of histamine accumulation

During storage of raw fresh fish, muscle tissue disintegrates by self-digestion, and then microorganisms invade the body of the fish from the surface and intestine of fish. As self-digestion progresses, a large amount of drip leaches out of the fish body and is used as a nutrient source for microbes existing in the surroundings. Generally, when the freshness of fish decreases (progression of putrefaction), spoilage components that release odors such as ammonia are generated, and it becomes possible to judge whether or not it is easy to eat. Since histamine-producing bacteria also invade the fish body with other microorganisms and contribute to decomposition of fish meat and produce histamine, the histamine content often correlates with the progression of decay. As mentioned above, it is highly likely that it will be contaminated by oceanic histamine-producing bacteria, after landing by terrestrial Enterobacteriaceae bacteria. Therefore, it is considered that the body of the fish is always contaminated by histamine-producing bacteria. However, whether histamine is actually accumulated in fish meat is influenced by the species and density of adhering bacteria. In the study that tracks the appearance of histamine-producing bacteria annually in the seawater of Tokyo Bay, the frequency of histamine-producing bacteria depends on seawater temperature, for instance histamine-producing bacteria was high in summer (Yoguchi et al. 1990a). Also, if other bacteria apart from histamine producers grow during storage in fish meat faster than histamine-producing bacteria, histamine accumulation may not be achieved due to eliminate cell count of histamine producers (Sato et al. 1994). Conversely, when special environments where only histamine-producing bacteria can grow are available, they accumulate a large amount of histamine even though they can be judged as edible in organoleptically. In addition, even in the same individual, histamine concentration is biased due to the attachment density of histamine producing bacteria and it is difficult to accurately grasp the appearance of histamine accumulation even if only a part of the fish body is examined (Yoguchi et al. 1990b, Tao et al. 2002, 2009). Thus, while the mechanism of histamine food poisoning is gradually being elucidated, many points remain unclear.

4.1.2 Relationships between freshness and histamine level

In the case of fresh fish, the progress of spoilage and the accumulation of histamine are considered to be correlated, but since the production of histamine is greatly influenced by the number of bacteria and the proliferation state of histamine producing bacteria, the mode of histamine accumulation varies greatly depending on the storage conditions of fish meat. Therefore, FAO/WHO biogenic amine expert committee also reported to CODEX that histamine production and the freshness of fresh fishes are not synonymous. As mentioned above, there were many examples in which the spoilage of fish meat did not correlate with the accumulation of histamine, and it is clear that recognition such as "the less histamine is fresh", and "the histamine content as a freshness index" are incorrect. The histamine content shows the risk of causing allergy poisoning, and it does not reflect freshness.

4.1.3 Relationships between fish species and histamine level

As histidine in fish meat is converted to histamine, fish species that have high free histidine content in muscle are more likely to have histamine accumulation risk, because the histamine producing bacteria can utilize free histidine in fish easily without proteolytic digestion of fish fresh. Generally marine migratory fish such as mackerel, sardine, saury, swordfish, tuna, etc., are known as the causative fish species of major allergy-like food poisoning. A summary of histidine contents and fish species is shown in Table 12.6. In the FAO/WHO biogenic amine expert committee's report, a list of fish species that caused allergy-like food poisoning and fish species with high histidine content was prepared, and their common name, scientific name, histidine content, and catch amount were also indicated. The fish species listed on the list are presumed to be easy to accumulate histamine, but only that the histidine content is high. If handling is not suitable, histamine accumulation can occur even in fish species with low histidine content. In fact, allergy-like food poisoning occurrs in salmon. Currently, there is no known fish species in which it is hard to produce histamine, despite the high free histidine content.

Inhibition of histamine accumulation in raw materials

As mentioned above, histamine-producing bacteria in fresh fish are ordinary *Photobacterium* spp. and Enterobacteriaceae. Since they are general gram negative, non-spore forming bacteria, suppression methods for these bacteria has the same procedures as that of general gram negative food poisoning bacteria. Although some histamine-producing bacteria, such as psychrophilic *Photobacterium* and *Morganella psychrotolerans*, can grow even in a low-temperature environment, it takes a minimum of 3 days at 5°C to accumulate a significant amount of histamine (Fig. 12.4). Therefore, operation of cold chain transport systems, even in the processing step, is effective for prevent histamine accumulation. In the case of large fish caught in longline fishery, such as tuna, the history from being taken on a longline to being collected onboard is ambiguous. Therefore, when the seawater temperature of the fishing ground is high, accumulation of histamine in dead individuals is concerned. In FDA research, when the temperature of the seawater is 31°C or less, it shows that there is less risk of histamine accumulation within 10 hours after the longline. To track the history and condition at the time of fishing, the possibility of avoiding accumulation risk by histamine increases. The FAO/WHO biogenic amine expert committee has also reported that histamine poisoning is a controllable risk with thorough HACCP, GAP, traceability, and others. Although the psychrophilic histamine producing bacteria can grow and produce histamine at a temperature less than 10°C within 3 days (Torido et al. 2012a,b), it should take 6 days or more to grow and produce histamine at 5°C, even in these bacteria. It indicates that the risk of histamine poisoning is reduced by applying basic GAP and HACCP system (EFSA 2011, FAO/WHO 2012).

4.2 Salted/Fermented seafoods

Accumulation of histamine is a problem even in fermented seafoods, especially fish sauce, but in many cases, since a large amount of salt is added to raw materials, there are no histamine-producing bacteria and bacteria that produce putrefactive odors which are problematic with fresh fish. In other words, such foods often accumulate histamine without causing putrid smell. Unlike those of fresh fish, histamine-producing bacteria of salted and fermented foods containing a large amount of common salt are halophilic lactic acid bacteria and salt resistant gram positive bacteria. Here, in recent years, the histamine accumulation mechanism will be explained taking a fish sauce whose production volume has increased dramatically.

a) Fish sauce

The history of fish sauce production is older than that of soy sauce (Ishige and Ruddle 1990). The fermentation mechanism of fish sauce involves the digestion of fish materials using self-digestion enzymes and microorganisms, resulting in the accumulation of chemical compounds that affect the taste and flavor, though mold starter, called Koji, is added to fish materials during fish sauce processing to improve the fishy smell and to reduce the fermentation period in the recent trend (Funatsu et al. 2000). Typical fermented process is shown in Fig. 12.5. The accumulation of amino acids and peptides, which are associated with the typical fish sauce taste, results from a self-digestion enzyme secreted by fish materials. During fermentation, bacterial counts, total nitrogen, and amino acids and peptides increase; hence, the pH decreases due to lactic acid accumulation. Halophilic lactic acid bacteria, mainly *T. halophilus*, are dominant during fermentation in a variety of Japanese fish sauces. Changes in the bacterial flora of Japanese

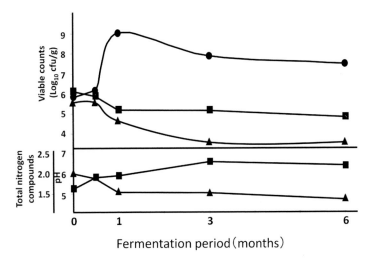

Figure 12.5 Changes in the viable counts of microorganisms and chemical compounds during Japanese fish sauce fermentation. In upper graph, •; total viable halophilic bacterial count, ▪; standard plate count, ▲; viable mold count (mainly *Aspergillus* sp.). Lower graph, ▪; total nitrogen compounds, ▲; pH.

fish sauce during fermentation have been studied previously (Fukui et al. 2012, Shozen et al. 2012a, Taira et al. 2007). The halophilic lactic acid bacteria play an important role in the accumulation of lactic acid during fish sauce fermentation, though they have little effect on the digestion of fish materials or the production of amino acids via enzymes. The basic quality of fish sauce is stabile as long as an appropriate salt concentration is used.

4.2.1 Mechanisms of histamine accumulation

The accumulation of biogenic amines, including histamine and tyramine, sometimes occurs during fish sauce production. In 2011, CODEX adopted a limit of 400 ppm for histamine content in fish sauce. The behavior of histamine-producing tetragenococci synchronizes that of the predominant bacteria occupied by tetragenococci (Fig. 12.6), because almost all histamine producer and dominant bacteria belong to the same bacterial genus, suggesting that the growth characteristics of both are the same in fish sauce mash. In the case of serious histamine accumulation in fish sauce, the viable counts are almost the same for histamine producers and total halophilic bacteria, due to the significant proportion of total viable counts by histamine-producing bacteria (Kimura et al. 2015). Simply, the numbers of histamine-producing bacteria reflect histamine content in fish sauce mash for the entire fermentation period (Kimura et al. 2015). The initial counts of histamine producers in fish sauce fermentation also affect histamine

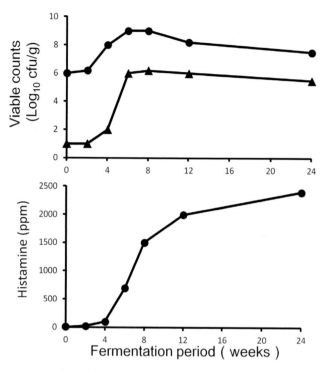

Figure 12.6 Changes in the viable counts of total halophilic and histamine-producing bacteria and histamine contents in the Japanese fish sauce accumulating a large amount of histamine. In upper graph, ●; total viable halophilic bacterial count (mainly *Tetragenococcus* sp.), ▲; viable histamine-producing bacterial count (also mainly *Tetragenococcus* sp.). Lower graph, ●; histamine content in the sample accumulating histamine.

accumulation, meaning that even a minor contamination of histamine producers may cause serious accumulation.

Control of histamine accumulation

Starter cultures

In the many countries which produce fish sauce, a starter culture has not been used to make the traditional fermented fishery products. However, it is obvious that starter cultures are useful when making fermented fishery products, because they can increase fermentation stability and the quality of products, and moreover inhibit histamine accumulation (Kimura et al. 2015, Shozen et al. 2012, Zaman et al. 2011). For example, it has been shown that using a bacterial starter can inhibit histamine accumulation during fish sauce fermentation (Fig. 12.7; Kimura et al. 2015). Of course, it is important to provide a suitable environment for the starter culture in

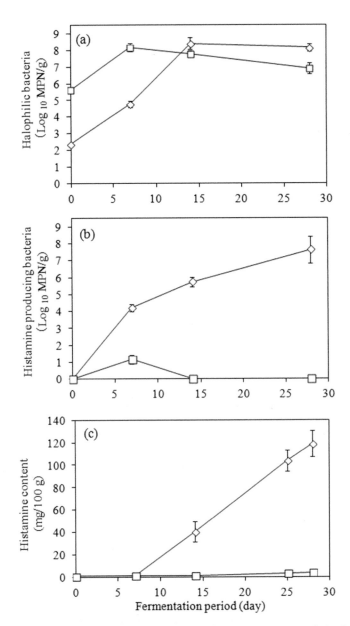

Figure 12.7 Effect of starter inoculation on suppression of histamine accumulation in the fish sauce mashes during fermentation. Addition of a starter (Sample A) can inhibit the growth of histamine-producing bacteria during fermentation; hence, in the sample B (non inoculation of starter) histamine accumulation occurred. Vertical bars indicate standard deviation (n = 3) of means. (a) Halophilic bacterial count. (b) Histamine producing bacterial count. (c) Histamine content. Symbols: □, NBRC12172 (10^5 MPN/g) as the Sample A; ◊, No inoculation as the Sample B.

fish sauce mash, such as nutritional factors, temperature, and other factors. Actually, additions of sub-material that affect nutritional help to starter cultures are effective in preventing histamine accumulation during fish sauce fermentation. Figure 12.8 shows that an effect of starter inoculation with or without sucrose addition on the accumulation of histamine in the fish sauce mashes during fermentation. It is obvious that supplementation of starter culture with sucrose is effective. It is likely that carbohydrate compounds, which are essential nutrition for growth of lactic acid bacteria, are not enough in the case of the fish sauce mashes (Shozen et al. 2012).

Reduction of histamine-producing bacteria at initial fermentation stage

As mentioned above, starter cultures are useful in preventing histamine accumulation in fish sauce fermentation. However, the initial contamination levels by histamine-producing bacteria in fermented products are considerable. To achieve steady fermentation, using a starter culture in combination with a reduction of initial counts of histamine producers is essential to the processing of fermented fishery products. The effects of sanitation and controlling fermentation on suppression of histamine accumulation in the fish sauce manufacturer are shown in Fig. 12.9 as a

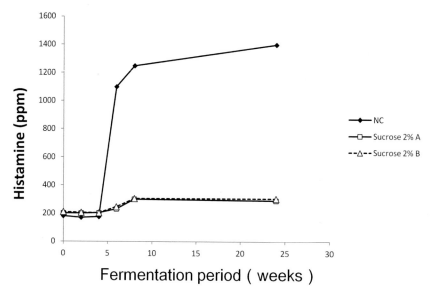

Figure 12.8 Effect of starter inoculation with or without sucrose addition on the accumulation of histamine in the fish sauce mashes during fermentation. Addition of a starter with sucrose (lot 1 and 2) can inhibit the accumulation of histamine during fermentation; hence in the sample inoculated with only starter culture accumulate histamine. Symbols: ♦, addition of only starter culture; □, addition of starter culture with sucrose (lot 1: 2% final concentration); and Δ, addition of starter culture with sucrose (lot 2: 2% final concentration).

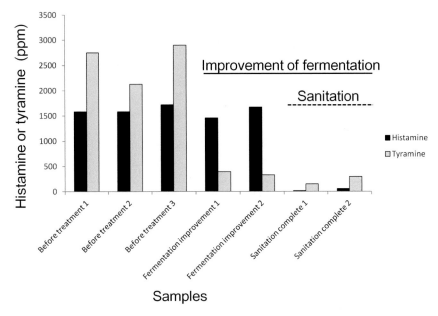

Figure 12.9 Effects of sanitation and controlling fermentation on suppression of histamine accumulation in the fish sauce manufacturer. Before sanitation and controlling fermentation, the products contain large amounts of amines. After controlling fermentation (addition of sucrose to the fish sauce mash), tyramine content was dramatically decreased and then sanitation affect decreasing histamine accumulation.

successive example. Before sanitation and controlling fermentation, the products contain large amounts of amines. After controlling fermentation (addition of sucrose to the fish sauce mash), tyramine content was dramatically decreased and then sanitation affects decreasing histamine accumulation (Satomi et al. 2014, Satomi 2016).

b) Other salted/fermented seafoods

Histamine-producing bacteria have been isolated from pickled seafood with rice bran, though there are few details known of their behavior during fermentation (Satomi et al. 2011, Yatsunami and Echigo 1991). Changes in the bacterial flora in mackerel pickles with rice bran have been studied (Kosaka et al. 2012), and staphylococci are predominant in the initial stages of fermentation. Then, the predominant organisms become tetragenococci in the samples, suggesting a transition of bacterial flora similar to that in fish sauce. In addition, since *T. halophilus* has been isolated as a histamine producer from sardine pickles with rice (Satomi et al. 2012), it is likely the primary histamine-producing bacteria. Basically, mechanisms and causative microorganisms in histamine accumulation of rice bran pickles are similar to the case of fish sauce fermentation. As a secondary causative agent, staphylococci are proposed as histamine producers in the products (Yatsunami and Echigo 1991), because there are some cases where these bacteria were isolated as histamine-producing bacteria. As mentioned, in high salinity, *Tetragenococcus* may be problematic due to their tolerance to salt; hence, in the comparatively low salt environment, staphylococci can also be isolated, according to the previous sources, such as dried fish, sausages, and fish pastes.

c) Other seafood

As major fish products related to histamine accumulation, salted and marinade fish is also known, for instance, soused herring in Europe. These products have impressive vinegar flavor, taste and texture, and the utilization of vinegar is effective for short-term preservation. In the case of salted and marinade mackerel, the risk of histamine accumulation is a concern, because mackerel is known as a fish that easily accumulates histamine. Although the low pH of vinegar broth is expected to prohibit bacterial growth and produce a preservative effect after packing, little information is available regarding the behavior of histamine-producing bacteria in these products during the production process and during storage after manufacture. Interestingly, *L. otakiensis* like strains have been isolated as histamine producers from salted and marinade mackerel in Japan (Hamaya et al. 2014, Satomi 2016). Since it has been regarded that major histamine producer is gram-negative bacteria in the pickled mackerel production owing to contamination from materials (Furutani et al. 2013), it should be studied further. The fact that histamine-producing *Lactobacillus* strains are present in the field where fishery products are manufactured is meaningful when considering the possibility of histamine accumulation in pickled type fishery products. To control histamine accumulation caused by *L. otakiensis* like species, which can grow under low temperature and low pH condition, further studies are needed.

5. Risk Management

5.1 Regulation levels of histamine

The regulation limits are set for an amount of histamine in many countries and organizations such as FDA, CODEX, and others. Table 12.7 summarizes the prescribed limits and sampling methods for histamine in each country and institution. These inspection methods require a large number of sampling inspections per lot, and it is expected that inspection will cost a lot, indicating that it is necessary to consider various situations, such as histamine content survey of products, toxicological analysis,

Table 12.7 List of prescribed limits for histamine in fish products set by various regulatory bodies around the world.

Regulatory bodies (country)	Products	Prescribed limits
CODEX	Fish and fish products	Histamine levels as indicators of decomposition and hygiene and handling. A maximum average level of not more than 100 ppm is considered satisfactory in relation to decomposition, while an upper limit of 200 ppm in any one sample is applied for hygiene and handling.
	Fish sauce	The product shall not contain more than 400 ppm of histamine in any sample unit tested.
FSANZ (Australia and New Zealand)	Fish and fish products	Maximum limit of 200 ppm for histamine.
FDA (USA)	Tuna and related fish	Guidelines for tuna and related fish establishing a 'defect action level' of 50 ppm in any sample (It is said to be indicative of spoilage and may mean that toxic levels are present in other samples). A separate toxicity level of 500 ppm is also given.
EU (European Union)	Fish and fish products	Fish species belonging to families known to contain large amounts of histidine (e.g., Scombridae, Clupeidae, etc.) in their tissues should be tested for the presence of histamine. Nine samples should be tested from each lot and the mean value should be 100 ppm or less. The lot is considered unsatisfactory if more than two samples give results of between 100 and 200 ppm, or if any sample gives a result of 200 ppm or more.
	Fermented products	Maximum limit of 200 ppm for histamine.

epidemiological survey, etc., for establishment of regulation limits and sampling plan in histamine.

5.2 Sampling plan

As mentioned above, standard values are set by each country and institution, but it is necessary to discuss the sampling method (sampling inspection method). This is due to the distribution of histamine in fishery products. As described in Fig. 12.10 (FAO/WHO 2012), the distribution of histamine in products lot is uneven. For example, if there are two foods with the same average histamine content, one is a liquid food and the other is a round of fish. Since liquid foods were mixed in a large tank once and then bottled, the distribution of histamine concentration in each bottle should show very close to the average value (distribution map is a very steep mountain shape). In the case of round-fish products, the amount of

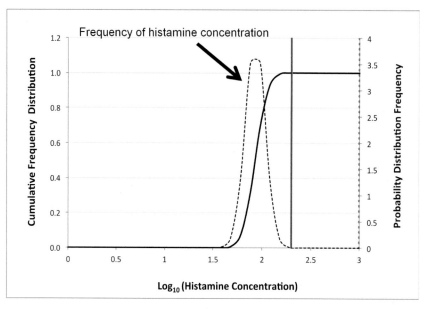

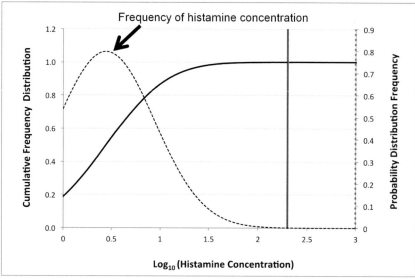

Figure 12.10 Difference in histamine concentration distribution in food. The histamine concentration distribution of the food that accumulates 200 ppm of histamine 1 in 10000 samples in both the upper and lower samples. The histogram concentration in the above figure is uniform in the above figure. In the lower figure the histamine content average value is low but the concentration distribution is wide. Dashed line: concentration distribution of histamine. Solid line: Distribution frequency of sample (1.0 in total) (modified from reference 4).

histamine in each individual is varied and it is contained in the same lot from one with very high histamine to the one not accumulating at all. Therefore, it shows a mountain shape with a smooth foot around the average value in a distribution map. In the former case, there is a very low probability that a high concentration of specimen exists which largely deviates from the mean value and causes food poisoning, but the latter means that a high histamine concentration sample is contained in the lot. Figure 12. 10 shows the histamine concentration distribution of food having histamine 200 mg/kg accumulated in 10,000 samples. In the above figure, although the average histamine concentration is high, the variation in histamine concentration is small, and it can be supposed that there are few specimens that are largely out of the mean. On the other hand, in the figure below, although the average value is low, it can be supposed that the distribution of histamine concentration is wide and there are specimens at concentrations far away from the average value. Both have the same probability that there are specimens that accumulate 200 mg/kg of histamine. In other words, risk cannot be correctly evaluated unless the number of sampling and analysis method of data are appropriate. As described above, it is desirable to establish a sampling method based on scientific evidence for marine products whose histograms differ greatly in the form of histamine for each processing mode.

References

Abe, K., Ohnishi, F., Yagi, K., Nakajima, T., Higuchi, T., Sano, M., Machida, M., Sarker, R.I. and Maloney, P.C. 2002. Plasmid-encoded asp operon confers a proton motive metabolic cycle catalyzed by an aspartate–alanine exchange reaction. J. Bacteriol. 184: 2906–2913.

Alekseeva, E.A., Prozorovskii, V.N. and Grebenshchikov, O.G. 1976. Amino acid sequence in tryptic peptides of maleylated β-polypeptide chain of histidine decarboxylase from *Micrococcus* sp. n. Biokhimiya 41: 1760–1765.

Alekseeva, E.A., Grebenshchikova, O.G. and Prozorovskii, V.N. 1986. Chemical fragmentation by o-iodosobenzoic acid of α-chain of histidine decarboxylase from *Micrococcus* sp. n. at tryptophan residues. Biokhimiya 51: 1235–1241.

Basby, M., Jeppesen, V. F. and Huss, H. H. 1998. Characterization of the microflora of lightly salted lumpfish (*Cyclopterus lumpus*) roe stored at 5C. J. Aquat. Food Prod. Technol. 7: 35–51.

Benachour, A., Frère, J., Flahaut, S., Novel, G. and Auffray, Y. 1997. Molecular analysis of the replication region of the theta-replicating plasmid pUCL287 from *Tetragenococcus* (*Pediococcus*) *halophilus* ATCC33315. Mol. Gen. Genet. 255: 504–13.

Beuzón, C.R., Chessa, D. and Casadesús, J. 2004. IS200: an old and still bacterial transposon. Int. Microbiol. 7: 3–12.

Bjornsdottir-Butler, K., McCarthy, S.A., Dunlap, P.V. and Benner, Jr. R.A. 2017. *Photobacterium angustum* and *Photobacterium kishitanii*, psychrotrophic high-level histamine-producing bacteria indigenous to tuna. Appl. Environ. Microbiol. 82: 2167–2176.

Brenner, D.J. and Farmer III, J.J. 2005. Family I. Enterobacteriaceae. pp. 587–850. *In*: Vos, P., Garrity, G., Jones, D., Krieg, N.R., Ludwig, W., Rainey, F.A., Schleifer, K-H. and Whitman, W. (eds.). Bergey's Manual of Systematic Bacteriology. Vol. 3. Springer, New York.

Burdychova, R. and Komprda, T. 2007. Biogenic amine-forming microbial communities in cheese. FEMS Microbiol. Lett. 276: 149–155.

Calles-Enríquez, M., Eriksen, B.H., Andersen, P.S., Rattray, F.P., Johansen, A.H., Fernández, M., Ladero, V. and Alvarez, M.A. 2010. Sequencing and transcriptional analysis of the *Streptococcus thermophilus* histamine biosynthesis gene cluster: factors that affect differential *hdcA* expression. Appl. Environ. Microbiol. 76: 6231–6238.

Chang, G.W. and Snell, E.E. 1968. Histidine decarboxylase of *Lactobacillus* 30a. II. Purification, substrate specificity, and stereospecificity. Biochemistry 7: 2005–2012.

Chang, S.F., Ayres, J.W. and Sandine, W.E. 1985. Analysis of cheese for histamine, tyramine, tryptamine, histidine, tyrosine, and tryptophane. J. Dairy Sci. 68: 2840–2846.

Codexalimentarius. 2014. http://www.codexalimentarius.org.

Collins, M.D., Williams, A.M. and Wallbanks, S. 1990. The phylogeny of *Aerococcus* and *Pediococcus* as determined by 16S rRNA sequence analysis: description of *Tetragenococcus* gen. nov. FEMS Microbiol. Lett. 70: 255–262.

Coton, E., Rollan, G.C. and Lonvaud-Funel, A. 1998. Histidine carboxylase of *Leuconostoc oenos* 9204: purification, kinetic properties, cloning and nucleotide sequence of the *hdc* gene. J. Appl. Microbiol. 84: 143–151.

Coton, E. and Coton, M. 2005. Multiplex PCR for colony direct detection of Gram-positive histamine and tyramine-producing bacteria. J. Microbiol. Methods 63: 296–304.

Coton, E. and Coton, M. 2009. Evidence of horizontal transfer as origin of strain to strain variation of the tyramine production trait in *Lactobacillus brevis*. Food Microbiol. 26: 52–57.

Coton, M., Romano, A., Spano, G., Ziegler, K., Vetrana, C., Desmarais, C., Lonvaud-Funel, A., Lucas, P. and Coton, E. 2010. Occurrence of biogenic amine-forming lactic acid bacteria in wine and cider. Food Microbiol. 27: 1078–1085.

De las Rivas, B., Rodríguez, H., Carrascosa, A.V. and Muñoz, R. 2008. Molecular cloning and functional characterization of a histidine decarboxylase from *Staphylococcus capitis*. J. Appl. Microbiol. 104: 194–203.

Dicks, L.M.T., Holzapfel, W.H., Satomi, M., Kimura, B. and Fujii, T. 2009. Genus *Tetragenococcus*. pp. 611–616. *In*: Vos, P., Garrity, G., Jones, D., Krieg, N.R., Ludwig, W., Rainey, F.A., Schleifer, K-H. and Whitman, W. (eds.). Bergey's Manual of Systematic Bacteriology, Vol. 3. Springer, New York.

EFSA. 2011. Scientific opinion on risk based control of biogenic amine formation in fermented foods. EFSA J. 9: 2393.

Emborg, J., Dalgaard, P. and Ahrens, P. 2006. *Morganella psychrotolerans* sp. nov., a histamine-producing bacterium isolated from various seafoods. Int. J. Syst. Evol. Microbiol. 56: 2473–2479.

Emborg, J. and Dalgaard, P. 2008. Modelling the effect of temperature, carbon dioxide, water activity and pH on growth and histamine formation by *Morganella psychrotolerans*. Int. J. Food Microbiol. 128: 226–233.

FAO, WHO. 2012. Hazard identification. *In*: FAO, WHO (eds.). Joint FAO/WHO Expert Meeting on the Public Health Risks of Histamine and Other Biogenic Amines from Fish and Fishery Products.

Farmer III, J.J. and Janda, J.M. 2005. Family I. Vibrionaceae. pp. 491–546. *In*: Vos, P., Garrity, G., Jones, D., Krieg, N.R., Ludwig, W., Rainey, F.A., Schleifer, K-H. and Whitman, W. (eds.). Bergey's Manual of Systematic Bacteriology. Vol. 3. Springer, New York.

Fujii, T., Kurihara, K. and Okuzumi, M. 1994. Viability and histidine decarboxylase activity of halophilic histamine-forming bacteria during frozen storage. J. Food Prot. 57: 611–613.

Fujii, T., Kimura, B. and Mizoi, M. 2008. Changes in chemical composition and microbial flora during fermentation of squid sauce of Tobishima Island. Yamawaki Studies of Arts and Sciences 46: 121–131. (in Japanese with an English abstract.)

Fukui, Y., Yoshida, M., Shozen, K., Funatsu, Y., Takano, T., Oikawa, H., Yano, Y. and Satomi, M. 2012. Bacterial communities in fish sauce mash culture-dependent and -independent methods. J. Gen. Appl. Microbiol. 58: 271–281.

Funatsu, Y., Sunago, R., Konagaya, S., Imai, T., Kawasaki, K. and Takeshima, F. 2000. A comparison of extractive components of a fish sauce prepared from frigate mackerel using soy sauce koji with those of Japanese-made fish sauce and soy sauce. Nippon Suisan Gakkaishi 66: 1036–1045. (in Japanese with an English abstract.)

Furutani, A., Matsubara, H., Ishikawa, T. and Satomi, M. 2013. Behavior of histamine-producing bacteria in shimesaba, raw mackerel salted and marinated in vinegar during processing and storage at various temperatures. Fish Sci. 79: 725–733.

Furutani, A., Harada, Y., Shozen, K., Yokoi, K., Saitou, M. and Satomi, M. 2014. Purification and properties of a histidine decarboxylase from *Staphylococcus epidermidis* TYH1 isolated from Japanese fish-miso. Fish Sci. 80: 93–101.

Gallagher, T., Rozwarski, D.A., Ernst, S.R. and Hackert, M.L. 1993. Refined structure of the pyruvoyl-dependent histidine decarboxylase from *Lactobacillus* 30a. J. Mol. Biol. 230: 516–528.

Garvie, E.I. 1986. Genus *Pediococcus* Claussen 1903, 68AL. pp. 1075–1079. *In*: Sneath, P.H.A., Mair, N.S., Sharpe, M.E. and Holt, J.G. (eds.). Bergey's Manual of Systematic Bacteriology. Vol. 2. Williams & Wilkins Co., Baltimore.

Giacomini, A., Squartini, A. and Nuti, M.P. 2000. Nucleotide sequence and analysis of plasmid pMD136 from *Pediococcus pentosaceus* FBB61 (ATCC43200) involved in pediocin A production. Plasmid 43: 111–122.

Guirard, B.M. and Snell, E.E. 1987. Purification and properties of pyridoxal-5'-phosphate-dependent histidine decarboxylases from *Klebsiella planticola* and *Enterobacter aerogenes*. J. Bacteriol. 169: 3963–3968.

Götz, F., Bannerman, T. and Schleifer, K-H. 2006. The genera *Staphylococcus* and *Micrococcus*. pp. 5–75. *In*: Dworkin, M., Falkow, S., Rosenberg, E., Scleifer, K-H. and Stackebrandt, E. (eds.). The Prokaryotes Volume 4: Bacteria: Firmicutes, Cyanobacteria. Springer, New York.

Hamaya, Y., Furutani, A., Fukui, Y., Yano, Y., Takewa, T. and Satomi, M. 2014. Growth characteristics of histamine-producing lactic acid bacteria isolated from marinade broth in shimesaba production. Nippon Suisan Gakkaishi 80: 956–964. (in Japanese with an English abstract.)

Hammes, W.P., Weiss, N. and Holzapfel, W. 1992. The genera *Lactobacillus* and *Carnobacterium*. pp. 1535–1594. *In*: Balows, A. et al. (eds.). The Prokaryotes 2nd Ed. Vol. 2. Springer Verlag, New York, NY.

Halasz, A., Barath, A., Simon-Sarkadi, L. and Holzapfel, W. 1994. Biogenic amines and their production by microorganisms in food. Trends Food Sci. Technol. 5: 42–49.

Harada, Y., Shozen, K.-I., Satomi, M. and Yokoi, K. 2008. Effect of citrate treatment on quality of fermented fish-miso prepared from small horse mackerel. Nippon Shokuhin Kagaku Kogaku Kaishi 55: 25–31. (in Japanese with an English abstract.)

Hernandez-Herrero, M.M., Roig-Sages, A.X., Rodrigez-Jerez, J.J. and Mora-Ventura, M.T. 1999. Halotolerant and halophilic histamine-forming bacteria isolated during the ripening of salted anchovies (*Engraulis encrasicholus*). J. Food Prot. 62: 509–514.

Hungerford, J.M. 2010. Scombroid poisoning: A review. Toxicon 56: 231–243.

Huynh, Q.K. and Snell, E.E. 1985. Pyruvoyl-dependent histidine decarboxylases. J. Biol. Chem. 260: 2798–2803.

Ibe, A., Tabata, S., Sadamasu, Y., Yasui, A., Shimoi, T., Endoh, M. and Saito, K. 2003. Production of tyramine in "Moromi" mash during soy sauce fermentation. Food Hyg. Saf. Sci. 44: 220–226. (in Japanese with an English abstract.)

International working group on the classification of staphylococcal cassette chromosome elements (IWG-SCC). 2009. Classification of staphylococcal cassette chromosome mec (SCCmec): guidelines for reporting novel SCCmec elements. Antimicrob. Agents Chemother. 53: 4961–4967.

Ishige, N. and Ruddle, K. 1990. Gyosho in Southeast Asia: A study of fermented aquatic products. Iwanami, Tokyo (in Japanese).

Ito, H., Hadioetomo, R.S., Nikkuni, S. and Okada, N. 1985a. Studies on lactic acid bacteria in fish sauces (part 1). Chemical composition and microflora of fish sauce. Rep. Natl. Food Res. Inst. 47: 23–30.

Ito, H., Hadioetomo, R.S., Nikkuni, S. and Okada, N. 1985b. Studies on lactic acid bacteria in fish sauces (part 2). Identification of salt-tolerance and acid-producing bacteria from fish sauces. Rep. Natl. Food Res. Inst. 47: 31–40.

Ito, T., Katayama, Y. and Hiramatsu, K. 1999. Cloning and nucleotide sequence determination of the entire mec DNA of pre-methicillin-resistant *Staphylococcus* aureus N315. Antimicrob. Agents Chemother. 43: 1449–1458.

Ito, T., Okuma, K., Ma, X.X., Yuzawa, H. and Hiramatsu, K. 2003. Insights on antibiotic resistance of *Staphylococcus aureus* from its whole genome: genomic island SCC. Drug Resist. Updat. 6: 41–52.

Joosten, H.M.L.J. and Northolt, M.D. 1989. Detection, growth, and amine-producing capacity of lactobacilli in cheese. Appl. Environ. Microbiol. 55: 2356–2359.

Justé, A., Lievens, B., Frans, I., Marsh, T.L., Klingeberg, M., Michiels, C.W. and Willems, K.A. 2008. Genetic and physiological diversity of *Tetragenococcus halophilus* strains isolated from sugar- and salt-rich environments. Microbiology 154: 2600–2610.

Justé, A., Van Trappen, S., Verreth, C., Cleenwerck, I., De Vos, P., Lievens, B., Willems, K.A. 2012. Characterization of *Tetragenococcus* strains from sugar thick juice reveals a novel species, *Tetragenococcus osmophilus* sp. nov., and divides *Tetragenococcus halophilus* into two subspecies, *T. halophilus* subsp. *halophilus* subsp. nov. and *T. halophilus* subsp. *flandriensis* subsp. nov. Int. J. Syst. Evol. Microbiol. 62: 129–137.

Kamath, A.V., Vaaler, G.L. and Snell, E.E. 1991. Pyridoxal phosphate-dependent histidine decarboxylases. Cloning, sequencing, and expression of genes from *Klebsiella planticola* and *Enterobacter aerogenes* and properties of the overexpressed enzymes. J. Biol. Chem. 266: 9432–9437.

Kandler, O., Weiss, N. 1986. Genus *Lactobacillus* Beijerinck 1901, 212AL. pp. 1209-1234. *In*: Sneath, P.H.A. et al. (eds.). Bergey's Manual of Systematic Bacteriology. Vol. 2. The William and Wilkins Co., Baltimore Md.

Kanki, M., Yoda, T., Tsukamoto, T. and Shibata, T. 2002. *Klebsiella pneumoniae* produces no histamine: *Raoultella planticola* and *Raoultella ornithinolytica* strains are histamine producers. Appl. Environ. Microbiol. 38: 3462–3466.

Kanki, M., Yoda, T., Ishibashi, M. and Tsukamoto, T. 2004. *Photobacterium phosphoreum* caused a histamine fish poisoning incident. Int. J. Food Microbiol. 92: 79–87.

Kanki, M., Yoda, T., Tsukamoto, T. and Baba, E. 2007. Histidine decarboxylases and their role in accumulation of histamine in tuna and dried saury. Appl. Environ. Microbiol. 73: 1467–1473.

Kantor, A., Montville, T.J., Mett, A. and Shapira, R. 1997. Molecular characterization of the replicon of the *Pediococcus pentosaceus* 43200 pediocin A plasmid pMD136. FEMS Microbiol. Lett. 151: 237–244.

Kimura, B., Hokimoto, S., Takahashi, H. and Fujii, T. 2000. *Photobacterium histaminum* Okuzumi et al. 1994 is a later subjective synonym of *Photobacterium damselae* subsp. *damselae* (Love et al. 1981) Smith et al. 1991. Int. J. Syst. Evol. Microbiol. 50: 1339–1342.

Kimura, B., Konagaya, Y. and Fujii, T. 2001. Histamine formation by *Tetragenococcus muriaticus*, a halophilic lactic acid bacterium isolated from fish sauce. Int. J. Food Microbiol. 70: 71–77.

Kimura, M., Furutani, A., Fukui, Y., Shibata, Y., Nei, D., Yano, Y. and Satomi, M. 2015. Isolation and identification of the causative bacterium of histamine accumulation during fish sauce fermentation and the suppression effect of inoculation with starter culture of lactic acid bacterium on the histamine accumulation in fish sauce processing. Nippon Suisan Gakkaishi 81: 97–106. (in Japanese with an English abstract.)

Kloos, W.E. and Shileifer, K.H. 1986. Genus IV. *Staphylococcus*. pp. 1013–1035. *In*: Brenner, D.J. et al. (eds.). Bergey's Manual of Systematic Bacteriology. Vol. 2, 2nd Edn. Springer, New York.

Kobayashi, K., Okuzumi, M. and Fujii, T. 1995. Microflora of fermented puffer fish ovaries in rice-bran "fugunoko nukazuke". Fish Sci. 61: 291–295.

Kobayashi, T., Kimura, B. and Fujii, T. 2000. Differentiation of *Tetragenococcus* populations occurring in products and manufacturing processes of puffer fish ovaries fermented with rice-bran. Int. J. Food Microbiol. 56: 211–218.

Konagaya, Y., Kimura, B., Ishida, M. and Fujii, T. 2002. Purification and properties of a histidine decarboxylase from *Tetragenococcus muriaticus*, a halophilic lactic acid bacterium. J. Appl. Microbiol. 92: 1136–1142.

Kosaka, Y., Satomi, M., Furutani, A. and Ooizumi, T. 2012. Microfloral and chemical changes during processing of heshiko produced by aging of salted mackerel with rice bran by means of conventional practice in Wakasa Bay area, Fukui, Japan. Fish. Sci. 78: 463–469.

Kuda, T., Miyamoto, H., Sakajiri, M., Ando, K. and Yano, T. 2001. Microflora of fish nukazuke made in Ishikawa, Japan. Nippon Suisan Gakkaishi 67: 296–301. (in Japanese with English abstract.)

Ladero, V., Calles-Enriquez, M., Fernández, M. andAlvarez, M.A. 2010. Toxicological effects of dietary biogenic amines. Curr. Nutr. Food Sci. 6: 145–156.

Landeta, G., De las Rivas, B., Carrascosa, A.V. and Munoz, R. 2007. Screening of biogenic amine production by coagulase-negative staphylococci isolated during industrial Spanish dry-cured ham processes. Meat Sci. 77: 556–561.

Landete, J.M., Ferrer, S. and Pardo, I. 2005. Which lactic acid bacteria are responsible for histamine production in wine? J. Appl. Microbiol. 99: 580–586.

Landete, J.M., De las Rivas, B., Marcobal, A. and Munoz, R. 2008. Updated molecular knowledge about histamine biosynthesis by bacteria. Crit. Rev. Food Sci. Nutr. 48: 697–714.

Lehane, L. and Olley, J. 2000. Histamine fish poisoning revisited. Int. J. Food Microbiol. 58: 1–37.

Le Jeune, C., Lonvaud-Funel, A., Ten Brink, B., Hofstra, H. and van der Vassen, J.M.B.M. 1995. Development of a detection system for histidine decarboxylating lactic acid bacteria based on DNA probes, PCR and activity test. J. Appl. Microbiol. 78: 316–326.

Lonvaud-Funel, A. and Joyeux, A. 1994. Histamine production by wine lactic acid bacteria. Isolation of a histamine-producing strain of *Leuconostoc oenos*. J. Appl. Microbiol. 77: 401–407.

Lonvaud-Funel, A. 2001. Biogenic amines in wines: role of lactic acid bacteria. FEMS Microbiol. Lett. 199: 9–13.

Lucas, P., Landete, J., Coton, M., Coton, E. and Lonvaud-Funel, A. 2003. The tyrosine decarboxylase operon of *Lactobacillus brevis* IOEB 9809: characterization and conservation in tyramine-producing bacteria. FEMS Microbiol. Lett. 229: 65–71.

Lucas, P.M., Wolken, W.A., Claisse, O., Lolkema, J.S. and Lonvaud-Funel, A. 2005. Histamine-producing pathway encoded on an unstable plasmid in *Lactobacillus hilgardii* 0006. Appl. Environ. Microbiol. 71: 1417–1424.

Lucas, P.M., Claisse, O. and Lonvaud-Funel, A. 2008. High frequency of histamine-producing bacteria in the enological environment and instability of the histidine decarboxylase production phenotype. Appl. Environ. Microbiol. 74: 811–817.

Martín, M.C., Fernández, M., Linares, D.M. and Alvarez, M.A. 2005. Sequencing, characterization and transcriptional analysis of the histidine decarboxylase operon of *Lactobacillus buchneri*. Microbiology 151: 1219–1228.

Miyaki, T. 1954. Scombroid fish poisoning. Food Sanit. Res. 12: 7–14.

Molenaar, D., Bosscher, J.S., ten Brink, B., Driessen, A.J. and Konings, W.N. 1993. Generation of a proton motive force by histidine decarboxylation and electrogenic histidine/histamine antiport in *Lactobacillus buchneri*. J. Bacteriol. 175: 2864–2870.

Mongkolthanaruk, W., Nagase, M., Kawai, Y., Tanigawa, K., Li, Y., Yamaguchi, T. and Aimi, T. 2012. Evaluation of histamine productivity of *Tetragenococcus halophilus* isolated from salted mackerel (saba-shiokara). Fish Sci. 78: 441–449.

Morii, H., Kasama, K. and Herrera-Espinoza, R. 2004. Activity of two histidine decarboxylases from *Photobacterium phosphoreum* at different temperatures, pHs, and NaCl concentrations. J. Food Prot. 67: 1736–1742.

Morii, H., Kasama, K. and Herrera-Espinoza, R. 2006. Cloning and sequencing of the histidine decarboxylase gene from *Photobacterium phosphoreum* and its functional expression in *Escherichia coli*. J. Food Prot. 69: 1768–1776.

Nakazato, M., Kobayashi, C., Yamajima, Y., Tateishi, Y., Kawai, Y. and Yasuda, K. 2002. Determination of volatile basic nitrogen (VBN) and non volatile amines in fish sauce. Ann. Rep. Tokyo. Metr. Res. Lab. P. H. 53: 95–100. (in Japanese with English abstract.)

Nei, D., Kawasaki, S., Inatsu, Y., Yamamoto, K. and Satomi, M. 2012. Effectiveness of gamma irradiation in the inactivation of histamine-producing bacteria. Food Control 28: 143–146.

Nicoloff, H. and Bringel, F. 2003. ISLpl1 is a functional IS30-related insertion element in *Lactobacillus plantarum* that is also found in other lactic acid bacteria. Appl. Environ. Microbiol. 69: 6032–6040.

Okuzumi, M., Okuda, S. and Awano, M. 1981. Isolation of psychrophilic and halophilic histamine-forming bacteria from Scomber japonicus. Nippon Suisan Gakkaishi 47: 1591–1598.

Okuzumi, M., Hiraishi, A., Kobayashi, T. and Fujii, T. 1994. *Photobacterium histaminum* sp. nov., a histamine-producing marine bacterium. Int. J. Syst. Bacteriol. 44: 631–636.

Prozorovski, V. and Jörnvall, H. 1974. Separation and characterisation of subunits of histidine decarboxylase from *Micrococcus* sp. n. Eur. J. Biochem. 42: 405–409.

Recsei, P.A. and Snell, E.E. 1970. Histidine decarboxylase of *Lactobacillus* 30a. VI. Mechanism of action and kinetic properties. Biochemistry 9: 1492–1497.

Recsei, P.A., Moore, W.M. and Snell, E.E. 1983. Pyruvoyl-dependent histidine decarboxylases from *Clostridium perfringens* and *Lactobacillus buchneri*. J. Biol. Chem. 258: 439–444.

Recsei, P.A. and Snell, E.E. 1984. Pyruvoyl enzymes. Annu. Rev. Biochem. 53: 357–387.

Röling, W.F.M. and van Verseveld, H.W. 1996. Characterization of *Tetragenococcus halophila* populations in Indonesian soy mash (Kecap) fermentation. Appl. Environ. Microbiol. 62: 1203–1207.

Sato, T., Fujii, T., Masuda, M. and Okuzumi, M. 1994. Changes in numbers of histamine-metabolic bacteria and histamine content during storage of common mackerel. Fish. Sci. 60: 299–302.

Sato, T., Kimura, B. and Fujii, T. 1995. Histamine contents and histamine-metabolizing bacterial flora of fish sauce during fermentation. J. Food Hyg. Soc. Jpn. 36: 763–768. (in Japanese with an English abstract.)

Satomi, M., Kimura, B., Mizoi, M., Sato, T. and Fujii, T. 1997. *Tetragenococcus muriaticus* sp. nov., a new moderately halophilic lactic acid bacterium isolated from fermented squid liver sauce. Int. J. Syst. Bacteriol. 47: 832–836.

Satomi, M., Furushita, M., Oikawa, H., Yoshikawa-Takahashi, M. and Yano, Y. 2008. Analysis of a 30 kbp plasmid encoding histidine decarboxylase gene in *Tetragenococcus halophilus* isolated from fish sauce. Int. J. Food Microbiol. 126: 202–209.

Satomi, M., Furushita, M., Oikawa, H. and Yano, Y. 2011. Diversity of plasmid encoding histidine decarboxylase gene in *Tetragenococcus* spp. isolated from Japanese fish sauce. Int. J. Food Microbiol. 148: 60–65.

Satomi, M., Mori-Koyanagi, M., Shozen, K., Furushita, M., Oikawa, H. and Yano, Y. 2012. Analysis of plasmids encoding histidine decarboxylase gene in *Tetragenococcus muriaticus* isolated from Japanese fermented seafoods. Fish. Sci. 78: 935–945.

Satomi, M., Shozen, K., Furutani, A., Fukui, Y., Kimura, M., Yasuike, M., Funatsu, Y. and Yano, Y. 2014. Analysis of plasmids encoding the tyrosine decarboxylase gene in *Tetragenococcus halophilus* isolated from fish sauce. Fish. Sci. 80: 849–858.

Satomi, M. 2016. Effect of histamine-producing bacteria on fermented fishery products. Food Sci. Technol. Res. 22: 1–21.

Shimizu, T., Ohtani, K., Hirakawa, H., Ohshima, K., Yamashita, A., Shiba, T., Ogasawara, N., Hattori, M., Kuhara, S. and Hayashi, H. 2002. Complete genome sequence of *Clostridium perfringens*, an anaerobic flesh-eater. Proc. Natl Acad. Sci. USA 99: 996–1001.

Shozen, K., Satomi, M., Yano, Y., Yoshida, M., Fukui, Y., Takano, T. and Funatsu, Y. 2012. Effect of sucrose and halophilic lactic acid bacterium *Tetragenococcus halophilus* on chemical characteristics and microbial proliferation during fish sauce fermentation. J. Food Safety 32: 389–398.

Silla Santos, M.H. 1996. Biogenic amines: their importance in foods. Int. J. Food Microbiol. 29: 213–231.

Silla Santos, M.H. 1998. Amino acid decarboxylase capability of microorganisms isolated in Spanish fermented meat products. Int. J. Food Microbiol. 39: 227–230.

Spano, G., Russo, P., Lonvaud-Funel, A., Lucas, P., Alexandre, H., Grandvalet, C., Coton, E., Coton, M., Barnavon, L., Bach, B., Rattray, F., Bunte, A., Magni, C., Ladero, V., Alvarez, M., Fernández, M., Lopez, P., de Palencia, P.F., Corbi, A., Trip, H. and Lolkema, J.S. 2010. Biogenic amines in fermented foods. Eur. J. Clin. Nutr. 64: S95–S100.

Stute, R., Petridis, K., Steinhart, H. and Biernoth, G. 2002. Biogenic amines in fish and soy sauces. Eur. Food Res.Technol. 215: 101–107.

Suzzi, G. and Gardini, F. 2003. Biogenic amines in dry fermented sausages: a review. Int. J. Food Microbiol. 88: 41–54.

Taira, W., Funatsu, Y., Satomi, M., Takano, T. and Abe, H. 2007. Changes in extractive components and microbial proliferation during fermentation of fish sauce from underutilized fish species and quality of final products. Fish. Sci. 73: 913–923.

Takahashi, H., Kimura, B., Yoshikawa, M. and Fujii, T. 2003. Cloning and sequencing of the histidine decarboxylase genes of gram-negative, histamine-producing bacteria and their application in detection and identification of these organisms in fish. Appl. Environ. Microbiol. 69: 2568–2579.

Takahashi, H., Ogai, M., Miya, S., Kuda, T. and Kimura, B. 2015. Effects of environmental factors on histamine production in the psychrophilic histamine-producing bacterium *Photobacterium iliopiscarium*. Food Control. 52: 39–42.

Tanase, S., Guirard, B.M. and Snell, E.E. 1985. Purification and properties of a pyridoxal 5′-phosphate-dependent histidine decarboxylase from *Morganella morganii* AM-15. J. Biol. Chem. 260: 6738–6746.

Tao, Z., Nakano, T., Yamaguchi, T. and SATO, M. 2002. Production and diffusion of histamine in the muscle of scombroid fishes. Fish. Sci. 68: 1394–1397.

Tao, Z., Nakano, T., Yamaguchi, T. and Sato, M. 2009. Formation and diffusion mechanism of histamine in the muscle of tuna fish. Food Control 20: 923–926.

Taylor, S.L. 1986. Histamine food poisoning: toxicology and clinical aspects. Crit. Rev. Toxicol. 17: 91–128.

Toda, M., Yamamoto, M., Uneyama, C. and Morikawa, K. 2009. Histamine food poisonings in Japan and other countries. Bull. Natl. Inst. Health Sci. 127: 31–38. (in Japanese with an English abstract.)

Tolmasky, M.E., Actis, L.A. and Crosa, J.H. 1995. A histidine decarboxylase gene encoded by the *Vibrio anguillarum* plasmid pJM1 is essential for virulence: histamine is a precursor in the biosynthesis of anguibactin. Mol. Microbiol. 15: 87–95.

Torido, Y., Takahashi, T., Kuda, T. and Kimura, B. 2012a. Analysis of the growth of histamine-producing bacteria and histamine accumulation in fish during storage at low temperatures. Food Control 26: 174–177.

Torido, Y., Oshima, C., Takahashi, T., Miya, S., Iwakawa, A., Kuda, T. and Kimura, B. 2012b. Distribution of psychrophilic and mesophilic histamine-producing bacteria in retailed fish in Japan. Food Control 46: 338–342.

Trip, H., Mulder, N.L., Rattray, F.P. and Lolkema, J.S. 2011. HdcB, a novel enzyme catalysing maturation of pyruvoyl-dependent histidine decarboxylase. Mol. Microbiol. 79: 861–871.

Tsai, Y-H., Lina, C-Y., Chien, L-T., Lee, T-M., Wei, C-I. and Hwang, D-F. 2006. Histamine contents of fermented fish products in Taiwan and isolation of histamine-forming bacteria. Food Chemistry 98: 64–70.

Tsai, Y-H., Chang, S-C. and Kung, H-F. 2007. Histamine contents and histamine-forming bacteria in natto products in Taiwan. Food Control 18: 1026–1030.

Udomsil, N., Rodtong, S., Tanasupawat, S. and Yongsawatdigul, J. 2010. Proteinase-producing halophilic lactic acid bacteria isolated from fish sauce fermentation and their ability to produce volatile compounds. Int. J. Food Microbiol. 141: 186–194.

Vanderslice, P., Copeland, W.C. and Robertus, J.D. 1986. Cloning and nucleotide sequence of wild type and a mutant histidine decarboxylase from *Lactobacillus* 30a. J. Biol. Chem. 261: 15186–15191.

van Poelje, P.D. and Snell, E.E. 1990. Pyruvoyl-dependent enzyme. Ann. Rev. Biochem. 59: 29–59.

Visciano, P., Schirone, M., Tofalo, R. and Suzzi, G. 2012. Biogenic amines in raw and processed seafood. Front. Microbiol. 3: 188.

Watanabe, K., Fujimoto, J., Tomii, Y., Sasamoto, M., Makino, H., Kudo, Y. and Okada, S. 2009. *Lactobacillus kisonensis* sp. nov., *Lactobacillus otakiensis* sp. nov., *Lactobacillus rapi* sp. nov. and *Lactobacillus sunkii* sp. nov., heterofermentative species isolated from sunki, a traditional Japanese pickle. Int. J. Syst. Evol. Microbiol. 59: 754–760.

Weiss, N. 1992. The genera *Pediococcus* and *Aerococcus*. pp. 1502–1507. *In*: Balows, A., Truper, H.G., Dworkin, M., Harder, W. and Schleifer K.H. (eds.). The Prokaryotes, 2nd Ed. Vol. 2. Springer Verlag, New York.

Yatsunami, K. and Echigo, T. 1991. Isolation of salt-tolerant histamine-forming bacteria from commercial rice-bran pickles of sardine. Nippon Suisan Gakkaishi 57: 1723–1728. (in Japanese with an English abstract.)

Yatsunami, K. and Echigo, M. 1993. Changes in the number of halotolerant histamine-forming bacteria and contents of nonvolatile amines in sardine meat with addition of NaCl. Bull. Jpn. Soc. Sci. Fish. 59: 123–127.

Yoguchi, R., Okuzumi, M. and Fujii, T. 1990a. Seasonal variation in numbers of mesophilic and halophilic histamine-forming bacteria in inshore of Tokyo Bay and Sagami Bay. Nippon Suisan Gakkaishi 56: 1467–1472.

Yoguchi, R., Okuzumi, M. and Fujii, T. 1990b. Seasonal variation in number of halophilic histamine-forming bacteria on marine fish. Nippon Suisan Gakkaishi 56: 1473–1479.

Yokoi, K., Harada, Y., Shozen, K., Satomi, M., Taketo, A. and Kodaira, K. 2011. Characterization of the histidine decarboxylase gene of *Staphylococcus epidermidis* TYH1 coded on the staphylococcal cassette chromosome. Gene 477: 32–41.

Zaman, M.Z., Bakara, F.A., Jinapa, S. and Bakarb, J. 2011. Novel starter cultures to inhibit biogenic amines accumulation during fish sauce fermentation. Int. J. Food Microbiol. 145: 84–91.

13

Marine Toxins Associated with Seafood-borne Illnesses

Toshiyuki Suzuki

1. Introduction

Fish and shellfish are delicious food, with important protein content, and provide nutrition resources considered healthy for humans. Nonetheless, they often accumulate marine toxins produced by microorganisms, mostly toxic dinoflagellates. Molecular structures of marine toxins are generally large and complicated, and several analogues including metabolites in fish and shellfish are present, making them one of the most challenging targets in testing seafood-borne illnesses. As marine toxins are generally stable under heating conditions, human poisoning by consumption of fish or shellfish contaminated with marine toxins cannot be prevented even when they are cooked, therefore monitoring of fish or shellfish is the most important measure to protect people from poisoning by marine toxins. In the present chapter, chemical nature of marine toxins, human symptoms associated with marine toxins, and seafood safety measures for marine toxins are described.

National Research Institute of Fisheries Science, 2-12-4 Fukuura, Kanazawa, Yokohama, Kanagawa 236-8648, Japan.
E-mail: tsuzuki@affrc.go.jp

2. Shellfish Toxins

Shellfish poisoning is a severe gastrointestinal or neuro toxic illness caused by consumption of filter feeding bivalves contaminated with marine toxins originating from the toxic dinoflagellates or diatom. Some predators of bivalves such as crabs and lobsters are also contaminated with marine toxins by feeding toxic bivalves. Based on the structures, shellfish toxins are classified into five groups in CODEX STAN 292-2008. The maximum level of these marine toxins in bivalve mollusks is shown in Table 13.1. These five toxin groups, except for domoic acid (DA), had been regulated by the mouse bioassay. Recently, instrumental methods for other toxin groups have been developed and used as an official testing method in many countries. As instrumental methods provide accurate quantification of toxin analogues with dissimilar potencies, knowledge of the relative toxicities is required for risk assessment and determining overall toxicity. The ratios between the toxicity of the analogues and that of a reference compound within the same toxin group are termed "Toxicity Equivalency Factors" (TEFs). TEFs of marine toxins associated with shellfish were recently summarized [1]. The TEFs recommended by the FAO/WHO expert group for each toxin group are listed in Table 13.2.

For the prevention of shellfish poisoning and to maintain the safe supply of major shellfish products, monitoring stations are established in the areas where shellfish are harvested. The amounts of toxins contained in shellfish are regularly monitored by the official testing methods, including mouse bioassay and other chemical or biochemical methods. In addition to the monitoring system for shellfish toxins, monitoring of toxic dinoflagellates responsible for contamination of shellfish with toxins has been conducted in many countries. This provides an early warning and risk assessment of the chances of toxicity developing in the shellfish.

2.1 Saxitoxin (STX) group

STX analogues are the causative toxins of paralytic shellfish poisoning (PSP), which is one of the oldest known marine poisoning in North America

Table 13.1 Maximum levels of marine toxins in CODEX STAN 292-2008.

Marine toxin groups	mg/kg of molluscus fresh
Saxitoxin (STX) group	≤ 0.8 mg (2HCl) of saxitoxin equivalent
Okadaic acid (OA) group	≤ 0.16 mg/kg of okadaic acid equivalent
Domoic acid (DA) group	≤ 20 mg/kg domoic acid
Brevetoxin (BTX) group	≤ 200 mouse units or equivalent
Azaspiracid (AZA) group	≤ 0.16 mg/kg

Table 13.2 TEFs recommended by EFSA and the FAO/WHO expert group.

Toxin groups	Analogues	TEF proposed by EFSA*1	Oshima relative toxicity values (MU/u mole)*2	TEF recommended by the expert group
OA group	OA	1.0		1.0
	DTX1	1.0		1.0
	DTX2	0.6		0.5
STX group	STX	1.00	1	1.0
	neoSTX	1.00	0.92	2.0
	GTX1	1.00	0.99	1.0
	GTX2	0.4	0.36	0.4
	GTX3	0.6	0.64	0.6
	GTX4	0.7	0.73	0.7
	GTX5	0.1	0.064	0.1
	GTX6	0.1		0.05
	C1	–	0.006	0.01
	C2	0.1	0.096	0.1
	C3	–	0.013	0.01
	C4	0.1	0.058	0.1
	dcSTX	1.0	0.51	0.5
	dcNeoSTX	0.4		0.2
	dcGTX2	0.2	0.15	0.2
	dcGTX3	0.4	0.38	0.4
AZA group	AZA1	1.0		1.0
	AZA2	1.8		0.7
	AZA3	1.4		0.5
	AZA4			
	AZA5			
	AZA6			0.7

* 1 European Food Safety Authority.
* 2 Ref.

and Europe. The first historical poisoning case occurred when a Russian expedition team consumed mussels in Alaska in 1790. Approximately hundred people died by PSP. PSP have been reported in North and South America, Europe, Africa, East and Southeast Asian countries, Australia, New Zealand, etc. Symptoms of human PSP intoxication vary from a slight tingling or numbness to complete respiratory paralysis [2]. In fatal cases, death is usually caused by respiratory paralysis within 2–12 hours of consumption of food contaminated with STX analogues. Neurotoxic symptoms of STX analogues are caused by the blockade of the voltage-gated sodium channel by STX analogues [2]. STX analogues act on the receptor site 1 of the sodium channel. It is estimated that consumption of between 456 and 12400 µg of STX per person results in fatal cases [2].

Figure 13.1 shows the structures of prominent STX [3] analogues. The structure of STX analogues is characterized with two guanidinium moieties and hydrated ketone on the C-12. STX analogues are crystalline water-soluble toxin, and are relatively stable in neutral and acidic conditions even with heating conditions. On the contrary, STX analogues are rapidly decomposed in alkaline conditions. STX analogues are classified into three subgroups based on the nature of side chain: carbamoyl toxins (R4 = $CONH_2$), N-sulfocarbamoyl toxins (R4 = $CONHSO_3^-$), and decarbamoyl toxins (R4 = H). Although they all bind to site 1 of sodium channels, different affinities of three groups result in different toxicities. The carbamoyl toxins are the most toxic, and the N-sulfocarbamoyl derivatives are the least toxic [4]. C1,2, GTX1-6, STX, and neoSTX are originally produced by toxic dinoflagellate *Alexandrium* and *Gymnodinium* spp [2]. Figure 13.2 shows chemical or enzymatic conversions of STX analogues occurring in shellfish and the environment, as reported by Oshima [5]. Due to the conversions of STX analogues in shellfish, toxin profiles of shellfish are generally more complicated than those of toxic dinoflagellates.

2.2 Okadaic acid (OA) group

OA [6] and its analogues dinophysistoxins (DTXs) are the causative toxins of diarrhetic shellfish poisoning (DSP) [7, 8]. DSP human symptoms are diarrhoea, nausea, vomiting, and abdominal pain, starting 30 min to a few

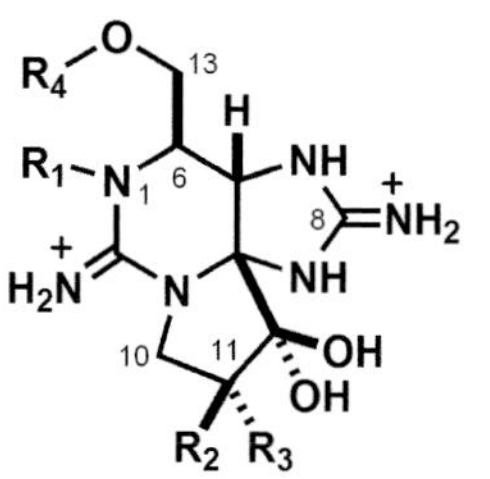

Toxins:	R1:	R2:	R3:	R4:	Toxicity (MU/µmol)	Net Charge
C1	H	H	OSO_3^-	$-CONHSO_3^-$	15	0
C2	H	OSO_3^-	H	$-CONHSO_3^-$	239	0
GTX1	OH	H	OSO_3^-	$-CONH_2$	2468	+1
GTX2	H	H	OSO_3^-	$-CONH_2$	892	+1
GTX3	H	OSO_3^-	H	$-CONH_2$	1584	+1
GTX4	OH	OSO_3^-	H	$-CONH_2$	1803	+1
GTX5	H	H	H	$-CONHSO_3^-$	160	+1
GTX6	OH	H	H	$-CONHSO_3^-$	180	+1
dcGTX2	H	H	OSO_3^-	-H	1617	+1
dcGTX3	H	OSO_3^-	H	-H	1872	+1
STX	H	H	H	$-CONH_2$	2483	+2
neoSTX	OH	H	H	$-CONH_2$	2295	+2
dcSTX	H	H	H	-H	1274	+2

Figure 13.1 Chemical structures of prominent saxitoxin analogues.

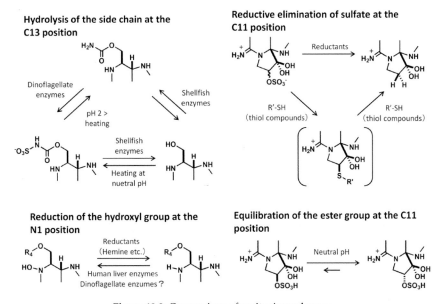

Figure 13.2 Conversions of saxitoxin analogues.

hours after ingestion and complete recovery occurs within 3 days [9]. It is estimated that consumption of 48 μg of OA or 38 μg of DTX1 per person results in diarrhea [10]. Along with PSP, DSP has been reported from almost all over the world, including North and South America, Europe, East Asian countries, and New Zealand, etc. OA/DTX analogues are produced by dinoflagellates that belong to the genera *Dinophysis* [11, 12, 13]. Although the dinoflagellate *Prorocentrum* is found to be a producer of OA/DTX analogues, involvement of bivalve contamination is unknown.

Figure 13.3 shows structures of OA/DTX analogues. OA is a lipophilic polyether compound of a C38 fatty acid. OA/DTX analogues are known as inhibitors of protein phosphatases (PPs) [14]. DTX2 is about half as toxic in the mouse bioassay, and has about half the affinity for PP2A as OA [15]. Molecular modeling studies indicated that 35(S)-methyl of DTX2 had an unfavorable interaction in the PP2A binding site [16], and this has been proposed as the reason for the reduced toxicity of DTX2 [17].

The C-1 carboxyl terminus and the C-7 hydroxyl group are both commonly modified by esterification. One group is the so-called "OA diol-esters", in which the carboxyl group of OA is conjugated to several different unsaturated diols to form allylic diol-esters. OA diol-esters are produced by *Prorocentrum* [18] and *Dinophysis* spp [19]. The other group is a complex mixture of 7-O-acyl ester derivatives of OA, DTX-1 and DTX-2 (also known as the "DTX-3" complex). 7-O-acyl ester derivatives can be formed as metabolites [20] of free toxins (OA, DTX-1, DTX-2) and several free

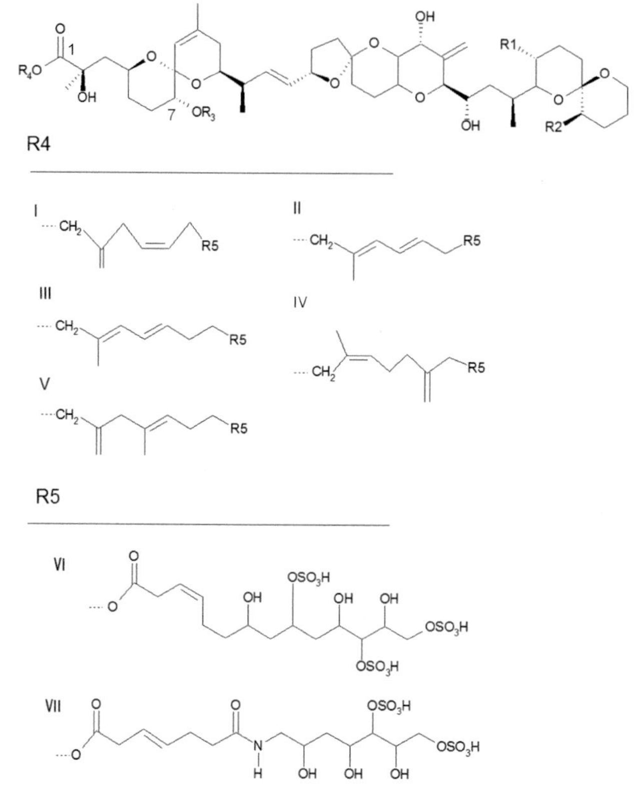

	R1	R2	R3	R4	R5	EMW
OA	CH_3	H	H	H	-	804.5
DTX1	CH_3	CH_3	H	H	-	818.5
DTX2	H	CH_3	H	H	-	804.5
DTX3	(H or CH_3)		Acyl	H	-	
OA-D7a	CH_3	H	H	I	OH	914.6
OA-D7b	CH_3	H	H	II	OH	914.6
OA-D8	CH_3	H	H	III	OH	928.6
OA-D9a	CH_3	H	H	IV	OH	942.6
OA-D9b	CH_3	H	H	V	OH	942.6
DTX4	CH_3	H	H	III	VI	1472.6
DTX5a	CH_3	H	H	II	VII	1392.6
DTX5b	CH_3	H	H	III	VII	1406.6

Figure 13.3 Structures of okadaic acid and dinophysistoxin analogues.

fatty acids in shellfish that have consumed toxic dinoflagellates. The most dominant fatty acids esterified to the C-7 position of OA, DTX1, and DTX2 are 16:0, 14:0, and 16:1. The activities of OA were generally decreased by acylation of the C-7 hydroxyl group [21]. The decrease was most significant in the mouse lethality, moderate in cytotoxicity, and only slight in the fluid accumulating potency in mouse intestinal loops. Diarrheagenicity measured by suckling mouse assays was affected little by the acylation.

2.3 Domoic acid (DA) group

DA [22] was first isolated from red algae *Chondria armota*, which was used as an anthelmintic in Tokunoshima Island, Japan. DA was later identified as the causative toxin of amnesic shellfish poisoning (ASP), which was recorded in 1987 in Prince Edward Island, Canada [23]. More than hundred people suffered from sickness, and at least three people died in the case in Prince Edward Island. The human symptoms of ASP include abdominal cramps, vomiting, disorientation, and memory loss (amnesia) [23]. Critical patients were old people aged more than 65. It is estimated that consumption between 135 and 295 mg per person results in fatal cases [24]. Bivalves are contaminated with DA by ingesting the causative diatom species *Pseudo-nitzschia multiseries, P. australis, P. seriata* [25]. DA is also suggested as the causative toxin of the mass mortalities of pelicans and cormorants in Monterey Bay, California.

Figure 13.4 shows structures of DA analogues. DA is a crystalline water-soluble acidic amino acid. Besides DA, nine isomers have been reported. DA is a potent neurotoxin and an excitatory neurotransmitter that binds to the kainate receptor proteins in neuronal cells [10]. It is thought that this activity causes amnesia. Other isomers are less toxic because they bind less strongly to the kainate receptor proteins, and they are not included in the regulation in many countries.

2.4 Brevetoxin (BTX) group

BTX analogues are lipophilic polyether compounds that are produced by the toxic dinoflagellate *Karenia brevis*. Blooms of *K. brevis*, called "red tide", which had been recorded since the mid-1800s in the Gulf of Mexico, occur regularly in this region. It had been reported that *K. brevis* caused massive fish kills, mollusk poisoning, and human food poisoning along the Florida coast in the Gulf of Mexico [26]. BTX-B (BTX2; also known as PbTx-2) was probably the first identified structure with a unique ladder-shaped polyether [26] (Fig. 13.5). The structure of BTX-A (BTX1; also known as PbTx-1) was later elucidated [27]. BTX-A and BTX-B are considered the parent algal toxins from which all other analogues are derived. A serious neurotoxic

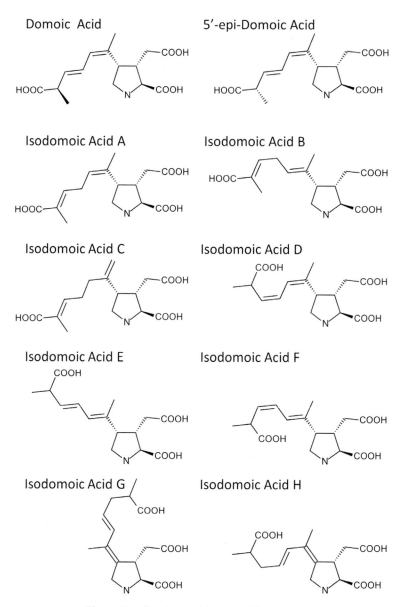

Figure 13.4 Structures of domoic acid analogues.

illness, neurotoxic shellfish poisoning (NSP), is caused by the ingestion of molluskan shellfish contaminated with BTX analogues. BTX analogues bind to site 5 of the voltage-gated sodium channels, causing them to open at normal cell resting membrane potentials. This leads to persistent activation

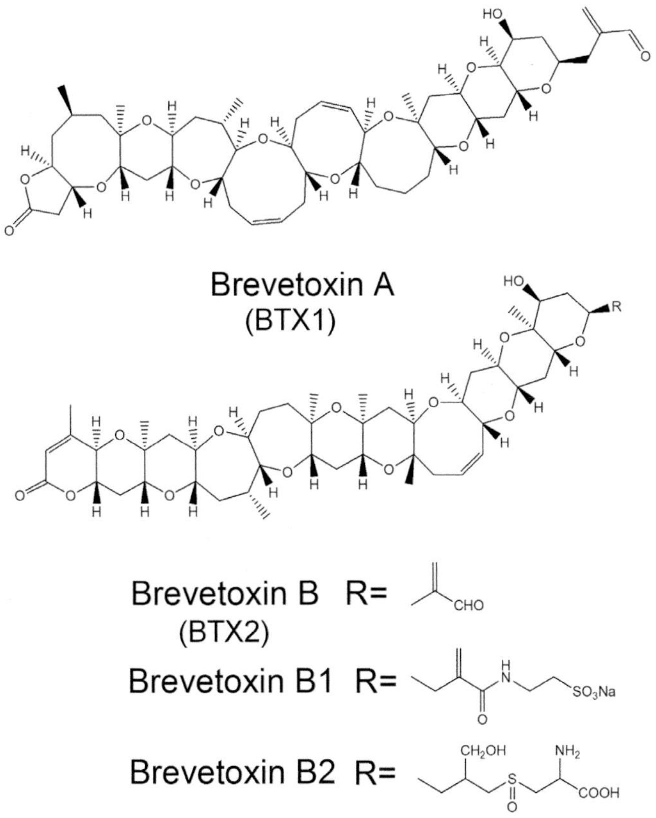

Brevetoxin A
(BTX1)

Brevetoxin B   R=

(BTX2)

Brevetoxin B1   R=

Brevetoxin B2   R=

Figure 13.5 Structures of brevetoxin analogues.

of neuronal, muscle, and cardiac cells [10]. This function is opposite to that caused by paralytic shellfish toxin STX analogues. The human symptoms of NSP are nausea, diarrhea, vomiting, abdominal pain, paresthesia, myalgia, ataxia, bradycardia, loss of coordination, vertigo, and mydriasis [10]. At least nine BTX analogues were isolated from *K. brevis*. There were a few fatal cases reported. Besides BTX-B produced by *K. brevis*, several bivalve metabolites (BTX-B1, BTX-B2, BTX-B3, BTX-B4, etc.) converted from BTX-B were reported [28, 29, 30, 31].

2.5 *Azaspiracid (AZA) group*

AZA analogues are the causative toxins of azaspiracid poisoning (AZP). In 1995, human poisonings were caused by the consumption of mussels, *Mytilus edulis*, harvested from Killary Harbor Ireland. The human symptoms are nausea, vomiting, severe diarrhea, and stomach cramps [10, 32]. The symptoms are similar to those caused by diarrhetic shellfish toxin OA/DTX analogues. However, diarrhetic shellfish toxins OA and DTX1 were hardly detected in the mussels implicated in the food poisoning case. Later, AZA1 [32] was identified with the structural elucidation as the causative toxin. It is estimated that consumption of AZAs between 23 and 86 μg per person results in illness [33].

The structures of ten prominent AZA analogues are shown in Fig. 13.6. AZA1, AZA2, and AZA3 are the most dominant toxins found in shellfish [32, 34]. Approximately 40 analogues, including stereoisomers, have been identified by MS/MS and NMR studies. The original reported structure [32] was later revised by the total synthesis of AZA1 [35]. AZA analogues are lipophilic polyether amino acids with rare structural features including a tri-spiro ring and an azaspiro ring assembly, together with a terminal carboxylic acid group. AZA analogues are easily soluble in aqueous methanol or acetone. AZA1 and AZA2 are produced by a small dinoflagellate *Azadinium spinosum* [36]. Many other AZA analogues are shellfish metabolites or chemically degraded artefact [37]. Although AZA analogues show similar symptoms to those by OA/DTX analogues, they do not inhibit protein phosphatases (PPs) [10]. AZA analogues induce pathological changes to the liver, pancreas, thymus, and spleen of mice. These pathological changes differ from those caused by diarrhetic shellfish toxins OA and DTX analogues [10].

2.6 Other lipophilic toxins, pectenotoxin (PTX) and yessotoxin (YTX) group

PTX and YTX analogues were included in diarrhetic shellfish toxin (DST) group due to their potent toxicity on the mouse bioassay used for the official testing method of diarrhetic shellfish poisoning (DSP). Although several toxicological data of PTX [38, 39, 40, 41] and YTX [42, 43] have been reported, symptoms of intoxication produced by PTX and YTX in humans are relatively unknown due to the fact that no human intoxication has been reported till date [10]. Animal studies indicate that PTX and YTX are much less toxic via the oral route, and that they do not induce diarrhea. Although PTX and YTX groups were eliminated from the definition of DST and regulation of marine toxins in CODEX STAN 292-2008, PTX and YTX analogues are regulated in some countries. PTX and YTX analogues are lipophilic polyether compounds (Figs. 13.7 and 13.8). YTX analogues resemble brevetoxin (BTX) and ciguatoxin (CTX) in having a ladder-shape polycyclic ether skeleton and an unsaturated side chain. PTX analogues

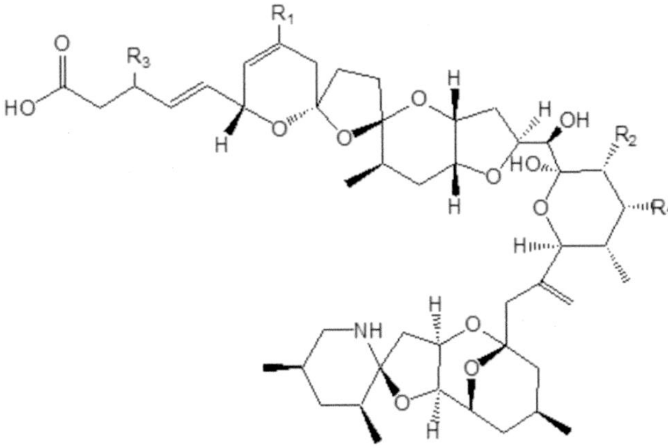

Toxins	R1	R2	R3	R4
AZA1	H	CH_3	H	H
AZA2	CH_3	CH_3	H	H
AZA3	H	H	H	H
AZA4	H	H	OH	H
AZA5	H	H	H	OH
AZA6	CH_3	H	H	H
AZA7	H	CH_3	OH	H
AZA8	H	CH_3	H	OH
AZA9	CH_3	H	OH	H
AZA10	CH_3	H	H	OH

Figure 13.6 Structures of azaspiracid analogues.

often coexist in bivalves contaminated with OA/DTX analogues, because both groups are produced by the same dinoflagellate *Dinophysis* spp [11, 12, 13]. YTX analogues are produced by dinoflagellate *Protoceratium reticulatum* [44], *Gymnodinium spinifera* [45], and *Lingulodinium polyedrum* [46].

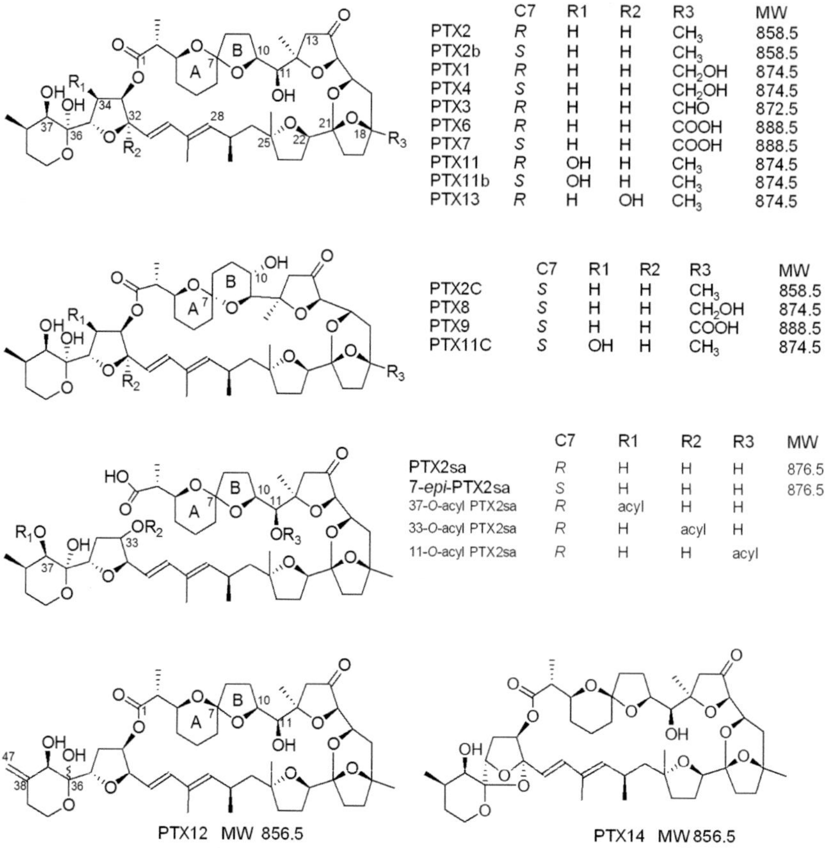

	C7	R1	R2	R3	MW
PTX2	R	H	H	CH₃	858.5
PTX2b	S	H	H	CH₃	858.5
PTX1	R	H	H	CH₂OH	874.5
PTX4	S	H	H	CH₂OH	874.5
PTX3	R	H	H	CHO	872.5
PTX6	R	H	H	COOH	888.5
PTX7	S	H	H	COOH	888.5
PTX11	R	OH	H	CH₃	874.5
PTX11b	S	OH	H	CH₃	874.5
PTX13	R	H	OH	CH₃	874.5

	C7	R1	R2	R3	MW
PTX2C	S	H	H	CH₃	858.5
PTX8	S	H	H	CH₂OH	874.5
PTX9	S	H	H	COOH	888.5
PTX11C	S	OH	H	CH₃	874.5

	C7	R1	R2	R3	MW
PTX2sa	R	H	H	H	876.5
7-*epi*-PTX2sa	S	H	H	H	876.5
37-O-acyl PTX2sa	R	acyl	H	H	
33-O-acyl PTX2sa	R	H	acyl	H	
11-O-acyl PTX2sa	R	H	H	acyl	

PTX12 MW 856.5

PTX14 MW 856.5

Figure 13.7 Structures of pectenotoxin analogues.

3. Palytoxin (PLTX) and Ovatoxin (OVTX)

PLTX is one of the most potent non-protein marine biotoxins with a long polyhydroxylated and partially unsaturated aliphatic backbone [47, 48], which was first isolated from the marine zoanthid *Palythoa toxica* in Hawaii (Fig. 13.9). PLTX is known as a potent inhibitor for ATPase Na⁺/K⁺ pump [49], a transmembrane enzyme that plays a role in maintaining the resting potential of nerve, muscle, and heart cells. Species of the benthic dinoflagellates of the genus *Ostreopsis* are known to produce PLTX analogues, ostreocin [50], and ovatoxin (OVTX) [51, 52]. Blooms of *Ostreopsis ovata* in the Mediterranean Sea have been causing serious human health problems by inhalation of aerosol and dermal contact [53]. It is reported that PLTX or PLTX-like compounds are causative toxins in Blue humphead parrot fish (*Scarus ovifrons*) in Japan, however PLTX and OVTX

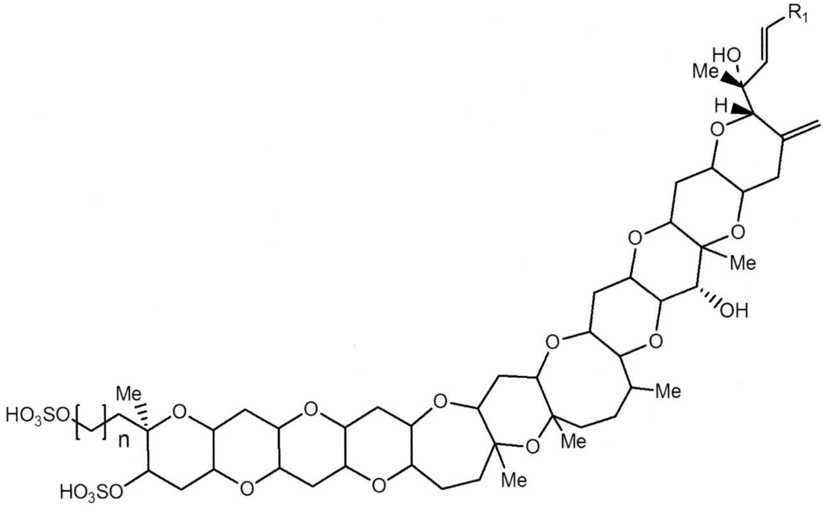

Toxin	n	m/z	R₁
YTX	1	1141	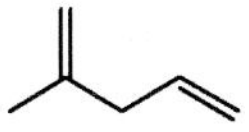
1-homoYTX	2	1155	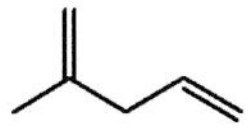
45-hydroxyYTX	1	1157	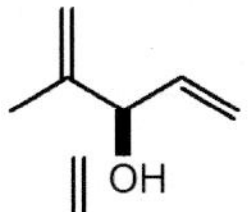
45-hydroxy-1-homoYTX	2	1171	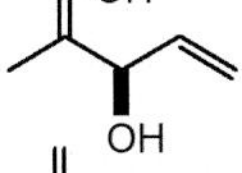
45,46,47-trinorYTX	1	1101	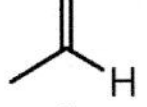
45,46,47-trinor-1-homoYTX	2	1115	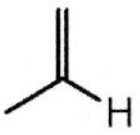

Figure 13.8 Structure of prominent yessotoxin analogues.

analogues were not detected in blue humphead parrot fish implicated in food poisoning case in Japan [54]. Due to co-occurrence with other seafood toxins, it has been difficult to assess the true risk of PLTX poisoning through seafood consumption in humans [53]. The European Food Safety Authority (EFSA) suggested a tentative maximum limit of 30 µg/kg of palytoxin equivalents in shellfish.

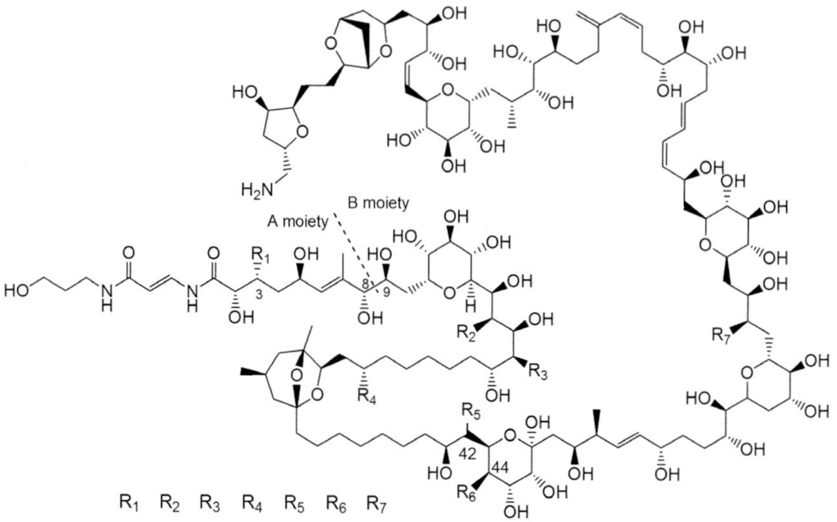

	R_1	R_2	R_3	R_4	R_5	R_6	R_7
Palytoxin	Me	OH	OH	Me	H	OH	OH
Ostreocin-d	H	OH	H	H	OH	H	OH
Ovatoxin-a	Me	H	OH	Me	OH	H	H

Figure 13.9 Structures of palytoxin analogues.

4. Tetrodotoxin (TTX)

TTX analogues are the causative toxins of puffer fish poisonings generally occurring in Asian countries [55]. Symptoms of human intoxication are the same as those caused by paralytic shellfish toxins STX analogues. Neurotoxic symptoms of TTX analogues are caused by the blockade of the voltage-gated sodium channel by TTX analogues as well as STX analogues. TTX occurs in a wide range of marine and terrestrial animals including frogs, newts, and salamanders. Aquaculture puffer fish in Japan is not contaminated with TTX analogues, suggesting that TTX in puffer fish could be extrinsic. Although marine bacteria (*Vibrio alginolyticus, Shewanella alga*, etc.) producing TTX were isolated from marine environment [55], the toxification mechanism of puffer fish is totally unambiguous. Recently, TTX was found in bivalves from New Zealand [56] and Europe [57], raising a discussion that TTX should be included in paralytic shellfish toxin group. As a safety measure for puffer fish poisoning, the sale and consumption of puffer fish is banned in Japan. However, some parts of some puffer fish (Table 13.3) are allowed to be eaten if the puffer fish is cooked by a licensed chef. Similar safety measures are taken in some countries.

Figure 13.10 shows the structures of prominent TTX analogues. The structure of TTX [58, 59, 60] is characterized with a guanidinium

Table 13.3 Edible puffer fish and the edible parts controlled by Ministry of Health, Labour and Welfare in Japan.

English name	Species	Parts Muscles	Skins	Testes
Grass puffer	*Takifugu niphobles*	o	–	–
Finepatterned puffer	*Takifugu poecilonotus*	o	–	–
Panther puffer	*Takifugu pardalis*	o	–	–
Vermiculated puffer	*Takifugu snyderi*	o	–	o
Genuine puffer or Purple puffer	*Takifugu porphyreus*	o	–	o
*1	*Takifugu obscurus*	o	–	o
Red-eyed puffer	*Takifugu chrysops*	o	–	o
Ocella puffer or Tiger puffer	*Takifugu rubripes*	o	o	o
Eyespot puffer	*Takifugu chinensis*	o	o	o
Striped puffer or Yellowfin puffer	*Takifugu xanthopterum*	o	o	o
Spottyback puffer	*Takifugu stictonotus*	o	–	o
Smooth-backed blowfish	*Lagocephalus inermis*	o	o	o
Brown-backed toadfish	*Laocephalus wheeleri*	o	o	o
Brown-backed toadfish	*Laocephalus gloveri*	o	o	o
Slackskinned puffer	*Sphoeroides pachygaster*	o	o	o
*1	*Takifugu flavidus*	o	–	–

o: Edible.
–: Not edible.
*1: Corresponding English name is not found.

moiety connected to a highly oxygenated carbon skeleton that possesses a 2,4-dioxaadamantane portion containing five hydroxyl groups. TTX analogues are water-soluble toxins and they are relatively stable in weak acidic conditions. On the contrary, TTX analogues are decomposed in alkaline or strong acidic conditions. TTX analogues are also unstable in neutral conditions if they are heated.

5. Ciguatoxin (CTX)

CTX analogues are the causative toxins of ciguatera fish poisoning (CFP). CFP characterizes the intoxication caused by consumption of fish from tropical and subtropical areas. It is reported that more than at least ten thousand people are estimated to suffer annually from CFP, which is the largest scale of food poisonings of nonbacterial origin [2]. CTXs are

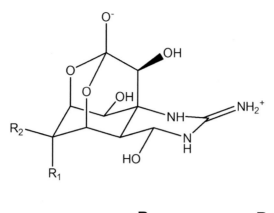

	R₁	R₂
TTX	OH	CH₂OH
6-*epi*TTX	CH₂OH	OH
11-deoxyTTX	OH	CH₃
11-oxoTTX	OH	CHO
11-norTTX-6(*R*)-ol	H	OH
11-norTTX-6(*S*)-ol	OH	H
Chiriquitoxin	OH	CH(OH)CH(NH₂)COOH
		R *S*

Figure 13.10 Structures of tetrodotoxin analogues.

produced by certain strains of the dinoflagellate *Gambierdiscus toxicus* [61, 62] and other *Gambierdiscus* spp. [63], and presumably accumulated in various kinds of reef fish through the food chain. Ingestion of affected fish leads to neurological (e.g., tingling, itching), gastrointestinal (e.g., vomiting, diarrhea, nausea), and cardiovascular (e.g., hypotension, bradycardia) disorders, which may last up to a month or more [2]. The most characteristic symptom is a disorder of temperature sensation. Specifically, touching cold water can induce pain similar to that of an electric shock. CTX analogues are mainly found in the Pacific, Caribbean, and Indian Ocean region, and they are classified as Pacific (P) [61, 62], Caribbean (C) [64], and Indian Ocean (I) CTX-group toxins. The structure of CTX from Indian Ocean has not been elucidated, as along with puffer fish, the sale and consumption of fish involved in CFP is banned Japan and some countries. Currently, there

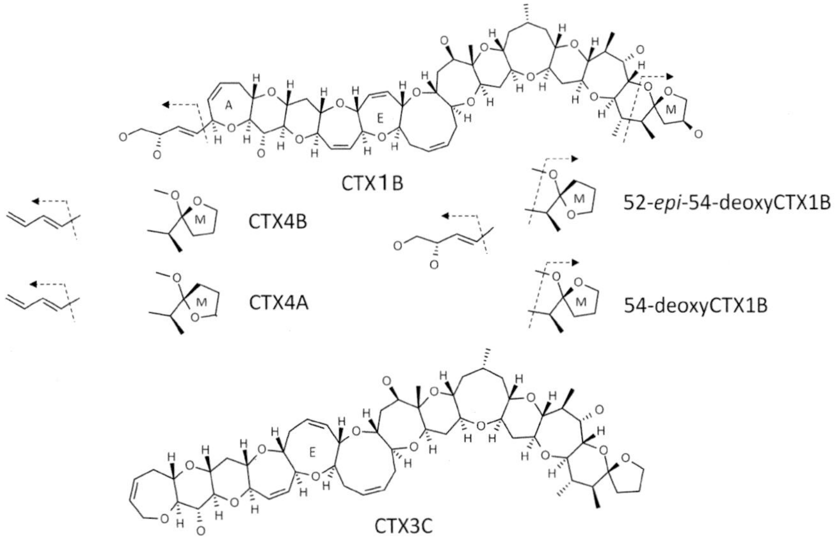

Figure 13.11 Structures of ciguatoxin analogues found in fish and *Gambierdiscus* in Pacific Ocean.

are no regulatory limits for CTX in fish, but a guidance level of 0.01 ppb CTX1B equivalent has been issued by the United States Food and Drug Administration (US FDA). This recommendation was based on a 10-fold reduction of the lowest concentration of CTX in meal remnants found to cause human illness.

Figure 13.11 shows the structures of prominent CTX analogues found from the Pacific region. CTX analogues are lipophilic polyether compounds. CTX analogues from Pacific region are divided into CTX1B [61, 65] type and CTX3C [62] type on the structures. It is thought that CTX1B, which is the most toxic compound in LD_{50} in mice, is a fish metabolite converted from CTX4A [65] or CTX4B [61] originating from *Gambierdiscus* spp. CTX analogues are characterized by their affinity binding to voltage sensitive sodium channels, causing them to open at normal cell resting membrane potentials [2]. This function is opposite to that caused by STX analogues and TTX. CTX analogues act at the same receptor site (site 5) of the sodium channel as brevetoxin (BTX). This results in an influx of sodium ions, cell depolarization, and the appearance of spontaneous action potentials in excitable cells. As a consequence of the increased sodium permeability, the plasma membrane is unable to maintain the internal environment of cells and volume control. These could result in several neurotoxin symptoms of CTX analogues.

References

1   Botana, L.M., Hess, P., Munday, R., Nathalie, A., DeGrasse, S.L., Feeley, M., Suzuki, T., van den Bergh, M., Fattori, V., Gamarro, E.G., Tritscher, A., Nakagwa, R. and Karunasagar, I. 2017. Derivation of toxicity equivalency factors for marine biotoxins associated with bivalve molluscs. Trends in Food Science and Technology 59: 15–24.

2   FAO (ed.). 2004. Marine Biotoxins. Rome, Italy (ISBN95–5–105129–1).

3   Schantz, E.J., Ghazarossian, V.E., Schnoes, H.K. and Strong, F.M. 1975. The structure of saxitoxin. J. Am. Chem. Soc. 97: 1238–1239.

4   Oshima, Y. 1995. Postcolumn derivatization liquid chromatographic method for paralytic shellfish toxins. J. AOAC Int. 78: 528–532.

5   Oshima, Y. 1995. Chemical and enzymatic transformation of paralytic shellfish toxins in marine organisms harmful marine algal blooms. pp. 475–480. *In*: Lassis, P., Arzul, G., Erard, E., Gentien, P. and Marcaillou, C. (eds.). Lavoisier Publ.

6   Tachibana, K., Scheuer, P.J., Tsukitani, Y., Kikuchi, H., Engen, D.V., Clardy, J., Gopichand, Y. and Schmitz, F.J. 1981. Okadaic acid, a cytotoxic polyether from two marine sponges of the genus *Halichondria*. J. Am. Chem. Soc. 103: 2469–2471.

7   Murata, M., Shimatani, M., Sugitani, H., Oshima, Y. and Yasumoto, T. 1982. Isolation and structural elucidation of the causative toxin of the diarrhetic shellfish poisoning. Bull. Japan Soc. Sci. Fish. 48: 549–552.

8   Yasumoto, T., Murata, M., Oshima, Y., Sano, M., Matsumoto, G.K. and Clardy, J. 1985. Diarrhetic shellfish toxins. Tetrahedron 4: 1019–1025.

9   Yasumoto, T., Oshima, Y. and Yamaguchi, M. 1978. Occurrence of a new type of shellfish poisoning in the Tohoku district. Bull. Japan. Soc. Sci. Fish. 44: 1249–1255.

10 FAO. 2011. Assessment and management of biotoxin risks in bivalve. Rome, Italy (ISBN 978-92-5-107003-1).

11 Lee, J.S., Igarashi, T., Fraga, S., Dahl, E., Hovgaard, P. and Yasumoto, T. 1989. Determination of diarrhetic shellfish toxins in various dinoflagellate species. J. Appl. Phycol. 1: 147–152.

12 Reguera, B. and Pizarro, G. 2008. Planktonic dinoflagellates which contain polyether toxins of the old "DSP complex". pp. 257–284. *In*: Botana, L.M. (ed.). Seafood and Freshwater Toxins. Pharmacology, Physiology, and Detection. CRC Press.

13 Suzuki, T., Miyazono, A., Baba, K., Sugawara, R. and Kamiyama, T. 2009. LC-MS/MS analysis of okadaic acid analogues and other lipophilic toxins in single-cell isolates of several *Dinophysis* species collected in Hokkaido, Japan. Harmful Algae 8: 233–238.

14 Bialojan, C. and Takai, A. 1988. Inhibitory effect of a marine-sponge toxin, okadaic acid, on protein phosphatases. Specificity and kinetics, Biochem. J. 256: 283–290.

15 Aune, T., Larsen, S., Aasen, J.A.B., Rehmann, N., Satake, M. and Hess, P. 2007. Relative toxicity of dinophysistoxin-2 (DTX-2) compared with okadaic acid, based on acute intraperitoneal toxicity in mice. Toxicon 49: 1–7.

16 Larsen, K., Petersen, D., Wilkins, A.L., Samdal, I.A., Sandvik, M., Rundberget, T., Goldstone, D., Arcus, V., Hovgaard, P., Rise, F., Rehmann, N., Hess, P. and Miles, C.O. 2007. Clarification of the C-35 stereochemistries of dinophysistoxin-1 and dinophysistoxin-2 and its consequences for binding to protein phosphatase. Chem. Res. Toxicol. 20: 868–875.

17 Huhn, J., Jeffrey, P.D., Larsen, K., Rundberget, T., Rise, F., Cox, N.R., Arcus, V., Shi, Y. and Miles, C.O. 2009. A structural basis for the reduced toxicity of dinophysistoxin-2. Chem. Res. Toxicol. 22: 1782–1786.

18 Yasumoto, T., Seino, N., Murakami, Y. and Murata, M. 1987. Toxins produced by benthic dinoflagellates. Biol. Bull. 172: 128–131.

19 Suzuki, T., Beuzenberg, V., Mackenzie, L. and Quilliam, M.A. 2004. Discovery of okadaic acid esters in the toxic dinoflagellate *Dinophysis acuta* from New Zealand using

liquid chromatography/tandem mass spectrometry. Rapid Commun. Mass Spectrom. 18: 1131–1138.

20 Suzuki, T., Ota, H. and Yamasaki, M. 1999. Direct evidence of transformation of dinophysistoxin-1 to 7-*O*-acyl-dinophysistoxin-1 (dinophysistoxin-3) in the scallop *Patinopecten yessoensis.* Toxicon 37: 187–198.

21 Yanagi, T., Murata, M., Torigoe, K. and Yasumoto, T. 1989. Biological activities of semisynthetic analogs of dinophysistoxin-3, the major diarrhetic shellfish toxin. Agric. Biol. Chem. 53: 525–529.

22 Takemoto, T., Daigo, K., Kondo, Y. and Kondo, K. 1966. Studies on the constituents of Chondira armota VIII, On the structure of domoic acid. Yakugaku Zasshi 86: 874–877.

23 Wright, J.L.C., Boyd, R.K., de Freitas, A.S.W., Falk, M., Foxall, R.A., Jamieson, W.D., Laycock, M.V., McCulloch, A.W., McInnes, A.G., Odense, P., Pathak, V.P., Quilliam, M.A., Ragan, M.A., Sim, P.G., Thibault, P., Walter, P.J.A., Gilgan, M., Richard, D.J.A. and Dewar, D. 1989. Identification of domoic acid, a neuroexcitatory amino acid, in toxic mussels from eastern Prince Edward Island. Can. J. Chem. 67: 481–490.

24 Todd, E.C.D. 1993. Domoic acid and amnesic shellfish poisoning—A review. J. Food Protection 56: 69–83.

25 Lelong, A., Hegaret, H., Soudant, P. and Bates, S.S. 2012. Pseudo-nitzschia (Bacillariophyceae) species, domoic acid and amnesic shellfish poisoning: revisiting previous paradigms. Phycologia 51: 168–216.

26 Lin, Y.Y., Risk, M., Ray, S.M., Engen, D.V., Clardy, J., Golik, J., James, J.C. and Nakanishi, K. 1981. Isolation and structure of brevetoxin B from the "red tide" dinoflagellate *Ptychodiscus brevis* (*Gymnodinium breve*). J. Am. Chem. Soc. 103: 6773–6775.

27 Shimizu, Y., Chou, H.N., Bando, H., Duyne, G.V. and Clardy, J. 1986. Structure of brevetoxin A (GB-1 toxin), the most potent toxin in the Florida red tide organism *Gymnodinium breve* (*Ptychodiscus brevis*). J. Am. Chem. Soc. 108: 514–515.

28 Nozawa, A., Tsuji, K. and Ishida, H. 2003. Implication of brevetoxin B1 and PbTx-3 in neurotoxic shellfish poisoning in New Zealand by isolation and quantitative determination with liquid chromatography-tandem mass spectrometry. Toxicon 42: 91–103.

29 Ishida, H., Nozawa, A., Nukaya, H., Rhodes, L., McNabb, P., Holland, P.T. and Tsuji, K. 2004. Confirmation of brevetoxin metabolism in cockle, Austrovenus stutchburyi, and greenshell mussel, *Perna canaliculus*, associated with New Zealand neurotoxic shellfish poisoning, by controlled exposure to Karenia brevis culture. Toxicon 43: 701–712.

30 Wang, Z., Plakas, S.M., El Said, K.R., Jester, E.L.E., Granade, H.R. and Dickey, R.W. 2004. LC/MS analysis of brevetoxin metabolites in the Eastern oyster (Crassostrea virginica). Toxicon 43: 455–465.

31 Abraham, A., Wang, Y., El Said, K.R. and Plakas, S.M. 2012. Characterization of brevetoxin metabolism in Karenia brevis bloom-exposed clams (Mercenaria sp.) by LC-MS/MS. Toxicon 60: 1030–1040.

32 Satake, M., Ofuji, K., Naoki, H., James, K.J., Furey, A., McMahon, T., Silke, J. and Yasumoto, T. 1998. Azaspiracid, a new marine toxin having unique spiro ring assemblies, isolated from Irish Mussels, *Mytilus edulis.* J. Am. Chem. Soc. 120: 9967–9968.

33 European Union/Sante et Consommateurs (EU/SANCO). 2001. Report of the meeting of the working group on toxicology of DSP and AZP, 21–23 May, 2001. Brussels.

34 Ofuji, K., Satake, M., McMahon, T., Silke, J., James, K.J., Naoki, H., Oshima, Y. and Yasumoto, T. 1999. Two analogues of azaspiracid isolated from mussels, *Mytilus edulis*, involved in human intoxication in Ireland. Nat. Toxins 7: 99–102.

35 Nicolaou, K.C., Vyskocil, S., Koftis, T.V., Yamada, Y.M.A., Ling, T., Chen, D.Y.K., Tang, W., Petrovic, G., Frederick, M.O., Li, Y. and Satake, M. 2004. Structural revision and total synthesis of azaspiracid-1, Part 1: Intelligence gathering and tentative proposal. Angew. Chem. Int. Ed. 43: 4312–4318.

36 Tillmann, U., Elbrachter, M., Krock, B., John, U. and Cembella, A. 2009. *Azadinium spinosum* gen. et sp. nov. (Dinophyceae) identified as a primary producer of azaspiracid toxins. Eur. J. Phycol. 44: 63–79.

37 Hess, P., McCarron, P., Krock, B., Kilcoyne, J. and Miles, C.O. 2014. Azaspiracids: Chemistry, biosynthesis, metabolism, and detection, seafood and freshwater toxins. Pharmacology, Physiology, and Detection, pp. 799–821.

38 Terao, K., Ito, E., Yanagi, T. and Yasumoto, T. 1986. Histopathological studies on experimental marine toxin poisoning. I. Ultrastructural changes in the small intestine and liver of suckling mice induced by dinophysistoxin-1 and pectenotoxin-1. Toxicon 24: 1141–1151.

39 Ito, E., Suzuki, T., Oshima, Y. and Yasumoto, T. 1998. Studies of diarrhetic activity on pectenotoxin-6 in the mouse and rat. Toxicon 51: 707–716.

40 Zhou, Z.H., Komiyama, M., Terao, K. and Shimada, Y. 1994. Effects of pectenotoxin-1 on liver cells *in vitro*. Nat. Toxins 2: 132–135.

41 Jung, J.H., Sim, C.S. and Lee, C.O. 1995. Cytotoxic compounds from the two-sponge association. J. Nat. Prod. 58: 1722–1726.

42 De la Rosa, L.A., Alfonso, A., Vilariño, N., Vieytes, M.R. and Botana, L.M. 2001. Modulation of cytosolic calcium levels of human lymphocytes by yessotoxin, a novel marine phycotoxin. Biochem. Pharmacol. 61: 827–833.

43 Ronzitti, G., Callegari, F., Malaguti, C. and Rossini, G.P. 2004. Selective disruption of the E-cadherin–catenin system by an algal toxin. Br. J. Cancer 90: 1100–1107.

44 Satake, M., MacKenzie, L. and Yasumoto, T. 1997. Identification of *Protoceratium reticulatum* as the biogenetic origin of yessotoxin. Nat. Toxins 5: 164–167.

45 Rhodes, L., McNabb, P., de Salas, M., Briggs, L., Beuzenberg, V. and Gladstone, M. 2006. Yessotoxin production by Gonyaulax spinifera. Harmful Algae 5: 148–155.

46 Paz, B., Riobo, P., Fernandez, M.L., Fraga, S. and Franco, J.M. 2004. Production and release of yessotoxins by the dinoflagellates *Protoceratium reticulatum* and *Lingulodinium polyedrum* in culture. Toxicon 44: 251–258.

47 Moore, R.E. and Bartolini, G. 1981. Structure of palytoxin. J. Am. Chem. Soc. 103: 2491–2494.

48 Uemura, D., Ueda, K., Hirata, Y., Naoki, H. and Iwashita, T. 1981. Further studies on palytoxin. II. structure of palytoxin. Tetrahedron Let. 22: 2781–2784.

49 Rossini, G.P. and Bigiani, A. 2011. Palytoxin action on the Naþ,Kþ-ATPase and the disruption of ion equilibria in biological systems. Toxicon 57: 429–439.

50 Ukena, T., Satake, M., Usami, M., Oshima, Y., Naoki, H., Fujita, T., Kan, Y. and Yasumoto, T. 2001. Structure elucidation of ostreocin D, a palytoxin analog isolated from the dinoflagellate Ostreopsis siamensis. Bioscience, Biotechnology, and Biochemistry 65: 2585–2588.

51 Ciminiello, P., Dell'Aversano, C., Fattorusso, E., Forino, M., Tartaglione, L., Grillo, C. and Melchiorre, N. 2008. Putative palytoxin and its new analogue, ovatoxin-a, in Ostreopsis ovata collected along the Ligurian coasts during the 2006 toxic outbreak. J. Am. Soc. Mass Spectrom. 19: 111–120.

52 Suzuki, T., Watanabe, R., Uchida, H., Matsushima, R., Nagai, H., Yasumoto, T., Yoshimatsu, T., Sato, S. and Adachi, M. 2012. LC-MS/MS analysis of novel ovatoxin isomers in several Ostreopsis strains collected in Japan. Harmful Algae 20: 81–91.

53 Deeds, J.R. and Schwartz, M.D. 2010. Human risk associated with palytoxin exposure. Toxicon 56: 150–162.

54 Suzuki, T., Watanabe, R., Matsushima, R., Ishihara, K., Uchida, H., Kikutsugi, S., Harada, T., Nagai, H., Adachi, M., Yasumoto, T. and Murata, M. 2013. LC-MS/MS analysis of palytoxin analogues in blue humphead parrotfish Scarus ovifrons causing human poisoning in Japan. Food Additives and Contaminants: Part A 30: 1358–1364.

55 Noguchi, T. and Arakawa, O. 2008. Tetrodotoxin—Distribution and accumulation in aquatic organisms, and cases of human intoxication. Mar. Drugs 6: 220–242.

56 McNabb, P.S., Taylor, D.I., Ogilvie, S.C., Wilkinson, L.A.A., Hamon, D.W.S. and Peake, B.M. 2014. First detection of tetrodotoxin in the bivalve paphies australis by liquid chromatography coupled to triple quadrupole mass spectrometry with and without precolumn reaction. J. AOAC. Int. 97: 325–333.

57 Turner, A.D., Powell, A., Schofield, A., Lees, D.N. and Baker-Austin, C. 2015. Detection of the puffer fish toxin tetrodotoxin in European bivalves, England, 2013 to 2014. Eurosurveillance 20: 2–8.

58 Tsuda, K., Ikuma, S., Kawamura, M., Tachikawa, R., Sakai, S., Tamura, C. and Amakasu, O. 1964. Tetrodotoxin. VII. On the structures of tetrodotoxin and its derivatives. Chem. Pharm. Bull. 12: 1357–1374.

59 Woodward, R.B. and Zanos, J. 1964. Gougoutas, the structure of tetrodotoxin. J. Am. Chem. Soc. 86: 5030–5030.

60 Goto, T., Kishi, Y., Takahashi, S. and Hirata, Y. 1965. Tetrodotoxin. Tetrahydron 21: 2059–2088.

61 Murata, M., Legrand, A.M., Ishibashi, Y. and Yasumoto, T. 1989. Structure of ciguatoxin and its congener. J. Am. Chem. Soc. 111: 8929–8931.

62 Satake, M., Murata, M. and Yasumoto, T. 1993. The structure of CTX3C, a ciguatoxin congener isolated from cultured *Gambierdiscus toxicus*. Tetrahedron Let. 34: 1975–1978.

63 Chinain, M., Darius, H.T., Ung, A., Cruchet, P., Wang, Z., Ponton, D., Laurent, D. and Pauillac, S. 2010. Growth and toxin production in the ciguatera-causing dinoflagellate *Gambierdiscus polynesiensis* (Dinophyceae) in culture. Toxicon 56: 739–750.

64 Lewis, R.J., Vernoux, J.P. and Brereton, I.M. 1998. Structure of caribbean ciguatoxin isolated from *Caranx latus*. J. Am. Chem. Soc. 120: 5914–5920.

65 Satake, M., Ishibashi, Y., Legrand, A.M. and Yasumoto, T. 1997. Isolation and structure of ciguatoxin-4A, a new ciguatoxin precursor, from cultures of dinoflagellate *Gambierdiscus toxicus* and parrotfish Scarus gibbus. Biosci. Biotech. Biochem. 60: 2103–2105.

14

Safety Management Systems (Programs) for Seafood

Dominic Kasujja, Bagenda

1. Historical Background

"Good morning America, there's glass in your baby food." It was the spring of 1971, and a woman from Connecticut had found glass in her baby's cereal. The contaminated food was a creamy white cereal for infants, Farina, produced by The Pillsbury Company. It had been contaminated by shattered glass that fell into a storage bin at the plant. According to a chapter in the book titled Societal Impact of Spaceflight (Nazzal J. R 2007), Pillsbury responded to the crisis by appointing a microbiologist Dr Howard E. Bauman to implement a secure product safety system to minimize the need for product recalls. Dr Bauman applied the principles of FMEA and Critical Control Points to all consumer food production at Pillsbury. In April 1971, in the first National Conference on Food Protection, Dr Bauman delivered a presentation in which he described critical control points and good manufacturing practices. These events accelerated the development of today's food safety management (FSM) systems.

So how did Dr Bauman come up with the idea of critical control points? He had earlier been part of a team tasked to develop safe food for the National Aeronautics and Space Administration (NASA) expeditions. That team included experts from the National Aeronautics and Space

Future University Hakodate, 116-2, Kameda-nakano, Hokkaido, 041-8655, Japan.
E-mail: bagenda@fun.ac.jp

Administration (NASA), the U.S. Army Natick Laboratories, and The Pillsbury Company. For NASA, it was crucial that there was a "zero risk" of food safety incidents during any of its space expeditions. However, food safety programs at the time depended on end product testing. For the experts to give NASA the guarantees it wanted, all food for the expedition would have to be subjected to destructive testing. This meant there would be no food for the expedition! A good solution was to be found in a protocol called the failure modes and effects analysis (FMEA) that the military engineers used to test reliability of electrical components. The protocol was successfully adapted to assess hazards and control measures in the production of food for space expeditions. NASA had the guarantees it needed, and Dr Bauman had learned the critical control points principle that he later applied to food safety at Pillsbury.

After Pillsbury's Farina incident, several highly publicized food poisoning incidents such as the "Bon Vivant", "Campbell", and "Stokley-Van Camp" cases occurred in the U.S. The Food and Drug Authority (FDA) sought a comprehensive program to protect the public from food poisoning. In September 1972, in response to a request from the FDA, Pillsbury's three-week course in Hazard Analysis and Critical Control Points (HACCP) was started. The 16 FDA inspectors who attended the course later developed the permanent low-acid canned food regulation. This was the first regulatory use of HACCP in the food industry. By 1974, Pillsbury had developed HACCP into a FSM system based on three principles: (1) analyzing for hazards, (2) determining critical control points, and (3) establishing monitoring procedures.

2. Developing a FSM System

2.1 What is a FSM system?

A food safety management (FSM) system is the sum of activities that a food business undertakes to ensure that its products do not compromise consumer safety. For all food businesses, having a comprehensive FSM system is critical for success, as such a system will detail internal regulations, and acceptable workplace practices. Moreover, good FSM systems detail contingency plans for potential crises, such as rapid traceability, product recalls, and effective communication protocols. A comprehensive FSM system can therefore be viewed as a business strategy to remain competitive in the market.

2.2 What does it take to have a comprehensive FSM system

To set up a comprehensive FSM system, a food business needs personnel with sufficient training on matters of food safety and hygiene. Such personnel should also have a clear understanding of the activities that are carried out in the business. Depending on the scale of the business and the food safety risks involved, a food business can set up a basic or advanced (detailed and requiring highly skilled staff) FSM system. In either case, an effective FSM system will control food safety hazards by stipulating the process flows, standard operating procedures, critical controls points, as well as corrective/preventive actions. The FSM system will also help manage the production process by tracking inputs/ingredients and rejects, as well as equipment performance. This enables real-time tracking of production and prediction of possible failures. However, even after a product has been packed and dispatched, the FSM system should store records, such as unique identifying numbers and shipping destinations. This enables quarantining of all ingredients used in that product in the event of a recall. It also enables rapid communication, thus managing the consequences of non-conforming products. Finally, an effective FSM system should support management of resources related to production. Such resources may include finances, client/supplier information, and costs of production. The inclusion of resource management functions in FSM systems provides evidence of commitment of managers to food safety.

2.3 FSM systems, GMPs, sanitation control programs and HACCP

Any food processed under unsanitary conditions is considered to be adulterated, and therefore harmful to consumers. Seafood processors and importers must ensure that all their products have been processed under conditions that conform to good sanitary conditions and practices. To achieve this, a comprehensive FSM system should have sanitation control procedures based on good manufacturing practices and a HACCP plan. In the U.S., conditions and practices that must be followed to avoid adulteration of food are often referred to as Good Manufacturing Practices (GMP).

GMP requirements of FSM systems are usually general, defining the minimum conditions necessary for the production of safe and wholesome foods. The GMP guidelines arose from sanitation concerns recorded during inspections of food establishments by the FDA. These concerns included safety of water, cleanliness of food contact surfaces, prevention of cross-contamination, hand and toilet hygiene, protection of food from adulteration, management of toxic substances in processing plants, management of employee health, and exclusion of pets (FCA Seafood HACCP Regulation, 21 CFR, Part 123.11). GMPs provide a hygienic context in which FSM systems can perform effectively. However, GMPs do articulate specific procedures that must be followed to achieve regulatory compliance.

FSM systems must include sanitation control programs to articulate procedures used to control or eliminate potential hazards due to the processing environment and personnel. These hazards are usually associated with the eight sanitation areas listed in the previous paragraph on GMPs. Procedures used to control or eliminate potential hazards due to the processing environment and personnel are referred to as sanitation standard operating procedures (SSOP). FSM systems should document these SSOP, how they are monitored, as well as assurances that unsanitary conditions are promptly corrected.

HACCP (Hazard Analysis and Critical Control Point) is a system that helps food businesses ensure the safety of their consumers. For most food businesses, having a FSM system based on the principles of HACCP is now a regulatory requirement. In the U.S., the Procedures for the Safe and Sanitary Processing and Importing of Fish and Fishery Products regulation (21 CFR 123) requires that HACCP plans are prepared for all fish and fishery products, if there is a risk of significant food safety hazards (FDA 2011). Moreover, a separate HACCP plan should be developed for each location where fish and fishery products are processed, as well as for each kind of fish or product processed at that location. Documentation and keeping of records is important, as operators of food businesses may be asked by regulators to justify why certain hazards were included or excluded from the HACCP plans. It should be noted that maintenance of sanitation monitoring programs is an essential prerequisite for the development of a HACCP program. Implementing a HACCP program can be looked at as a three-step process.

The first step of implementing a HACCP program is developing detailed process flow diagrams for each product. It is important that the product, its packaging, and distribution methods are clearly described. This information will be important in determining the significance of hazards associated with the product. It is also important that the intended use and consumers are identified. Process flow diagrams should be detailed enough to explain how products move from receipt of raw materials and ingredients to distribution.

The second step of implementing a HACCP program is identifying potential hazards associated with the product or processing. Hazards could be related to the fish species being processed or the process being used. If the hazard is significant (i.e., if not controlled, safety of the consumer would be significantly undermined), processing activities to control or eliminate the hazard must be identified. These processing activities are referred to as critical control points (CCPs). Examples of CCPs are washing raw material in iced chlorinated water, or refrigerated storage of final products.

The third step of implementing a HACCP program is identifying values that can serve as critical limits. Examples of such values are active chlorine concentrations and storage room temperatures. It is important that these values are monitored regularly to ensure that no product is exposed to conditions outside the critical limits. It is prudent to document all corrective actions (i.e., what employees should do if processing violates the critical limits). Since consumer safety is at stake, all monitoring records should be stored for future reference. The HACCP plans should also be regularly verified to confirm that critical controls are being implemented and each significant hazard is sufficiently controlled.

3. Assessing Performance of FSM Systems (FSMS-DI Tool)

3.1 Introduction to the FSMS-DI tool

For any food business, the cost of maintaining a FSM system will be a major concern. For some food businesses, a basic (low cost) FSM system may be sufficient to achieve good quality products with an acceptable food safety status. It is crucial that food businesses continually assess the performance of their FSM systems by checking benefits against the effort (costs) required to maintain the system. Several tools have been developed to achieve this. This section focuses on the "food safety management system diagnostic instrument" (FSMS-DI) as a tool for assessing performance of food safety management systems (Luning et al. 2008). The FSMS-DI analyzes activities (control and assurance) and contextual factors that could impact the performance of the FSM system implemented by a food business. The tool also analyzes system outputs based on external (e.g., third party audits) and internal indicators (e.g., design of microbiological sampling plans) of performance.

The overall assumption of the FSMS-DI tool is that food businesses that are set in high food safety risk contexts (high risk products, processes, or conditions) require advanced FSM systems (based on precise scientific information) in order to ensure good quality products with an acceptable food safety status. If the food business is operating in a low food safety risk context, even a basic FSM system will result in good quality products with an acceptable food safety status. However, if the context changes to high risk, such an FSM system must be upgraded to a more advanced system.

3.2 Assessing risk levels of a food business context

For any food business, the food safety risk context is a factor of the product, processes, organization, and the food chain environment. The company may have a low, moderate, or high food safety risk context. If food safety of the product or processes can be easily compromised, the food business is operating in a high-risk context. If the company management has no capacity for appropriate decision-making (organization), the food business is operating in a high-risk context. Capacity may include competency of staff, management commitment, formalized protocols, and sufficient information systems. The context of the FSM system may be assessed as high risk if the company has limited control of the food chain environment. Control of the food chain environment may include the ability to negotiate requirements with suppliers, and the ability to dictate how customers use the product. If a food business can determine whether its FSM system is operating in a low, moderate, or high-risk context, decisions can be made to reduce the risk or change to a more advanced FSM system in order to achieve good quality products with an acceptable food safety status.

3.3 Assessing core control activities

Assessment of core control activities analyzes design and operation of preventive measures, intervention processes, and monitoring systems. Core control activities of an FSM system may be low, basic, average, or advanced. Core control activities in a system may be assessed as advanced if they are designed and operated on the basis of critical analysis of specific scientific information. If design and operation of control activities is not possible, or no information is available, the system is assessed as low. In average systems, control activities may be guided by government regulations or best practices. Depending on the food safety risk context, it may be possible for a food business to achieve good quality products with an acceptable food safety status with average core control activities.

3.4 Assessing assurance activities

Assessment of core assurance activities analyzes the processes of setting system requirements, validation, verification, and documentation. Core assurance activities of an FSM system may be low, basic, average, or advanced. If they are advanced, it means that system requirements, their validation, verification, and documentation are characterized by systematic and critical analysis of specific scientific information. On the other hand, average assurance activities may be limited to regular reporting, and expert support. If assurance activities are limited to scarcely reported checks or problem-driven reports, then they may be assessed as basic. In some situations, such basic core assurance activities are sufficient to achieve acceptable food safety status of a product.

3.5 Assessing FSMS outputs

Assessment FSM system outputs gives a perspective of the hygiene, safety, and quality of the products. External and internal indicators are used to assess systems outputs. System outputs may be considered low if the food business has no information on the hygiene, safety, and quality of the product. If there is ad hoc sampling, or evidence of remarks from audits or inspections, the system outputs may be assessed as poor. Moderate system outputs are characterized by regular sampling, evaluations based on several criteria, and limited food safety problems. On the other hand, good systems will have systematic evaluations of products, and inspection (or audit) records with no food safety problems.

4. Getting a FSM System Recognized by a Benchmarked Standard

4.1 Introduction to food safety standards and benchmarking

As food safety gains importance around the world, regulators and consumers are requiring more assurances that food is safe. Food businesses must assure stakeholders that products have been processed to the highest possible safety standards. Certification of an FSM system by a recognized third party is one way to assure stakeholders of the quality of product safety.

The certification of FSM systems has been a food industry priority for a while. The turn of the century was marked by major international food safety incidents like BSE, and dioxin. To restore consumer confidence in the industry, regulatory bodies of countries like Brazil, the Netherlands, Ireland, and Australia developed national standards for food safety management systems. In addition to these national standards, many commercial standards were gaining prominence. These included the BRC food safety standard of 1998 (published by the British Retail Consortium), International Food Standard (published by French and German retailers), and Safe Quality Food Standard (developed in Australia and imported to the U.S. by the Food Marketing Institute). The proliferation of standards and certification schemes resulted in audit fatigue, as food businesses were put under pressure to apply multiple, and often incompatible, standards to their food safety management systems.

To solve this challenge, top executives of major retailers and manufacturers decided to harmonize food safety standards. This is how the Global Food Safety Initiative (GFSI) was founded in May 2000. The goal of GFSI is simple and compelling: once certified, it is recognized everywhere. GFSI measures existing food safety standards or certification schemes against food safety criteria specified in the GFSI Guidance Document. With GFSI, anyone can know how food safety schemes and standards measure up against each other. It is easier for businesses to confidently select a food safety scheme from the GFSI list of recognized schemes (available at the GFSI webpage), because GFSI is recognized internationally. It is important to note that GFSI is not a standard or food safety management scheme. It is a collaboration of representatives from the entire food chain to recognize standards that meet certain criteria. An international trade association, the Consumer Goods Forum (CGF), manages GFSI. Detailed information about GFSI can be accessed at www.mygfsi.com.

For the food industry, the formation of GFSI was a welcome development. GFSI reduced food safety risks by providing a neutral forum in which global food safety management systems could be contrasted against each other to make informed choices. A chef in Japan could decide if trout imported from Brazil met the minimum food safety criteria for his restaurant, as long as both actors used GFSI recognized standards. This reduced redundancy and improved efficiency in protecting consumers from hazards. The formation of GFSI provided motivation to create consistent and effective global food safety systems, as well as an international networking platform for all actors in the food industry.

4.2 Requirements of GFSI benchmarked standards

For any food safety standard to be recognized by the GFSI, it will have all of the following three components: requirements on the control food safety hazards during processing (HACCP and pre-requisite programs); requirements related to management of processes (e.g., food safety manuals, policies, documentation); requirements related to good hygiene (e.g., good agricultural practice, good manufacturing practice). The GFSI has classified the entire food chain into sectors with unique food safety requirements. Recognized FSM systems or schemes must therefore specify a sector for which they are applicable. Food businesses dealing in seafood will most probably be classified into EI (Processing of animal perishable products; Production of animal products including fish and seafood) or AII (Farming of fish). This section offers an overview of requirements related to seafood safety, but details can be found in the GFSI guidance document.

4.2.1 Requirements related to control of hazards (HACCP)

Schemes that are recognized by GFSI require food businesses to implement HACCP programs in a systematic, comprehensive, and thorough manner. The implemented HACCP programs should cover food allergens, and be specific to a product (or product category), process, or location. Results from hazard analysis performed under such programs should demonstrate food safety management, and the programs should be capable of adapting to change. GFSI recognized schemes also require that implemented HACCP programs include documented standard operating practices and work instructions.

4.2.2 Requirements related to management of processes

For a scheme to be recognized by GFSI, it must require the food business to have a FSM system that is well documented, implemented, and regularly reviewed for possible improvements. The food business must be able to show its food policy statement and measurable objectives to demonstrate commitment to food safety. Other management strategies that recognized schemes require of the food business include clear organizational structures (who does what), evidence of sufficient allocation of resources to food safety, secure storage of food safety documents so as to meet legal requirements, internal audits, protocols to identify and control products that do not conform to food safety requirements, as well as evidence that all externally sourced materials are traceable, and compliant with food safety standards or regulations.

4.2.3 Requirements related to good hygiene practice

GFSI recognized schemes require the food business to demonstrate implementation of good manufacturing practice or good agricultural practice. Such implementation guides decisions on processing environments, equipment, inputs, outputs (waste or by-products), and personnel. It is important that in all decisions related to the above aspects, the risk of product contamination is minimized. It is important to mention that recognized schemes have strict requirements regarding use of chemicals such as drugs (in aquaculture), cleaning materials, lubricants, and pest control agents.

4.3 Certification and audit procedures; The case of SQF

Developed in Australia in the late 1990's, the Safe Quality Food (SQF) standard is currently one of the GFSI recognized schemes for certifying FSM systems. The scheme has two standards: SQF 1000 for primary producers (fish farmers) and SQF 2000 for food processors. The SQF Institute supports a comprehensive website that provides detailed information about this standard (www.sqfi.com).

To have your FSM system certified under SQF, on-site training is provided and a designated SQF practitioner with prior HACCP training is appointed. A certification body then performs a gap assessment of your FSM system to decide whether the system needs to be improved. Improvements must be completed before the FSM system is certified.

Certified FSM systems must be regularly audited. The auditing process takes 2 to 3 days, and includes a document review, as well as a site assessment. All non-conformances identified during an audit are classified as minor, major, or critical. Depending on the gravity of non-conformances, an audited FSM system will receive an A, B, or C certification. FSM systems with an A or B are audited annually, while those with a C are audited again after a period of 6 months.

5. Conclusions

FSM systems must continuously adapt to the ever-increasing demands and challenges associated with advancement of humanity. As such, these systems should always be considered transient, and actors in the food industry must continue to seek new developments in this field. It is hoped that this chapter provides a simplified but current overview of the important aspects of developing comprehensive FSM systems.

References

FDA. 2011. Fish and Fishery Products Hazards and Controls Guidance. Fourth Edition. [PDF]. Available online at: https://www.fda.gov/downloads/food/guidanceregulation/ucm251970.pdf last accessed on June 10, 2017.

Jennifer Ross-Nazzal. 2007. From Farm to Fork: How space food standards impacted the food industry and changed food safety standards. *In*: Dick, S.J. and Launius, R.D. (eds.). Societal Impact of Spaceflight. NASA, pp. 220.

Luning, P.A., Bango, L., Kussaga, J., Rovira, J. and Marcelis, W.J. 2008. Comprehensive analysis and differentiated assessment of food safety control systems: a diagnostic instrument. Trends in Food Science & Technology 19: 522–34.

15

International Control of Seafood Hazards

Regulation and Authorizes

M.A. Khaleque,[1],* M. Mostafizur Rahman[2] and Sabina Yeasmin[3]

1. Introduction

Fish is an important source of dietary protein and nutrients in many parts of the globe. It includes freshwater, coastal, and deep-sea fishing. Consumers are aware of the health risks of eating seafood and freshwater fish that could be associated with both environmental and public health problems due to contamination. Seafood safety is a major health concern as a substantial number of seafood-borne diseases are transmitted by contaminated water. The primary causes of diseases are the presence of biological (bacteria, virus, parasites) and chemical (biotoxins such as ciguatoxin) components which utilize reservoirs of non-biodegradable polymers that pollute seafood habitats [1, 2]. However, most hazards related to contamination

[1] Department of Biochemistry & Microbiology, North South University, Bashundhara, Dhaka-1229, Bangladesh.
[2] National Institute of Diabetes and Digestive and Kidney Diseases, Bethesda, Maryland, USA.
[3] Department of Genetic Engineering & Biotechnology, University of Dhaka, Dhaka-1000, Bangladesh.
* Corresponding author

due biological and chemical hazards can be controlled by applying Good Manufacturing Practice (GMP), Good Hygiene Practice (GHP), and a well implemented HACCP programme [3]. Measures to prevent growth of pathogenic microorganisms during distribution and storage, such as temperature control, are in place. Storage in a controlled temperature can prevent the growth of most pathogens, but it is still recommended not to prolong shelf-life of seafood, as that can cause outbreaks. It has been noted that a number of problems in the processing of certain seafood persist despite practicing GMP, GHP, and HACCP principles. For example, coming up with effective methods to control heat-stable bio-toxins has been difficult. Furthermore, consumption of raw fish in some parts of the world makes it challenging to eliminate seafood-borne illness. Although biotoxins are mostly formed in fishes from warm tropical areas, scrombotoxins (biogenic amine) are formed in fish post-mortem, especially when these are stored at elevated temperatures (above 5°C). About 12% seafood-borne illnesses are caused by bacteria such as *Clostridium, Eschericia, Salmonella, Staphylococcus, Vibrio,* and *Bacillus* spp. Furthermore, aquatic birds also spread salmonellae and other human pathogens [4, 5]. Shellfish is the main type of seafood product responsible for outbreaks due to bacteria, viruses, and marine algae. Molluskan shellfish accumulate these pathogenic agents during their filter-feeding metabolic process. Although seafood-borne illnesses are not a problem for a specific area, but worldwide, the highest occurrences of parasitic infections due to seafood are from Asia. According to a 1995 data by WHO, 40 million people suffer from such infections, with 10% of world population at risk [6, 7].

The main hazards from seafood are due to pathogenic bacteria, biogenic amines, parasites, viruses, biotoxins, non-biodegradable chemicals, such as, heavy metals and pesticides [7]. These biohazards mentioned here are a cause of safety concern both during pre-harvest and processing. Pre-harvest pathogenic bacteria contamination poses a threat when seafood is consumed raw, and contamination during processing poses a threat to ready-to-eat seafood products [8]. Biotoxins are mainly observed in warm-water fishes as pre-harvest contamination. Biotoxins are magnified in the upper trophic levels; therefore, larger fish contain more toxin than smaller fish. More than 400 species of fish are carriers of these toxins [9, 10, 11]. The source of chemical hazards is mainly from commercial aquaculture or polluted coastal areas. In coastal regions where raw-fish is mostly consumed, for example, sushi and sashimi in Japan are predominantly susceptible to seafood-borne illness due to inadequately cooked fish, as cooking at high temperatures could easily eliminate most bacterial pathogens and heat-labile marine toxins [11]. Similarly, countries with poor knowledge of seafood hazards and disposal of human and animal waste into water are a serious problem. Histamine, a biogenic amine, is mostly present in scomboid species of fish, and becomes dangerous when the fishes are handled abusively [12]. High concentration of marine algae toxins, such as saxitoxin, brevetoxin, bromoic acid, ocadoic acid in shellfish are associated with numerous seafood-borne illness outbreaks, such as Paralytic Shellfish Poisoning (PSP), Diarrhetic Shellfish Poisoning (DSP), Neurotoxic Shellfish Poisoning (NSO), and Amnesic Shellfish Poisoning (ASP). All the algal toxins are heat stable, and presence of large number of algae in shellfish could cause illness if consumed, either cooked or raw [13]. Knowledge of seafood hazards, monitoring seafood harvest, storage, packaging and distribution conditions can help assess and reduce hazards related to the different fish and seafood consumption. Thus, control of seafood hazards is necessary for public health safety [4, 5, 14, 15].

2. Scientific Standards to Control Hazards in Seafood Industry

Seafood, any form of marine life regarded as food by humans, encompasses a vast array of sea life that includes various genera, species, and phyla, such as mollusca (e.g., clams and oysters), arthropoda (e.g., crabs and crayfish), and chordata (e.g., finfish). All together more than 350 marine species are being consumed by humans in various parts of the globe. There is a regular consumption of heterogeneous species of seafood, so this assortment indicates a wide range of sensory attributes, product forms, and preparations that are unique to seafood. Seafood requires some particular safety concerns that arise both from the essential characteristics and the environmental pollutants from where the seafood is being harvested. As an example, for some species, their harvest location and season is important for food safety issues. In addition, handling and processing at harvest as well as storage and distribution constitute significant factors to enhance or reduce the risk of seafood-borne illness [16, 17]. Certain seafood may present hazards to public health because of their unique trait. First, the diversity of aquatic environment and its condition, such as different types of habitats, for example, freshwater, saltwater, estuarine water, tropical, polar, in-shore, off-shore, pristine, and polluted water. Presence of harmful substances, both biological and chemical, in aquatic environment in the harvesting area is a major concern for seafood safety. For example, methyl mercury in various fish in a specific aquatic environment or contamination of *Vibrio vulnificus* in raw molluskan shellfish can cause seafood-borne illness [18, 19, 20]. Second, the wide expansion of aquaculture production seems to be similar to land-based muscle food as an opportunity for environmental pollution, so therapeutic agents and human pathogens can be present as a consequence. Third, all these concerns are further complicated by an increasing dependence on seafood production in international waters where surveillance is less. Besides that, factors that increase the rate of risk in seafood as food-borne diseases relate to handling, distribution, and preparation. A unique and notable characteristic of seafood consumption is that a significant portion can be consumed live (oysters, mussels, and clams), raw (e.g., sushi), or cooked to a rare state (e.g., cod and mahi-mahi). In some parts of the globe, the consumption of non-muscle components such as eyes, eggs, and viscera pose unique risks. Occasionally recreational fishermen ignore marine advisories that prohibit harvest of molluskan shellfish from non-approved waters, thus exposing consumers to potentially contaminated toxic shellfish [21, 22, 23]. The concern with seafood safety also focuses on ready-to-eat products, for instance, processed crab meat does have a terminal heat step that destroys most food-borne pathogens, including *Listeria*. The terminal heat precedes the meat removal step, which is traditionally done manually by hand, and thus increases the risk. With respect to cold-smoked fish, this product does not have a lethal heating step, therefore other parameters, such as salt concentration, become important risk minimization steps [24, 25].

There is a need to develop specific strategies to address the unique challenges presented by aquaculture production of seafood. In 1995, the Food and Drug Administration (FDA) issued the Procedures for the Safe and Sanitary Processing and Importing of Fish and Fishery Products; Final Rule also known as the seafood HACCP rule [26]. This rule is a significant objective to apply on the processing sector which could increase the risk of seafood contamination and is more tractable than harvesting. The Seafood HACCP is periodically updated and available at the U.S. Food and Drug administration website (https://www.fda.gov/food/guidanceregulation/haccp/ucm2006764.htm). International trade is expected to increase in response to efforts by various industrialized nations to supplement their dwindling domestic seafood resources. Supply is becoming the most significant issue in the world of seafood commerce. The anticipated significant shortfalls for the next decade may result in the reduced availability of seafood and elevated prices in industrialized countries, while serious shortages could occur in regions of the world that are dependent on subsistence fisheries. Imports to the United States exceed 80 percent for certain popular seafood products. FDA recently estimated that over 8,500 importing firms are subject to surveillance in accordance with the seafood HACCP rule [27]. FDA and the U.S. Environmental Protection Agency (EPA) have set up different sustenance well-being criteria that address the inborn way of fish (e.g., scombro toxicity) or attributes of the earth from which it was reaped (e.g., incapacitated shellfish poison, methyl mercury, fecal coliforms). Among these controls are microbiological criteria related to particular microorganisms, for example, Salmonella and *Clostridium botulinum*, and with item classifications, for example, prepared to-eat items and molluskan shellfish [3, 4, 5, 27].

The Seafood HACCP framework varies from the meat and poultry industry in that the inspections are not performed on a constant location premise. It is hard to legitimize due to a wide assortment of species, variable sources, and differences in shapes and sizes. Fish and seafood safety concerns are not ruled by any single pathogen or contaminant. The FDA has been in charge of building up a broad control of sustenance security criteria—sorted as resistances, activity levels, and rules—with the basic motivation of ensuring well-being through adherence to GMPs. As an example, the resilience for methyl mercury content in fish (1.0 ppm) depends on the level essential for shopper well-being, and the naming prerequisite for sulfite buildups (10 ppm) depends on the lower furthest reaches of investigative ability [28, 29]. Chemical hazards that are not of ecological source (i.e., biogenic amines, for example, histamine) require an alternate control procedure. Hoisted biogenic amine levels in some finfish, for example, mackerel and mahi-mahi, are delivered according to the development of certain indigenous microbes in fish. FDA has standardized an activity level of 50 ppm histamine in any consumable bit of the fish [30].

The Seafood HACCP has been acclaimed as a proper, science-based, nourishment well-being confirmation framework [26], in spite of the fact that it has not yet been generally connected in the sustenance business. For a few gatherings, execution of HACCP raises worries about increased government oversight of sustenance handling [3]. For instance, a report issued by the General Accounting Office (GAO 2001) proposed that FDA's oversight of fish firms did not adequately secure customers against food-borne maladies [31]. Notwithstanding these debates, latest reports recommend that HACCP has assumed a part in lessening a portion of the country's notifiable food-borne illnesses [32]. However, HACCP has had a very distinct impact on the seafood industry, primarily through enhanced awareness and understanding of potential seafood safety hazards from pre-harvest to processing, and distribution through preparation and consumption. Given the differences inside the seafood industry, FDA issued an extraordinary guide, the Fish and Fisheries Products Hazards and Control Guide, usually known as "the Guide", to help execute HACCP in the seafood industry [30].

2.1 Current safety criteria for seafood

The current seafood safety criteria include mechanisms and guidelines that ensure proper harvest, storage, distribution, and consumption. These criteria have been developed using the latest scientific knowledge, available technologies, and international export and import regulations. They are amended constantly to keep up with the supply and demand of aquaculture products around the world to attain public health objectives. Seafood safety criterion provides harvest, aquaculture business, as well as consumption of the vast variety of seafood to be safe [21, 26]. In order to allow the fishery products in the markets, certain microbiological and legal criteria have to be maintained. In the United States of America, FDA is the prime institution moderating the different rules and regulations required to be followed to ensure the safety of fish and fish products, under the provisions of the Public Health Service Act and Federal Food, Drug and Cosmetic (FD&C) Act. The regulation responsible for warranting the procedures involving the safety and sanitary processing and importing of fish and fishery products, along with developing and implementing Hazard Analysis Critical Control Point (HACCP) system came to be active from December 18, 1997 [26]. This FDA program includes a widespread compilation of research, inspection methodologies, compliance, enforcement, and outreach policies that affect the safety of fish and fishery products through the means of "Fish and Fisheries Products Hazards and Controls Guide", which identify the hazards that could be associated with the fishery products, and formulate regulatory strategies to subjugate them. The U.S. Environmental Protection Agency (EPA) has also been involved in framing similar strategies that could further act as criteria to address the intrinsic nature of the seafood or the marine environment that might have toxic effects. As a whole, seafood that is harvested or cultured domestically or is imported can maintain its diversity and is safe to consume [33].

2.2 Microbiological criteria for seafood safety

Microbiological criteria for seafood are associated with numerous microbes that possess the risk to cause seafood-borne illness. For example, both Salmonella and Clostridium are liable for producing toxins commonly reported in ready-to-eat seafood products and molluskan shellfish [33]. The microbiological criteria for seafood safety are divided into Food Safety Criteria (FSC) and Process Hygiene Criteria (PHC). Food Business Operators (FBOs) comply with the food safety regulation guide based on HACCP principles and GHP. Fishery industries go through several microbiological testings to validate and verify whether the criteria are met to identify food safety issues that arise with harvesting, cultivating, extracting, importing, storing, and distributing seafood. The FSC are usually applied throughout the shelf-life of a particular seafood, which if not met with the requirements of the FBO, can be taken off the market voluntarily [29, 33]. PHC are applied until the end of the manufacturing process, and failures are usually taken care of before the product reaches the market.

Table 15.1 Microbiological criteria to ensure safe food.

Process hygiene criteria		Food safety criteria				Food descriptor
					Listeria monocytogenes	Ready-to-eat (RTE) foods (e.g., smoked salmon, cooked crab meat)
				Salmonella		Cooked crustaceans & molluskan shellfish
			E. coli	*Salmonella*		Live bivalve mollusks and live echinoderms, tunicates and gastropods
		Histamine				Fishery products from fish species associated with a high amount of histidine
		Histamine				Fishery products which have undergone enzyme maturation in brine, manufactured from fish species associated with high amounts of histidine
Coagulase-positive Staphylococci	*E. coli*					Shelled & shucked products of cooked crustaceans & molluskan shellfish

2.3 Health and hygiene criteria for seafood safety

In most countries seafood safety insurance entails a legal framework for export and import of fishery products. The countries follow certain guidelines that are based on the general public health and hygiene regulations, which refer to the fact that the exporting country must maintain strict health and hygiene requirements following the seafood safety criteria. The Food and Agriculture Organization of the United Nations (FAO) provides guidelines on the proper mannerisms of fishing, approved vessels, and registered farms to prevent and eliminate illegal, unreported, and unregulated fishing. Fishery and aquaculture products to be exported require proper safety certificates. Eco-labeled seafood products are also required to have quality and safety certificates following particular certification schemes [34]. Among the standard rules and regulations, the vital steps in ensuring seafood safety are as follows.

(a) **Hygiene:** This criterion includes health standards of the fisheries along with contaminants and packaging and storing of products at controlled temperatures, following the guidelines of HACCP.

(b) **Traceability and labeling:** It allows the importers to ascertain the harvesting and cultivation of the fishery products. Labels help buyers consider choosing foods that have been harvested and cultivated in a more sustainable way. Allergens are also included in labels according to the latest guideline of HACCP to confirm food-borne illnesses.

(c) **Contaminants:** Contaminants such as heavy metals lead, cadmium, mercury; dioxins, and PCP can taint fishery products at any stage of processing. Thus, their presence should be tested before shipping to prevent outbreak of seafood-borne illnesses.

(d) **Microbiological contamination:** Safety standards must be maintained by detecting microbiological contamination at an early stage of processing the seafood. Microbiological contamination can be prevented by conducting the advised procedures during harvesting and producing seafood.

2.4 The HACCP system for seafood safety

Hazard Analysis Critical Control Point or HACCP was initiated to maintain the safety of seafood [26, 27]. It is necessary to ensure consumer safety regarding seafood, as the fish industry faces new challenges every now and then with issues, such as processing, storage, and distribution of seafood and its products, similar to any other food safety standard [35]. The concept of HACCP not only boosts the confidence of the consumers, but it also plays a role in motivating the developing countries who would export seafood. There are some critical control points (CCP's) that needs to be identified for undertaking preventive measures by the processor, who is defined in the HACCP rule as any person engaged in commercial, custom, or institutional processing of fish or fishery products, either in the United States or in a foreign country; and persons engaged in the production of foods that are to be used in market or consumer tests are also included [27, 28]. HACCP plan is not only applied for the safety of seafood, but also for the safety of other products such as poultry, meat, dairy products, and so on, but it differs that government inspections of seafood are not usually conducted continuously, rather it is more of an onsite basis inspection [33]. This is because when it comes to seafood, looking for a single pathogen or contaminant would not be of any help. The inspection of seafood, be it imported or domestic, is done to ensure high quality of the product rather than the safety [33].

2.5 Application of seafood safety criteria

In the United States, the FDA ensures the safety of seafood, and the responsibility of developing a widespread list of regulation critical to seafood safety [27, 29]. The FDA also prepares guidelines for tolerance, and action levels [33]. The National Advisory Committee on Microbiological

Criteria for Foods (NACMCF) at USDA, an advisory body to federal food safety agencies, specifically addressed the issue of microbial criteria with the following statement: the use of microbiological testing is seldom an effective means of monitoring CCPs because of the time required to obtain results. In most instances, monitoring of CCPs can best be accomplished through the use of physical and chemical tests and through visual observations [36]. The FDA, EPA, and USDA established boundaries for chemicals, but direct tests for chemical contaminant is often impractical as it is time consuming, expensive, and the parameters for the testing are not rigid and can vary due to different geographical locations [33]. No fish may be harvested from an area that is closed to commercial fishing by foreign, federal, state, or local authorities; and no fish may be harvested from an area that is under a consumption advisory by federal, state, or local regulatory authority based on a determination by the authority that fish harvested from the waters are reasonably likely to contain contaminants above the federal tolerances, action levels, or guidance levels [33]. This is a precaution to avoid culturing or harvesting in highly contaminated environments and to ensure the safety of seafood and its products. There are some chemical contaminants that do not originate from the environment, that is, they are not of environmental origin, such as histamine which is a biogenic amine. According to the FDA, the limit of histamine in any edible portion is 50 ppm. Now it is impractical, expensive, and tedious to measure the histamine level in each fish, and thus an appropriate alternate method is to review the harvest records of each lot of the fish, and these records include the time and temperature [33]. According to the Scientific Criteria to Ensure Safe Food, the harvest records must include the following information:

(1) Icing on-board the harvest vessel;
(2) method of capture;
(3) date and time of landing;
(4) estimated time of death;
(5) method of cooling;
(6) date and time cooling began;
(7) sea and air temperature if exposure temperatures exceeds 83° F [28.3°C];
(8) adequacy of ice during on-board holding.

As early as 1980, there was an international drive towards reforming fish inspection systems to move away from end-product sampling and inspection into preventative Hazard Analysis Critical Control Point (HACCP)-based safety and quality systems. These preventative approaches are thus required.

1. Fish products are prepared/processed in certified plants and establishments. The certification process requires that the plant meets

minimal requirements in terms of layout, design and construction, hygiene and sanitation;

2. The industry takes responsibility in fish safety control and implements HACCP-based in-plant quality control programmes;

3. A regulatory competent authority is in charge of certifying fish plants and establishments, approving and monitoring HACCP-based in-plant quality control programmes, and certifying fish and fishery products before distribution;

4. National surveillance programmes of the harvesting areas should be in place to control the threats of biotoxins and other biological and chemical pollutants; and

5. For export, an additional control can be exercised by the importing party and involves an audit of the national control system of the exporting country to ensure that it meets the requirements of the importing country. This should lead to the signing of mutual recognition agreements between trading countries.

3. Public Health and Economic Feasibility of Seafood Safety

HACCP has been marked as a proper, science-based, food safety assurance system by Food science community, even though it has not been universally applied in the food industry yet. Some groups have been concerned that while trying to implement HACCP, for instance, an issue was reported by the General Accounting Office, which suggested that FDA's surveillance of seafood firms did not adequately protect consumers against food-borne diseases [3, 31]. Latest reports indicate that some of the countries told about food-borne illnesses have been observed to have reduced illnesses due to implementation of HACCP [37, 38]. HACCP has had a very prominent impact on the seafood industry, basically through raising awareness and understanding the hazards caused by seafood safety from production to processing, through preparation, and eventually consumption. Ever since HACCP rule has come into play, industry personnel are being trained, which has played an integral role in seafood safety. In addition to this, areas of interest are better identified, so that further government supervision and educational programs can be carried out. However, technical innovations are also as important as carrying out educational programs, for example, the endeavor to decrease illness caused by consuming raw oysters. Food-borne illness due to consumption of raw oysters is still a crucial flaw. The main culprit of this is the pathogenic bacterium *V. vulnificus*. An infection caused by these bacteria is comparatively rare, and usually affects consumers who have pre-existing liver diseases or are immune-deficient. The fatality rate is high [19]. In addition to consumer education programs, the oyster companies have included new technologies such as pre- or post-harvest high hydrostatic pressure to decrease such infections [39]. The adjustability of this approach displays a regulatory shift from initiating a specific standard, to requiring that processors will choose and authorize technologies necessary and proper to their particular operations. The strategies chosen should in fact have measurable and improved outcomes through increased ability to perform by having specific procedures to reduce *V. vulnificus* in raw oysters, and through a decrease in food-borne illnesses [39, 40]. FDA and state regulatory agencies use an essentially zero tolerance as the limit for performance standard for *V. vulnificus* in oysters meant for raw consumption. This measure recognized that some post-harvest treatments can be used on raw oysters for the purpose of seafood safety. Treated oysters may not only be not liable to a public advisory, but may also have an attached product label, such as "processed for added safety". The decision to allow or manifest the use of particular product labels or statements belongs to individual state authorities. Eventually, use of recent *Vibrio* risk assessments [34] might support the manifestation of science-based microbiological performance standards for *V. vulnificus* that assure a minimum level of public health protection, in addition to allowing flexibility and new methods in the use of post-harvest treatments.

One of the highlights of the current HACCP-based system is that industries are highly involved in determining proper seafood safety control strategies to identify hazards. While there is scope for an increased level of participation by the industry, most seafood processors still want to be advised by the FDA to make their decisions and carry out their processes. Since the seafood industry is diverse, the FDA identified that specialized guidance has to be provided in order to help the companies implement HACCP rule.

If and when concern arises about seafood safety, third party processing authorities may be called in to validate or verify seafood-processing methods, or get suggestions of different methods. For the time being, there is no FDA guideline that defines the attributes of processing authorities, or how the verifications or validations need to be in order for a HACCP plan to be accepted by the FDA, in which case the processing authorities can be experts from academic institutions, private consulting firms, or elsewhere. In particular, the validation of modern, rapid microbiological methods, and the design of appropriate sampling plans need adequate FDA guidance. As the U.S. is increasing their dependence on seafood imports, the worries about current HACCP governance for seafood safety are increasing. The FDA even created a new rule where every seafood processor under international commerce is considered equal, and therefore allows each country to be independently marked on their ability to provide seafood safety. Countries will require competent authorities, responsible criteria, and standards in order to achieve high marks. However, some nation's effort to downgrade others in their commercial performance is less attacking, and is seen as a trade barrier. The FDA should focus more on addressing specific issues raised by some countries or products as it's a necessity. For the time-being however, The Codex Alimentarius (FAO/WHO) provides some cooperation in respect to national authorities.

The committee would suggest the FDA to make clear the intent of the Guide to international partners and ensure the highest priority is given to the application of the Guide, confirming equal standards once the international food safety commerce is reached. A greater precaution must also be taken when screening seafood products at point of entry. The committee understands that limited checking of products does not go hand in hand with the preventive concept of HACCP. As a result, a better regulation rule should be placed that ensures in-depth inspection of foreign seafood products at the point of entries. Creating a seafood safety exchange program, which will provide training and research opportunities to investigate common concerns will be beneficial. Common hazards such as *Listeria* in fresh seafood, *Salmonella*, and methyl mercury tolerances could be investigated, and rules for the Best Aquaculture Practices can be created. It can become of similar importance as Good Manufacturing Practices and Good Agricultural Practices for produce and other land-based crops. It can even become an international prerequisite for aquaculture around the globe.

4. Regulations Governing Seafood Safety and Quality

There are several criteria of regulations are involved in the governance of fishery and seafood quality and safety. The following section particularly elaborates on these points and mentions the importance of these subjects.

4.1 Seafood popularity on international context

In recent years, seafood has gained enormous popularity; and has earned a huge financial support on international context in the last three years. The market shares have raised their values twenty times between the years 1968 to 1995, from the count of 2.2 billion US dollar to 46.95 billion US dollars. International trade accounts for about 30 percent of the fish and seafood harvested from developing countries, where it contributes about 50 percent of local fisheries [41, 42]. The table (derived from FAO) shows the import and export commodities worldwide and selected trading regions, with a three-year average from 1993 to 1995. (A) It includes fish, fresh or frozen dried, salted and smoked, canned and oiled meals. (B) It is rounded up, hence individuals are not added. (C) It represents the trading activities of 202 countries. (D) It represents the list of low income food deficit countries. (E) It includes Austria, Denmark, Belgium, Finland, France, Germany,

Table 15.2 Imports and exports of fishery commodities.

Area of world	Imports (b)		Exports (b)	
	US$ billions	Percent	US$ billions	Percent
World (c)	50.562	100.0	46.954	100.0
Africa	.863	1.7	1.995	4.3
North America	8.139	16.1	6.907	14.7
South America	.521	1.0	4.371	9.3
Asia	22.139	43.8	16.643	36.5
Europe	18.064	35.7	13.573	28.9
Oceania	.550	1.1	1.571	3.4
Former USSR	.286	1.0	1.895	4.1
Low-income food deficit countries (d)	1.829	3.6	8.395	17.9
European community (e)	17.045	33.7	10.002	21.3
NAFTA countries (f)	7.873	15.6	5.988	12.8
Japan	16.060	31.8	.741	1.6
Rest of world	9.58	19.0	30.223	64.4

Greece, Iceland, Italy, Netherlands, Luxembourg, Portugal, Spain, Sweden, and United Kingdom. (F) It represents USA, Canada, and Mexico [43, 44].

However, seafood safety protocols are implied throughout the world to enhance the safety issues regarding chemical analysis of total export. Several multinational conferences took place, and one of them entitled 'Seafood Safety: New findings and innovation Challenges' was focused on the environment contaminants present in the seafood. In early 1980, an international drive caused the newly established system involving inspection of fisheries to start following the HACCP [43, 45].

4.2 Food chain approach to food safety

Three fundamental points should be maintained for the quality and safety for seafood and fishery products. These involve Communication, Assessment, and Management. The risk management could be resolved by placing a science-based risk assessment authority. Improvement of harvesting, processing, and distribution, which consists of the therapeutants, aquaculture, and animal feed could build a better network of tracing techniques. These would allow the fishery and seafood quality and the standard harmonized with the development, and enhance wider consumption of the processes, such as inclusion of scientifically proven standards on a global agreement. The further development of an equivalent food chain system which attains the same levels of protection as the seafood-borne illness would ensure food safety globally [42, 45]. Increased emphasis on risk assessment and preventive measures throughout the entire food chain would help achieve and maintain food safety internationally. The fishery sector would invest more for the development and betterment of aquaculture product quality and safety, for example HACCP to aid the food safety on traditional measures and quality management on the basis of control and regulation [26, 45]. Along with HACCP, FAO plays major roles on the approach of seafood with effective, safe, healthy, and nutritious ingredients shared among the entire world's food chain. This includes food processing sector and government control services with an enabling policy and regulatory environment with specified and proper rules and standards on local, national, and international levels [43]. Preventive steps proposed on food safety on international context include:

(a) Processing of the fish and seafood into certified plants to keep it cost effective and hygienic.

(b) Implementations of HACCP by the industries on plant quality control programs.

(c) Strictly regulate the hazards produced by biotoxins and other pollutants with the help of national surveillance programs.

(d) Mutual settlements between the trading partners should be adjusted, ensuring food safety and cost effectiveness.

4.3 European union: rules and regulations for seafood safety

Currently the biggest fish and seafood import service is done by the European Union (EU) countries. The same import rule is applied by all the EU countries, as the fish and seafood products are consistent with hygiene, consumer food safety, as well as the animal health status concerns [35, 46]. All non-EU countries are now settling to negotiate with partner EU countries for proper EC safe seafood certifications and attain subsequent standard of fish and seafood quality. The food safety in EU is led by the European Commission's Directorate General for Health and Consumers (SANCO). The basis of the import rule involves the understanding of the main principles of the EU food law which should ensure the safety and quality of food import [47]. An official Certification is necessary for the imports of the seafood products into the EU. This is the prerequisite for any non-EU country to be able to get authorized and export it to any of the EU countries. The European Food Safety Authority (EFSA) has legal powers to enforce all the factors, including public health, hygiene, animal health concerns, and aquaculture quality to the national competent authority for all the bilateral negotiations involving quality and safe seafood import [44, 46, 47]. In brief, the eligibility criteria are as follows,

(a) Competent authority is required to be in control of the production chain in the exporting countries.

(b) The veterinary service should be able to ensure proper enforcement for the health hazard controls, for example, observation of the fisheries with eggs and gametes.

(c) Assure proper hygiene and public health. The legislation concerning proper hygiene should build their approaches on the vessel's structure, good landing sites, the processes to operate, processing establishments, storage and freezing standards.

(d) Close monitoring of seafood products. First, the production zone should work on excluding the contaminations with the chemicals bio-toxins leading to poisoning of the marine shellfish. Secondly, able to categorize fishery and seafood precisely, for example, bivalve mollusks, including clams and mussels, sea urchins from echinoderms, conchs and marine snails from the category of marine gastropods should be kept under specific conditions.

For the requirements mentioned above, recognition should be done by the veterinary Office for the confirmation of the compliances by all non-EU countries. This creates a confidence between the competent authority of any non-EU countries and the EU commission for food safety.

4.3.1 Border control

Fish and seafood products entering into a country must encounter a safety and quality standard check, as it differs widely from one market to another. These differences are dealt with regulations, standards, and control procedures, including controls at the border where seafood products can be rejected, quarantined, destroyed, or put in detainment while further decisions are taken by testing if it meets the import regulations. Border control concerning seafood import is essential to identify and evaluate the known or estimated hazards to subsequent population in the region. The food safety concerns involve:

(a) Biological hazards (parasites, pathogenic bacteria, and virus)

(b) Chemical hazards (radioactive particles, heavy metals, pesticide and drug residues, natural toxins, food decomposition, unapproved food or color additives, and food allergens)

(c) Physical hazards (glass, wood, plastics, non-biodegradable pollutants).

The EU has established directive in place for food safety checks on products entering the EU from all non-EU countries [48]. This FAO Directive dictates that all fish or animal derived products imported into the EU from non-EU countries must be checked at an approved Border Inspection Post

(BIP) to validate their conformity with EU legislation. At these BIPs, there are three principal types of food quality and safety checks done on all food and fishery consignments,

(a) **Documentary:** All consignments go through a documentary check. This requires checking that the relevant veterinary documentation is in place, for example, a properly completed health certificate.

(b) **Identity:** A thorough check is performed on all consignments to validate that it matches the described documentation (also see Table 15.4) and is properly categorized under correct identity.

(c) **Physical:** A physical check is performed nearly on all consignments, even though majority of the food products abide by proper import rules. It includes an examination of the consignment to make sure that it poses no risk to public health.

There are approved border control inspection posts established for the import of the fisheries from the non-EU countries. The physical check frequency depends on the risk profile of the fishery and seafood products, and the results found from the recent checks. The consignments that don't match the EU food safety standards are destroyed and removed in 60 days [48, 49].

4.3.2 The rapid alert system

Once a food quality and safety problem with any consignment has been determined at the border, the EU member state has an obligation to inform all the other EU member nations. This is accomplished by the Rapid Alert System of the European Union. The Rapid Alert System for Food and Feed (RASFF) was established in 1979, to administer food and feed control

Table 15.3 Consignment checks at European Union borders.

Check on some of the packages to confirm that the stamps, official marks, and health marks identifying the country and establishment of origin are present and comply with those on the certificate or document.	**Consignments that do not arrive in containers**
Documentary and identity checks for all consignments. Some may not need to be opened in order to complete an identity check given that official seals have been used in the country of dispatch and the seal numbers are clearly recorded in official veterinary certification.	**Consignments that arrive in containers with official seals**
If official seals have not been used, or there is doubt over whether the seal number was recorded by the certifying veterinarian, the container would need to be opened and a check made on the packages therein to ensure that the stamps, health marks and other marks identifying the country and establishment of origin are present and conform to those on the certificate or document.	**Consignments that arrive in containers with no official seals**

authorities to the European Union with an efficacious tool to exchange appropriate information on measures taken in response to serious risks regarding food quality and safety concerns [50] among EU member nations. The legal basis of the RASFF is to be found in Article 50 of Regulation (EC) N° 178/2002. RASFF authorizes exchange of information between its members, which include the EU-28 national food safety authorities, Commission, EFSA (European Food Safety Authority), ESA, Liechtenstein, Norway, Switzerland, and Iceland. RASFF provides a round-the-clock service to ensure that urgent notifications are sent, received, and responded to in an organized and efficient manner [47]. Article 50.3 of the Regulation states some additional measures for when a RASFF notification is required. Fundamentally, the EU member states are required to immediately inform the Commission under the rapid alert system of:

1. any measure they utilize to regulate the placing or removal of food or feed from the market in order to protect human health and requiring rapid action;

2. any recommendation or agreement with professional operators which is directed at stopping, restricting, or putting forward specific conditions on the placing on the marker or the subsequent use of food or feed on account of a serious risk to human health;

3. any rejection, in concern to a direct or indirect risk to human health, of a batch, container, or cargo of food by a proficient authority at a border control post.

The Regulation establishes varied requirements for members of the network and the procedure for transmission of the different types of notifications. The RASFF notification system has three levels, such as, alert, information, and border rejection, and following confirmation of notification to all the member nations [51]. A RASFF news refers to any information concerning the safety of food or feed which has not been reported as an alert, information, or border rejection notification, but which has been deemed of interest to the control authorities of the EU member states. The Commission is obligated to notify a third country where it is known that a product subjected to an alert notification is present, and therefore enables the country to take necessary actions. Alam (2013) states that 'the country of origin of the product is not always where the hazard originated' (p. 401).

4.4 Seafood safety in the United States of America

Fish and seafood import in the United States is mainly governed by three agencies: Food and Drug Administration (FDA), the Fish and Wildlife Service (FWS), and the National Marine Fisheries Service (NMFS). Here the retailers and wholesalers are responsible for the safety and true labeling of products following the guidelines provided by the FDA. FDA's automated operational import support system (OASIS) strictly reviews each import [45]. Seafood may contain many harmful substances, such as Salmonella, Clostridium, Vibrio species, parasites, marine toxins (paralytic, neurotoxic, and diarrheic shellfish poisoning), environmental contaminants, and heavy metals. FDA set an innocuous level of microorganisms in raw seafood products [29, 30, 45]. If a product adulterant exceeds the FDA approved level, it will automatically detain the product and it will not be imported. FDA controls the quality and safety of oysters, mussels, and clams under the National Shellfish Sanitation Program. This program ensures that the shellfish are obtained from harvesting waters. Samples are tested to check the level of microbial contaminants, and depending on the report it is rejected or accepted for future shipments.

Since marine life survive in water, in order to harvest quality seafood it is important to monitor the quality of the water. Thus, the US Environmental Protection Agency (EPA) carries out many programs, such as Federal Water Pollution Control Act, Marine Protection, Research and Sanctuaries Act. These programs regulate all waste in navigable waters of the United States [52]. It also monitors the environmental contaminants, chemical toxins, and metals, and this data facilitates to assess the toxicity of adulterants in seafood. Post harvesting seafood, it is processed for distribution to retailers. Some US states share this responsibility with the agriculture department. Shellfish and blue crap are processed mainly in southeastern US. The common seafood processing contraventions are adulteration or mislabeling, but these quality problems are solved by the frequent inspection and educating the direct consent. Other problems, such as excess use of sulfite agents to prevent melanosis of crustaceans, could cause health risk for asthmatic group. The HACCP safety criteria for seafood are different from other food products, mainly because it is not dominated by a single contaminant. However, with the widespread application of many safety and quality regulation programs, there are more chances of successfully protecting public health [5, 19, 26].

4.5 Seafood safety in Canada

In North America, Canada is the second largest economy related to fishery and seafood. The government of Canada and the Atlantic provinces have initiated a fund to focus on growing opportunities and increasing the value to high quality fish and seafood products [53]. The Canadian framework for fish and seafood safety is focused on consumer safety through product packaging and labeling requirements on all imported fish and foreign consumers of Canadian fishery products.

The Canadian food inspection agency (CFIA), which was set up through the Canadian food inspection agency act in 1997, is responsible

Table 15.4 CFIA bacteriological guidelines for fish and fish products.

Table of bacteriological levels with organism, product and acceptance numbers						
Criteria for action	**M/g**	**M/g**	**Acceptance number (c)**	**Number of sample units**	**Product type**	**Test organism**
Reject if 2 or more units exceed m, or if any unit exceeds M	40	4	1	5	Cooked or ready-to-eat products	Escherichia coli
Reject if 2 or more units exceed m, or if any unit exceeds M	330/100 g	230/100 g	1	5	Raw bivalve mollusks	Escherichia coli
Reject if 3 or more units exceed m, or if any unit exceeds M	40	4	2	5	All other types	Escherichia coli
Reject if 2 or more units exceed m, or if any unit exceeds M	10000	1000	1	5	All types	Coagulase-positive Staphylococci
Reject if Salmonella spp. is detected	–	–	Absent in each 25 g sample or in pooled samples of 125 g.	5	All types	Salmonella spp.
Reject if Vibrio cholerae is detected	–	–	Absent in each 25 g sample or in pooled samples of 125 g	5	Cooked or ready-to-eat products	Vibrio cholerae
Reject if any unit exceeds m	–	100	0	5	Live oyster	Vibrio parahaemolyticus

M—number of bacteria per gram separating acceptable from marginally acceptable samples, c—number of samples that may exceed this number of bacteria per gram, M—no sample can exceed this number of bacteria per gram.

Table 15.5 CFIA bacteriological guidelines for fish and fish products.

Policy on listeria monocytogenes in ready-to-eat foods		
Action level	**Laboratory method to be applied**	**Product type/category**
Detected	Presence/absence in 125 g (MFHPB-30) on 5 sample units of 25 g each	RTE Fish products in which the growth of L. monocytogenes can occur and could exceed 100 CFU/g before the end of the stated shelf-life. Includes all products that do not fall in either below-mentioned product types. (Equivalent to category 1 foods in the HC listeria policy)
> 100 CFU/g	Enumeration in 50 g (MFLP-74) on 5 sample units of 10 g each	RTE fish products in which the growth of L. monocytogenes can occur, but is limited to levels no greater than 100 CFU/g over the course of their stated shelf-life. RTE products that have a refrigerated shelf-life of 5 days or less fall under this category. Other products require validation data demonstrating growth cannot exceed 100 CFU/g. (Equivalent to category 2A foods in the HC listeria policy)
> 100 CFU/g	Enumeration in 50 g (MFLP-74) on 5 sample units of 10 g each	RTE fish products in which growth of L. monocytogenes cannot occur over the course of the stated shelf-life. Products with the following characteristics fall under this category: • products that are frozen, or • have a pH < 4.4 regardless of the a_w, or • have an a_w < 0.92 regardless of the pH, or • have a pH < 5.0 and an a_w < 0.94. For products that don't meet the above characteristics, validation data demonstrating the absence of growth is required. (Equivalent to category 2B foods in the HC listeria policy)

for the administration and enforcement of all federal legislation related to food inspection, agricultural inputs, and animal and plant health [54]. The legislation covered includes the fish inspection act, the fish inspection regulations, as well as the consumer packaging and labeling to control the safety, quality, and integrity of fish and fishery products. CFIA controls product monitoring and other verification activities executed by domestic processes and licensed importers which will provide reasonable assurance to meet Canadian regulatory requirements. The analysis of all fish or fishery products is conducted in accordance with approved methods as legislated in the health Canada compendium of analytical methods.

4.6 Seafood safety in Japan

In Asia, Japan is one of the largest consumers of fish and seafood products, such as shrimp, tuna, Nile perch, oyster, squid, salmon, seaweed, and many more. In Japan, both raw and processed seafood go through a different set of rules governing its harvest and processing; and has to be approved by local authority for hygiene and safety. The Food Safety Committee was established under a Food Safety Basic Law in the public health policy sector of Japan. Perishable products such as fish and other seafood items are strictly regulated, and are only allowed to be imported under specified procedures of 'The Food Sanitation Law' [55].

4.6.1 Food sanitation inspection in Japan

In Japan, the Food Sanitation law supervises the utilization of pesticide, veterinary drugs in food, food additives, and residue in consumer products. Food products which exceed the maximum residue limit (MRL) cannot be imported in Japan. For some years now, the Japanese government applied HACCP based food control and regulations [56]. It provides the guidelines to manage the hygienic handling of fish, seafood processing, ways of storage and transport, for example, the buyers of fishes such as Nile perch have long term business deals with the suppliers in Uganda, Kenya, and Tanzania. These countries provide service according to HACCP (Hazard analysis and critical control points) and EU (European Union) standards. Another example, shrimp, which is a popular seafood choice among the Japanese society, retailers often set their own quality standards along with The Food Sanitation Law. The North Sea shrimp is imported only if the bacterial count is less than 1,000 per gram, since it's consumed raw. Food hygiene is a major concern to ensure consumer's safety, as lack of food hygiene leads to severe food poisoning [4, 8, 12].

4.6.2 Japanese agriculture standard

In Japan, another legal instrument that is used in assuring seafood safety is Japanese Agriculture Standard (JAS). Fish and seafood products undergo 3 screening processes:

1st Screening—The distribution and measurement of sample species should be less than a certain Regulation value.

2nd Screening—The fishery industries along with the government monitors the levels of radioactive substances such as Germanium, using detectors in fishery products.

3rd Screening—Fishery product inspection is based on the certification issued by Japanese Government.

One of the main Japanese laws that ensures seafood safety is the 'Quarantine Law'. It restricts the entry of seafood products from various areas of interest where disease outbreaks were reported previously. There are about hundred quarantine stations at nearly all Japanese seaports and airports. All retailers apply the regulatory safety criteria, and often a company-specific food quality standards to ensure overall public food safety. For example, the guideline intended for oysters for raw consumption cannot exceed 100 *Vibrio parahaemolyticus* MPN/g of oysters in any of the five sample units [57].

5. Conclusions: Control Systems for Fish and Seafood Safety

From food safety perspective, higher risks may arise from freshwater fish, aquaculture products, and recreational fishing in certain coastal areas and polluted waters, whereas the risk in commercial marine fishing is significantly lower. Pre-harvest contamination is nearly impossible to eradicate as ecosystems cannot be modified easily. Biological agents, such as, pathogenic bacteria, parasites, viruses, biotoxins will always be there, while chemical agents and other pollutants can be controlled at an expense. Tables 15.6 and 15.7 demonstrated an action level based on contaminant levels in wet weight of muscle tissue which does not apply to other types of tissues (e.g., roe, organ meat) or products (e.g., oil), unless otherwise specified in the product type description. For dehydrated or concentrated products, the action level applies to the food that is rehydrated or reconstituted to its original form. Mercury has consumption advisory for certain types of fish in order to maintain mercury exposure within safe levels. Fish protein is a standardized product that is defined in B.21.027 of the food and drug regulations [58, 59]. The first plausible step would be effective monitoring of fishing and aquaculture areas for contamination that can be controlled. The safety of seafood products depends on its origin, microbiological ecology of the seafood product, local processing practices, and traditional preparation practices before consumption [60]. Despite the effectiveness of HACCP, special attention needs to be diverted to, (a) consumption of raw molluskan shellfish, due to its high concentration of Vibrio spp. even after all preventative measures have been impacted, (b) presence of heat stable biotoxins in fish products, and (c) growth of *Listeria monocytogenes* and similar microorganisms in lightly preserved (less than 5% NaCl) fish products [23, 25, 27].

As early as 1980, there was an international drive towards reforming fish inspection systems to move away from end-product sampling and inspection into preventative Hazard Analysis Critical Control Point (HACCP)-based safety and quality systems. These preventative approaches are thus required.

Table 15.6 Guidelines for chemical contaminants and biotoxins in fish and fishery products.

(a) Action level	Product type	Contaminants
1 ppm	Swordfish, shark, escolar, orange roughy, marlin; tuna in fresh and frozen form	Mercury
0.5 ppm	All other fish, including canned tuna	Mercury
3.5 ppm	Fish protein	Arsenic
0.5 ppm	Fish protein	Lead
150 ppm	Fish protein	Fluoride
20 ppt *under review*	All fish	2,3,7,8 TCDD (dioxin)
5.0 ppm	All fish	DDT and metabolites (DDD and DDE)
2.0 ppm *under review*	All fish	PCB
1.0 ppm	Dried cod	Piperonyl butoxide
0.1 ppm	All fish	Agricultural chemicals and derivatives
(b) Action level	Product type	Toxins
20 mg/100 g	Enzyme ripened products (e.g., anchovies, anchovy paste, fish sauce)	Histamine
10 mg/100 g	All other fish products (e.g., canned or fresh or frozen tuna, mackerel, mahi-mahi)	Histamine
80 µg/100 g	Bivalve molluskan shellfish (edible tissue)	Saxitoxins (PSP)
20 µg/g	Bivalve molluskan shellfish (edible tissue)	Domoic acid (ASP)
0.2 µg/g (under review) 1 mg/kg (under review)	Bivalve molluskan shellfish (edible tissue) Bivalve molluskan shellfish digestive tissue	Okadaic acid (OA) + DTX1 + DTX2 + OA esters + DTX1 esters + DTX2 esters (DSP)
0.2 µg/g 1 mg/kg	Bivalve molluskan shellfish (edible tissue) Bivalve molluskan shellfish digestive tissue	Pectenotoxins: PTX-1, PTX-2, PTX-3, PTX-4, PTX-6 and PTX-11

a. Fish products are prepared/processed in certified plants and establishments. The certification process requires that the plant meets minimal requirements in terms of layout, design and construction, hygiene and sanitation;

b. The industry takes responsibility in fish safety control and implements HACCP-based in-plant quality control programmes;

Table 15.7 Action levels for additives which may have naturally occurring background levels.

Action level	Amount permitted to be added	Background levels	Product type	Additive
> 15 ppm (see note 2)	Not permitted	15 ppm (see note 2)	All fish and fish products (except marine mammal meat)	Nitrites
> 15 ppm (see note 2)	Not permitted	15 ppm (see note 2)	All fish and fish products	Nitrates
> 10 ppm	Not permitted	10 ppm	Clams (raw and canned)	Sulphites
> 2.10%	0.5%	1.60%	Shrimp (frozen, raw, cooked and canned)	Phosphates
> 1.47%	Not permitted	1.47%	Scallops (raw)	Phosphates
> 1.87%	0.5%	1.37%	Frozen fish fillets, frozen minced fish, canned seafood	Phosphates
> 2.20%	0.5%	1.70%	Crab (frozen, raw, canned and cooked)	Phosphates
> 1.97%	0.5%	1.47%	Lobster (frozen, raw, canned and cooked)	Phosphates
> 1.50%	0.5%	1.00%	Surf clams (frozen, raw and cooked)	Phosphates
> 1.47%	0.1%	1.37%	A blend of prepared fish and prepared meat	Phosphates
> 1.60%	0.5%	1.10%	Squid	Phosphates

c. A regulatory competent authority is in charge of certifying fish plants and establishments, approving and monitoring HACCP-based in-plant quality control programmes and certifying fish and fishery products before distribution;

d. Where necessary, national surveillance programmes of the harvesting areas should be in place to control the threats of biotoxins and other biological and chemical pollutants; and

e. For export, an additional control can be exercised by the importing party and involves an audit of the national control system of the exporting country to ensure that it meets the requirements of the importing country. This should lead to the signing of mutual recognition agreements between trading countries.

Finally, to ensure seafood safety, more focus should be put into informing consumers of the risks associated, and limiting shelf-life of preserved fish [22].

References

1   Vandermeersch, G., Lourenço, H.M., Alvarez-Muñoz, D., Cunha, S., Diogène, J., Cano-Sancho, G. et al. 2015. Environmental contaminants of emerging concern in seafood—European database on contaminant levels. Environ. Res. 2015 Nov. 143(Pt B): 29–45. [Medline].

2   Cano-Sancho, G., Sioen, I., Vandermeersch, G., Jacobs, S., Robbens, J., Nadal, M. et al. 2015. Integrated risk index for seafood contaminants (IRISC): Pilot study in five European countries. Environ. Res. 2015 Nov. 143(Pt B): 109–15. [Medline].

3   Wallace, C. and Williams, T. 2001. Pre-requisites: a help or a hindrance to HACCP? Food Cont. 2001 Jun. 12(4): 235–40.

4   Trienekens, J. and Zuurbier, P. 2008. Quality and safety standards in the food industry, developments and challenges. Int. J. Prod. Eco. 2008 May 113(1): 107–222.

5   Huss, H.H., Reilly, A. and Embarek, P.K.B. 2000. Prevention and control hazards in seafood. Food Cont. 2000 Apr. 11(2): 149–156.

6   Anon, Event. Seafood safety - new findings and innovation challenges - Research & amp; Innovation - European Commission. Available at: http://ec.europa.eu/research/index.cfm?pg=events&eventcode=1FEADAC1-02B9-AD69-9E9AD350A4589DD3.

7   Mowry, J.B., Spyker, D.A., Brooks, D.E., McMillan, N. and Schauben, J.L. 2014 Annual report of the american association of poison control centers' National Poison Data System (NPDS): 32nd Annual Report. Clin. Toxicol. (Phila) 2015 Dec. 53(10): 962–1147. [Medline].

8   Levine, K., Yavelak, M., Luchansky, J.B., Porto-Fett, A.C.S. and Chapman, B. 2017. Consumer perceptions of the safety of ready-to-eat foods in retail food store settings. J. Food Prot. 2017 Aug. 80(8): 1364–77.

9   Scoging. 1998. Marine biotoxins. J. Appl. Micro. 1998 Jul. 84(S1): 415–505.

10 Nicolas, J., Hendriksen, P.J.M., Gerssen, A., Bovee, T.F.H., Ivonne, M. and Rietjens, C.M. 2014. Marine neurotoxins: State of the art, bottlenecks, and perspectives for mode of action based methods of detection in seafood. Mol. Nutr. Food Res. 2014 Jan. 58(1): 87–100.

11 Todd, E.C. 1994. Emerging diseases associated with seafood toxins and water-borne agents. Ann. N. Y. Acad. Sci. 1994 Dec. 740: 77–94.

12 Feng, C., Teuber, S. and Gershwin, M.E. 2016. Histamine (Scombroid) fish poisoning: a comprehensive review. Clin. Rev. Allergy Immunol. 2016 Feb. 50(1): 64–9.

13 Arvanitoyannis, I.S., Kotsanopoulos, K.V. and Papadopoulow, A. 2014. Rapid detection of chemical hazards (toxins, dioxins, and PCBs) in seafood. Crit. Rev. Food Sci. Nutr. 54(11): 1473–528.

14 Hussain, M.A., Saputra, T., Szabo, E.A. and Nelan, B. 2017. An overview of seafood supply, food safety and regulation in New South Wales, Australia. Foods. 2017 Jul. 6(7): E52.

15 Hassoun, A. and Karoui, R. 2017. Quality evaluation of fish and other seafood by traditional and nondestructive instrumental methods: Advantages and limitations. Crit. Rev. Food Sci. Nutr. 2017 Jun. 57(9): 1976–1998.

16 Sioen, I., Henauw, S.D., Camp, J.V., Volatier, J.L. and Leblanc, J.C. 2009. Comparison of the nutritional-toxicological conflict related to seafood consumption in different regions worldwide. Reg. Toxic. Pharm. 2009 Nov. 55(2): 219–228.

17 Dalgaard, P. 2000. Fresh and lightly preserved seafood. pp. 110–139. *In*: Man, C.M.D. and Jones, A.A. (eds.). Shelf Life Evaluation of Foods. 2nd Edition. Aspen Publishing Inc. Gaithersburg, Maryland, USA.

18 Sumner, J. and Ross, T. 2002. A semi-quantitative seafood safety risk assessment. Int. J. Food Micro. 2002 Jul. 77(1-2): 55–59.

19 Han, F., Walker, R.D., Janes, M.E., Prinyawiwatkul, W. and Ge, B. 2007. Antimicrobial susceptibilities of Vibrio parahaemolyticus and Vibrio vulnificus isolates from Louisiana Gulf and retail raw oysters. Appl. Env. Micro. 2007 Nov. 73(1): 7096–98.

20 Srogi, K. 2007. Mercury content of hair in different populations relative to fish consumption. Rev. Environ. Contam. Toxicol. 189: 107–30.

21 Ababouch, L. 2006. Assuring fish safety and quality in international fish trade. Mar. Poll. Bull. 53(10-12): 561–568.

22 Fleming, L.E., Broad, K., Clement, A., Dewally, E., Elmir, S., Knap, A., Pomponi, S.A., Smith, S., Gabriele, H.S. and Walsh, P. 2006. Oceans and human health: Emerging public health risks in the marine environment. Mar. Poll. Bull. 53(10-12): 545–560.

23 James, K.J., Carey, B., O'Halloran, J., van Pelt, F.N. and Skarabakova, Z. 2010. Shellfish toxicity: human health implications of marine algal toxins. Epideomiol Infect. 2010 Jul. 138(7): 927–40.

24 Mejlholm, O. and Dalgaard, P. 2007. Modeling and predicting the growth boundary of Listeria monocytogenes in lightly preserved seafood. J. Food Prot. 2007 Jan. 70(1): 70–84.

25 Dalgaard, P. and Jorgensend, L.V. 1998. Predicted and observed growth of Listeria monocytogenes in seafood challenge tests and in naturally contaminated cold-smoked salmon. Int. J. Food Microbiol. 1998 Mar. 40(1-2): 105–15.

26 Seafood HACCP. https://www.fda.gov/food/guidanceregulation/haccp/ucm2006764.htm.

27 Panisello, P.J. and Quantick, P.C. 2001. Technical barriers to Hazard Analysis Critical Control Point (HACCP). Food Cont. 2001 Apr. 12(3): 165–173.

28 Srogi, K. 2007. Mercury content of hair in different populations relative to fish consumption. Rev. Environ. Contam. Toxicol. 189: 107–30.

29 Ahmed, F.E., Hattis, D., Wolke, R.E. and Steinman, D. 1993. Risk assessment and management of chemical contaminants in fishery products consumed in the USA. J. Appl. Toxic. 1993 Nov. 13(6): 395–410.

30 U.S. FDA Center for Food Safety and Applied Nutrition (CFSAN). 2001. Guideline. https://www.fda.gov/ICECI/EnforcementActions/EnforcementStory/EnforcementStoryArchive/ucm106655.htm.

31 U.S. Government Accountability Office. https://www.gao.gov/cghome/2001/pascg.html.

32 CDC. 2002. https://www.cdc.gov/mmwr/preview/mmwrhtml/mm5153a1.htm.

33 The National Academies of Sciences, Engineering and Medicine. Consensus Study Reprot. Scientific Criteria to Ensure Safe Food 2003. https://www.nap.edu/read/10690/chapter/1.

34 FAO/WHO [Food and Agriculture Organization of the United Nations/World Health Organization]. 2011. Risk assessment of Vibrio parahaemolyticus in seafood: Interpretative summary and Technical report. Microbiological Risk Assessment Series No. 16. Rome. 193pp.

35 Tzouros, N.E. and Arvanitoyannis, I.S. 2001. Agricultural produces: synopsis of employed quality control methods for the authentication of foods and application of chemometrics for the classification of foods according to their variety or geographical origin. Crit. Rev. Food Sci. Nutr. 2001 May 41(4): 287–319.

36 USDA Food Safety and Inspection Service. NACMCF Reports and Recommendations. 2013–2015. https://www.fsis.usda.gov/wps/portal/fsis/topics/regulations/advisory-committees/nacmcf-reports/nacmcf-reports.

37 Cusato, S., Gameiro, A.H., Sant'ana, A.S., Corassin, C.H. and Cruz, A.G., de Oliveira, C.A. 2014. Assessing the costs involved in the implementation of GMP and HACCP in a small dairy factory. Qual. Assur. Safety Crops Foods 6(2): 135–139.

38 Tomasevic, I., Dodevska, M., Simic, M., Raicevic, S., Matovic, S., Matovic, V. and Djekic, I. 2017. The use and control of nitrites in Serbian meat industry and the influence of mandatory HACCP implementation. Meat Sci. 2017 Jul. 134: 76–78.

39 Ye, M., Huang, Y., Gurtler, J.B., Niemira, B.A., Sites, J.E. and Chen, H. 2013. Effects of pre- or post-processing storage conditions on high-hydrostatic pressure inactivation of Vibrio parahaemolyticus and Vibrio vulnificus in oysters 2013 May 163(2-3): 146–52.

40 Ronholm, J., Lau, F. and Banerjee, S.K. 2016. Emerging seafood preservation techniques to extend freshness and minimize Vibrio contamination. Front Microbiol. 2016 Mar. 22(7): 350.

41 Olson, J., Clay, P.M. and da Silva, P.P. 2014. Putting the seafood in sustainable food systems. Mar. Pol. 2014 Jan. 43: 104–111.

42 Smith, M.D., Roheim, C.A., Crowder, L.B., Halpen, B.S., Turnipseen, M. et al. Sustainability and global seafood.

43 Anon. 2013. http://www.fao.org/docrep/008/y5924e/y5924e06.htm.

44 Anon. 2017. https://ec.europa.eu/food/sites/food/files/safety/docs/ia_trade_import-cond-fish_en.pdf.

45 FDA Fish and Fishery Products Hazards and Controls Guidance. 2011. Fourth edition. Available at: https://www.fda.gov/downloads/food/guidanceregulation/ucm251970.pdf.

46 EC safe Seafood. 2017. Available at: http://www.ecsafeseafood.eu/.

47 European Food Safety Authority. Available: at http://www.efsa.europa.eu/en/publications.

48 Abagouch, L., Gandini, G. and Ryder, J. 2005. *In:* Causes of detentions and rejections in international fish trade. FAO xiv: 110.

49 Summer, J., Ross, T. and Ababouch, L. 2004. *In:* Application of risk assessment in the fish industry. FAO ix: 78.

50 EU RASFF. 2015. Available at: https://ec.europa.eu/food/safety/rasff_en.

51 Alam, S. 2013. Bangladesh in the rapid alert system for food and feed notifications in the period 2000–2012: A review. Veterinarni Medicina 58(8): 399–404.

52 EPA Clean Water Act. 1972. Available at: https://www.epa.gov/laws-regulations/summary-clean-water-act.

53 Foley, P. 2013. National government responses to Marine stewardship council fisheries certification: Insights from Atlantic Canada. New Pol. Econ. 2013 Apr. 18(2): 228–307.

54 Bietiot, H.P. and Kolakowski, B. 2012. Risk assessment and risk management at the Canadian Food Inspection Agency (CFIA): A perspective on the monitoring of foods for chemical residues. Drug Test Anal. 2012 Aug. 4 Suppl. 1: 50–8.

55 Japan Ministry of Health, Labour and Welfare. The food sanitation law is available at: https://www.jetro.go.jp/ext_images/en/reports/regulations/pdf/foodext2010e.pdf.

56 USDA Global Agriculture Information Network Report 2009 is available at: https://gain.fas.usda.gov/recent%20gain%20publications/food%20and%20agricultural%20import%20regulations%20and%20standards%20-%20narrative_tokyo_japan_8-19-2009.pdf.

57 Mahmoud, B. 2009. Effect of X-ray treatments on inoculated Escherichia coli O157: H7, Salmonella enterica, Shigella flexneri and Vibrio parahaemolyticus in ready-to-eat shrimp. Food Microbiology 26(8): 860–864.

58 Coppes, Z., Pavlisko, A. and de Vecchi, S. 2008. Texture measurements in fish and fish products. J. Aqua. Food Prod. Tech. 2008 Oct. 11(1): 89–105.

59 Cunningham, C. and Robledo, J.F. 2015. Molluscan immunology. Fish & ShellFish Immunol. 2015 Sep. 46(1): 1–160.

60 Institute of Medicine (US) Committee on Evaluation of the Safety of Fishery Products; Ahmed, F.E. (ed.). Seafood Safety. Washington (DC): National Academies Press; 1991. 8, Seafood Surveillance and Control Programs. Available from: https://www.ncbi.nlm.nih.gov/books/NBK235724/.

Index